The Ecology of Soil

BIOLOGY OF HABITATS SERIES

The Biology of Agroecosystems
Nicola Randall, Barbara Smith

The Biology of Caves and Other Subterranean Habitats
Second Edition
David C. Culver, Tanja Pipan

The Biology of Urban Environments
Philip James

The Biology of Grasslands
Brian J. Wilsey

The Biology of Mediterranean-Type Ecosystems
Karen J. Esler, Anna L. Jacobsen, R. Brandon Pratt

The Biology of Coral Reefs
Second Edition
Charles Sheppard, Simon Davy, Graham Pilling, Nicholas Graham

The Biology of Lakes and Ponds
Third Edition
Christer Brönmark, Lars-Anders Hansson

The Ecology of Soil

From communities to ecosystems

Richard D. Bardgett

OXFORD
UNIVERSITY PRESS

OXFORD
UNIVERSITY PRESS

Great Clarendon Street, Oxford, OX2 6DP,
United Kingdom

Oxford University Press is a department of the University of Oxford.
It furthers the University's objective of excellence in research, scholarship,
and education by publishing worldwide. Oxford is a registered trade mark of
Oxford University Press in the UK and in certain other countries.

Published in the United States of America by Oxford University Press
198 Madison Avenue, New York, NY 10016, United States of America

British Library Cataloguing in Publication Data
Data available

Library of Congress Control Number: 2025940541

ISBN 9780198817031
ISBN 9780198817048 (pbk.)

DOI: 10.1093/9780191858710.001.0001

Printed and bound by
CPI Group (UK) Ltd, Croydon, CR0 4YY

The manufacturer's authorised representative in the EU for product safety is
Oxford University Press España S.A. of Parque Empresarial San Fernando de Henares,
Avenida de Castilla, 2 – 28830 Madrid (www.oup.es/en or product.safety@oup.com).
OUP España S.A. also acts as importer into Spain of products made by the manufacturer.

In memory of Juliet Frankland, John Whittaker, Gregor Yeates, and Diana Wall, for guiding me in my exploration of the ecology of soil.

Preface

I bequeath myself to the dirt, to grow from the grass I love,
If you want me again, look for me under your boot-soles.
Walt Whitman (1819–1892)

This book began its life as a second edition of *The Biology of Soil*, which was published by Oxford University Press 20 years ago in 2005. As I started writing, however, it fast became clear that *The Biology of Soil* needed much more than an update; rather, it needed a major overhaul. In fact, it required a new book to capture the remarkable growth of soil ecology research over the last 20 years: *The Ecology of Soil: From communities to ecosystems.* There are inevitable similarities between this book and *The Biology of Soil*, and their goals are similar: to demonstrate how and why soil life matters at the community and ecosystem scale by shaping the structure and function of terrestrial ecosystems and their response to global change. But there is much that is new in *The Ecology of Soil* reflecting the rapid growth of the field since the early 2000s. This includes new knowledge on the diversity and distribution of soil life across a range of spatial and temporal scales, made possible, in part, by an increased accessibility of molecular tools for characterising below-ground communities. It also includes advances in our understanding of the major roles that soil organisms and their interactions with each other and plants play in shaping ecosystems, and how soil biodiversity and its many functional roles are modified by global change, especially land use and climate change. There is also a completely new chapter on soil biota and climate change, and on soil biota and global sustainability challenges, reflecting a growing awareness that soil ecology is central to meeting sustainability challenges. Finally, there is a new chapter that highlights some challenges that I think will occupy soil ecologists over the next decade and perhaps beyond.

Anyone who has written a book will know that it takes much time and effort and, in my case, needed to be fitted in around many competing demands of a full-time academic job and other responsibilities. This book was no different. It took me 8 years to write, which is much longer than the publisher or I intended. But hopefully this has given rise to a richer book: it allowed me to capture recent developments in what is an incredibly fast-moving field, but also to place these new discoveries in the context of foundational work by the pioneers of the field. Indeed, the growth in soil ecology research over the last two decades has been remarkable and never ceases to amaze me. Over this time, soil ecology has not only become a central theme in terrestrial ecology, but also it has gained more attention from policy makers, farmers, foresters, land use planners, gardeners, the public, and even artists, who increasingly recognise the importance of soil life for a healthy planet. This is in stark contrast to when I first penned *The Biology of Soil*. Then, soil ecology was a minority topic, often considered to be of minor relevance to understanding the ecology of terrestrial ecosystems and how they respond to global change. But, fortunately, things have changed, and the rich diversity of soil ecology knowledge included in this book, spanning the foundational to new and derived from many parts of the world, hopefully reflects this. Put simply, soil ecology is booming and shows no signs of slowing. I hope this book captures this excitement and inspires the reader to delve deeper into the ecology of soil.

Many people helped me write this book. Ian Sherman deserves a special mention, as he first persuaded me to write the book, as he did for *The Biology of Soil.* He also kept me on track when other commitments pulled me away. Put simply, if it wasn't for Ian, I wouldn't have written this book in the first place, or *The Biology of Soil*, and for this I am very grateful. I am also grateful to Adam Breivik who guided me through the final stages of the book, and the many colleagues who have read chapters of the book, providing valuable and constructive reviews, including Stefan Giessen, Anton Potapov, Jeppe Kristensen, Manuel Delgado-Baquerizo, Arthur Broadbent, Tancredi Caruso, Dina in't Zant, Lena Neuenkamp, Michael Bahn, Carlos Guerra, Bill Grayson, Franciska De Vries, and David Smedley. A special mention should also go to Adam Bilton who proof checked the entire book, but also had the unenviable task of checking through all the references that I had compiled manually, much to the bewilderment of my colleagues who advised me to use reference software.

I also want to thank my wife, Jill, and three children, Alice, Lucy, and Marianne, who have encouraged me to keep going with writing the book, even if they don't know it, and my many colleagues who kindly provided me with advice, images, and figures. It has been quite a task, especially given the growing pressures of my job. But I hope my efforts are worthwhile, either as an introduction to the fascinating world of soil ecology to those not familiar or as a comprehensive, up-to-date synthesis for those who already have an interest in the ecology of soil.

When I first decided to write *The Biology of Soil* in the early 2000s, I was a young ecologist just beginning my own journey into the ecology of soil. I recall mentioning to a senior colleague at the time that I was planning to write the book. He asked: Is this really a good idea? I took this to mean that I would be better spending my time writing research papers and applying for grants, because books generally don't carry much kudos in the scientific world, at least in the UK. I couldn't disagree more with this view, and if I look back over my career, it is my books, including this one, that I am most proud of. Writing books has helped me keep abreast of the field and served to generate new knowledge. From this has come ideas.

I hope you enjoy reading *The Ecology of Soil* as much as enjoyed writing it, and that it inspires the reader to delve even deeper into the fascinating below-ground world.

Richard D. Bardgett
April 2025
Warton, northern England

Contents

CHAPTER 1

The soil environment

> **Key messages**
>
> - Within any landscape there is an incredible range of soil types resulting from variation in soil-forming factors, namely parent material, climate, topography, time, and biota.
> - Soil-forming factors operate simultaneously, usually with a hierarchy of importance, with one or two of them being pre-eminent in soil development at a particular location.
> - Soil-forming factors determine the physical and chemical nature of soils, which in turn strongly shapes the structure and activities of soil biota.
> - Soil biota both influence and are affected by the properties of soil, and the roles of soil biota and the factors that regulate their structure and activities vary from soil to soil.

Soil forms a thin mantle over the Earth's surface and acts as the interface between the atmosphere and lithosphere, the outermost shell of Earth. It is a multiphase system, consisting of minerals, plant roots, water, gases, and organic matter at various stages of decay. The soil is also a habitat for a staggering diversity of organisms. Recent estimates suggest that upward of 59% of species on Earth live in soil, making it the most biodiverse habitat on Earth (Anthony et al. 2023, Bardgett 2023). Most of these species are microorganisms (fungi, bacteria, archaea, and protists), but soil is also home to a host of animals, including nematodes, springtails, mites, earthworms, ants, spiders, and moles (Decaëns et al. 2006). These organisms not only use the soil as a habitat and source of energy, but also contribute to its formation, influencing a soil's physical and chemical properties as well as the diversity, composition, and productivity of the vegetation that grows on it (Bardgett and van der Putten 2014). Indeed, along with vegetation, biota, including soil organisms, is one of the five soil-forming factors according to Jenny (1941), namely parent material, climate, biota, relief, and time, which were first identified by the Russian soil scientist Dokuchaev (1883) to explain why the type of soil found in one place differs to that in another. To understand what controls the abundance and activities of these soil organisms, and the spatial and temporal variability in soil biological communities, first requires an understanding of the physical and chemical nature of the soil matrix in which they live. This chapter considers the factors responsible for regulating soil formation, and hence the diversity of soils in the landscape, and discusses key properties of the soil environment that most strongly influence soil biota, leading to variability in soil biological communities across a range of spatial and temporal scales.

Soil formation

To understand the properties of soils that influence the biota that dwell there, it is important to first consider some of the factors of soil formation that explain the variation in soils and soil properties within a landscape, or from one region to another. One of the most fascinating features of the terrestrial world is its tremendous variety in landform, reflecting not only a diversity of geological processes that have occurred over millions of years, but also biological processes and more recent influences of humans, which can be substantial. Within any landscape there is similarly an incredible range of soil types, resulting from variation in the role of the different soil-forming factors (Figure 1.1) (Jenny 1941). As will be discussed later, these soil-forming

The Ecology of Soil. Richard D. Bardgett, Oxford University Press. © Richard D. Bardgett (2025). DOI: 10.1093/9780191858710.003.0001

factors act simultaneously to influence the development of a particular soil, although at any given location their hierarchy of influence varies; in some situations, climate might be dominant, whereas in others it might be parent material. Further, vegetation has a dual role in soil formation: as a dependent variable, shaped for example by climate, but with reciprocal effects on soil properties, and as a true soil-forming factor, where soil formation is a direct function of vegetation.

A particular combination of soil-forming factors leads to the development of a unique soil type, with a relatively predictable series of horizons (layers) that constitute the soil profile (Figure 1.2). Of most interest to the soil ecologist are those horizons that are at or close to the soil surface; this is where most microbes and animals live, and where most root growth and nutrient recycling occurs. These horizons are referred to as the surface organic O horizon, derived from decomposing plant and animal material, and the uppermost A horizon, composed largely of mineral material, but also intimately mixed with organic matter derived from the surface O horizon. Importantly, soil ecologists are also concerned with the plant litter lying directly on the soil surface, deposited during the previous annual cycle of plant growth. This layer, referred to as the L layer, is often overlooked or even discarded in soil sampling regimes, but it can be the most biologically active and functionally important zone of the soil profile. Soil ecologists are becoming increasingly interested in subsurface soil, in the underlying mineral B and deeper C horizon, the latter being intermixed with the parent material. Indeed, soils can reach considerable depths (> 25 m in some cases), especially in tropical and subtropical regions where soil formation has been ongoing for millions of years. There is also growing evidence that a significant portion of the soil microbial biomass is in subsurface horizons (Fierer et al. 2003, Eliers et al. 2012), where it can play critical roles in the processes of weathering and biogeochemical cycling.

Whilst soil profiles vary greatly across landscapes, they can be classified into groupings based on a set of unifying soil properties and soil-forming characteristics, each having a unique set of ecosystem properties. For example, on free-draining, sandy parent material, in cold and wet climates, usually beneath heathland and coniferous forests, podzolic soils develop (Spodosols in

Figure 1.1 A collage of a variety of soil types from different parts of the world. Images by Richard Bardgett.

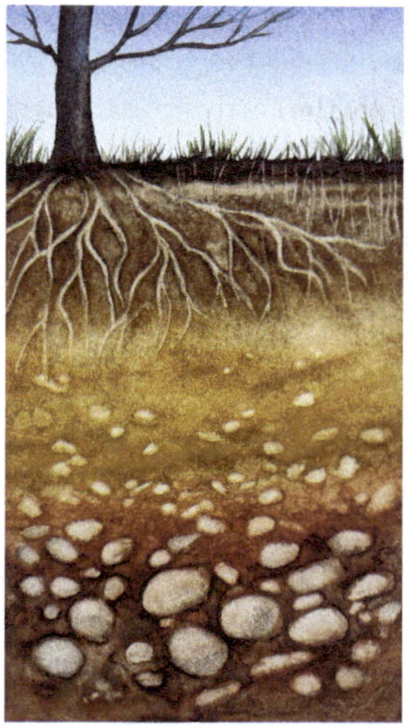

Figure 1.2 Schematic representation of a soil profile, showing major surface and subsurface horizons. Image by Jill Colquhoun Bardgett.

U.S. Department of Agriculture [USDA] soil taxonomy) (Figure 1.3). These Podzols, formed by podzolisation (Box 1.1), have a deep, acidic surface O horizon, referred to as mor humus. They are subject to heavy leaching (Box 1.1) and are characterised by low rates of decomposition and plant nutrient availability, resulting in low plant productivity. Typically, the microbial biomass of O horizon of a Podzol is dominated by fungi (rather than bacteria) and the fauna are characterised by high numbers of microarthropods (mites and Collembola), but an absence of earthworms. In contrast, on calcium (Ca)-rich, clayey parent material, typically beneath grassland and deciduous forest, Brown Earths (Inceptisols in USDA terminology) are often found (Figure 1.4). These soils have a mull humus form that is often mildly acidic, due to leaching of base cations (e.g., Ca_2^+) down the soil profile. The mull horizon is characterised by intimate mixing of the surface organic and uppermost mineral A horizon, resulting from high abundance and activity of soil biota, especially earthworms, leading to high rates of decomposition, nutrient availability, and plant growth. A total of 12 major soil groupings, termed soil orders, have been distinguished by the USDA Soil Taxonomy (Table 1.1), each with a unique set of ecosystem characteristics. Further details on the classification of soils can be found in general soil science textbooks by White (2005) and Weil and Brady (2017).

The soil-forming factors

As already noted, within most landscapes there are a tremendous variety of soil types of varying physical and chemical make-up. The soil-forming factors were originally identified by the Russian soil scientist Dokuchaev (1883), who argued that the type of soil and arrangement of its horizons within any place was 'always and everywhere a mere function' of a particular combination of soil-forming factors. The soil-forming factors were later formalised by Jenny (1941) in his classic book *Factors of Soil Formation*, and they are now central to our understanding of the variability in soils at the level of the landscape and the individual soil profile. As such, they deserve further consideration as the central forces responsible for creating variety in soil conditions and hence variations in the habitat of the soil biota. The biota themselves, along with vegetation, constitute one of the main soil-forming factors; both can act as important determinants of soil formation and profile development. This section summarises some of the important aspects of the five soil-forming factors. It is important to stress, however, that whilst soil-forming

Figure 1.3 Podzolic (Spodosol in USDA taxonomy) soil with deep O horizon (mor) and characteristic bleached E eluvial horizon above the red, depositional Bs or spodic horizon. Image by Otto Ehrmann.

factors are considered individually, they operate simultaneously in nature, usually with a hierarchy of importance, with one or two of them being pre-eminent in soil development at a particular location.

Parent material

Geological processes that act over millions of years determine the variation and distribution of parent materials from which soils develop. Soils are formed from the weathering of either consolidated rock *in situ* or from unconsolidated deposits—derived from erosion of consolidated rock—that have been transported by water, ice, wind, or gravity. The mineralogical composition of these deposits varies tremendously. For example, the mineralogy of igneous rocks, formed by solidification of molten magma in, or on, the Earth's crust, ranges from base-rich basalts (basic lava) with high amounts of Ca and magnesium (Mg), to acidic rhyolites (acid lava) which contain high amounts of silica (Si) and low amounts of Ca and Mg. Rocks of intermediate base status, such as andesites, also commonly occur. Some parent materials form soils enriched in toxic elements. For example, soils formed on black shales, developed in shallow, stagnant waters, can be enriched in heavy metals such as copper (Cu), cadmium (Cd), and lead (Pb). Also, soils developed on rocks containing metal ores can be enriched in heavy metals, and serpentine soils, derived from ultramafic rocks of very low silica content (e.g., serpentinite), contain high concentrations of nickel and chromium, which can be toxic to soil organisms and plants.

Parent material also varies greatly in grain size, the main determinant of soil texture, or the relative proportions of sand, silt, and clay. Soil texture in turn strongly influences many other soil properties that are key to its functioning, such as the ability of a soil to retain cations (cation exchange capacity) and moisture, and the infiltration and movement of water through the soil profile. Such variation in the mineralogy of rocks therefore strongly influences the type of soils that form and the character of the vegetation that they support (Figure 1.5). Soils formed from weathering of basic lava, for example, or on clay-rich marine or lake sediments, limestone, or glacial till, tend to be relatively rich in minerals such as Ca, Mg, and potassium (K) and fine textured (clayey), and, therefore, they have a high ability to retain moisture and cations of importance to plant nutrition (e.g., NH_4^+, Ca_2^+, Mg_2^+). These soils are typically fertile brown earths (Alfisols) with biologically active mull humus. In contrast, soils formed on coarse-textured parent materials, such as sandstones and acid volcanic rocks (e.g., granites and rhyolites), are low in Ca and Mg, coarse textured (sandy), and, hence, freely drained with low cation retention and water-holding capacity. The soils that typically develop on such rocks are strongly leached, are nutrient poor, have a mor humus, and often dry out quickly during periods of drought due

Box 1.1 Pedogenic processes

Weathering is caused by the action of a range of forces that combine to soften and break up rock into smaller particles that become the parent material of soil. Weathering can occur by physical, chemical, or biological means, and usually by a combination of the three. Physical weathering occurs mainly through the action of water, wind, and changes in temperature, which progressively break down rock into finer particles. Chemical weathering involves the decomposition of minerals by a range of processes including solution, hydration, oxidation, and hydrolysis. Biological weathering occurs under the influence of organisms, for example roots which penetrate and crack open rocks. Organisms such as lichens also produce organic acids which erode the rock surface.

Leaching refers to the downward movement of materials in soil solution, usually from one soil horizon to another. The mobility of elements depends on their solubility in water, the effect of pH on that solubility, and the rate of water percolation through the soil.

Podzolisation involves the leaching of aluminium (Al) and iron (Fe) from upper soil horizons and their deposition deeper in the soil. These elements are relatively immobile in soil, largely occurring as insoluble hydroxides. Their leaching, however, can be enhanced by the formation of soluble organo-metal complexes, or chelates, which are mobile in percolating water. The most active complexing agents are organic acids and polyphenols, which are especially abundant in decomposing litter of coniferous trees and ericaceous plants. Removal of Fe from the upper soil horizon leads to the formation of an eluvial E horizon, which is bleached in appearance; deposition of Fe deeper in the soil forms a characteristic Bs or spodic horizon, of orange-red colour. The presence of the spodic horizon is a diagnostic feature of the USDA soil taxonomy order Spodosol, or the Podzol soil group in UK terminology.

Lessivage refers to the translocation of clay down the soil profile and its deposition in oriented films (argillans) on ped faces and pore walls. This process gradually forms a subsurface horizon of clay accumulation, which is commonly referred to as an argillic, or Bt horizon. The end product of this process is the formation of an Argillic brown earth (UK soil group), or the USDA soil taxonomy order Alfisol. Lessivage occurs under conditions that favour deflocculation of clay minerals. This depends mainly on factors such as clay type and the presence of electrolytes such Na^+, which cause clay minerals to deflocculate and hence move in water. In acid soils, soluble organic compounds can produce hydrophilic films around clay particles which enhance their movement in percolating water. Clay minerals can also be destabilised by physical impact, for example by raindrops, cultivation, and frost action. Clay movement is unlikely to occur in calcareous soils with high amounts of exchangeable Ca^{2+} ions, which leads to flocculation of clay.

Gleying is the dominant biological process that occurs in hydromorphic soils that are characterised by waterlogging for significant periods of time, due to either impeded drainage or a seasonal water table that rises into the subsoil. Gleying is evidenced by the presence of a gleyed horizon that is predominately blue-grey or blue-green in colour. This coloration results from the microbial reduction of ferric (Fe^{3+}) to ferrous (Fe^{2+}) iron which occurs under anaerobic conditions, leading to the solution and depletion of Fe from the soil horizon. Mottling is common in gleyed soils due to re-oxidation and precipitation of Fe in better aerated zones, especially around plant roots and in larger soil pores. This process is evidenced by patches of orange-red colour, which are surrounded by the predominately blue-grey soil matrix. In UK soil taxonomy, soils are classified as being either Ground-water gley soils or Surface-water gley soils, the former resulting from a high water table and the latter from impeded drainage.

to their low water-holding capacity. Finally, soils formed from silt-rich parent material, such as the deposits of large rivers or estuaries, or wind-blown dust generally make for fertile soils; they drain more freely than clay, but being less heavy than clay are more susceptible to the erosive forces of wind and water.

Climate

Historically, climate has been considered pre-eminent in soil formation, due largely to the striking associations that exist, on continental scales, among regional climate, vegetation type, and associated soils. Indeed, these broad climatic associations led to the development in Russia of one

Figure 1.4 A typical brown earth soil with mull horizon, under grassland. Note the lack of horizon differentiation caused by intense biological activity and the mixing of organic matter with mineral material from the A horizon. Image by Otto Ehrmann.

of the first soil classification systems, the zonal concept of soils (Dokuchaev 1879). This system identified so-called zonal soils that are influenced over time more so by regional climate than by any other soil-forming factor. Whilst climate may be pre-eminent in soil development at continental scales, for example across Russia and Australia, it is arguably not as important in areas such as the subtropics and tropics where land surfaces are much older and more eroded, or in younger landscapes such as northern Europe where most soils are developed on recent, glacial deposits; here, the factors of topography and parent material are of major importance.

Effects of climate on soil development are largely due to temperature and precipitation, which vary considerably across climatic zones. These factors strongly govern rates of chemical reactions and the growth and activities of the soil biota, which in turn affect soil-forming processes of mineral weathering and decomposition of organic material. Effects of temperature on soil biological activity are well known; it is generally accepted that there is an approximate doubling of microbial activity and enzyme-catalysed reaction rates in soil for each 10°C rise in temperature, up to around 30–35°C. Above this temperature, however, most enzyme-catalysed reactions decline markedly, as proteins and membranes become denatured. Some microbes can live at extreme temperatures; for example, microbial activity has been detected in tundra soils frozen down to −39°C (Panikov et al. 2006), and thermophilic soil bacteria of the genus *Geobacillus* exhibit optimal growth at temperatures between 50 to 70°C (Santana and Gonzalez 2015). Cold-tolerant microorganisms are called pychrophiles, whereas

Table 1.1 USDA soil taxonomy orders

Order	Brief description
Entisols	Recently formed, lacking diagnostic horizons, often on unconsolidated sediments or rock
Vertisols	Clay-rich soils that swell when wet and shrink when dry, and typically of high base status
Andisols	Volcanic soils that are rich in amorphous colloidal materials, including allophane
Inceptisols	Young, poorly developed soils without contrasting subsurface horizons
Aridosols	Dry soils from deserts, with little surface organic matter and slow rates of development
Mollisols	Dark and deep surface mull horizon of high organic matter content and base rich
Gelisols	Soils of cold regions with permafrost close to the surface
Spodosols	Acid soils with humus and Fe-rich B horizons found in coniferous forests and heathland
Alfisols	Moderately leached but base-rich soils with a clay-rich subsurface B horizon
Ultisols	Highly weathered soils of the humid tropics depleted in base cations, with a clay B horizon
Oxisols	Heavily weathered Fe- and Al-rich soils found in humid tropics
Histosols	Organic hydromorphic soils (peats) found in bogs, moors, and blanket peats

microbes that live in extremely hot temperatures are called thermophiles.

Effects of temperature and precipitation on soil formation are especially marked at high altitudes and latitudes. For example, soil organic matter (SOM) content generally increases with increasing elevation, commonly reaching a peak in montane forests (Körner 1999), before declining at higher altitudes above the treeline where organic matter content of soils reaches almost zero in non-vegetated substrates in the upper alpine zone. This increase in SOM content up towards montane forest areas is largely due to declines in temperature combined with greater precipitation that reduces microbial activity and rates of decomposition. Similarly, in high latitude regions of Europe, extensive peatlands have developed in areas where the combined effects of high rainfall, low temperature, and minimal evapotranspiration have led to anaerobic conditions (waterlogging) and the retardation of decomposition of organic matter. The consequence of this has been accumulation of masses of peat (blanket peats), especially in topographically uniform areas where drainage is reduced (Figure 1.6). Dramatic shifts in the physiology and productivity of dominant plants also occur along altitudinal and latitudinal gradients (Díaz et al. 1998, Mayor et al. 2017),

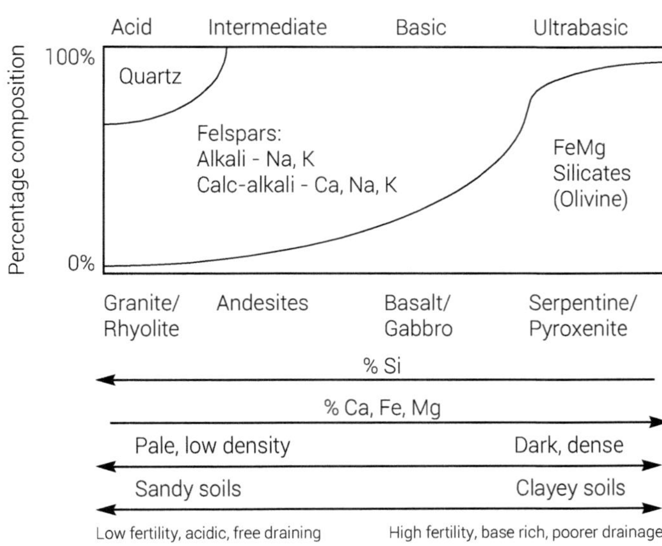

Figure 1.5 Schematic classification of igneous rocks and their resulting soils. From Bardgett (2005) by permission of Oxford University Press.

Figure 1.6 Blanket peat at Moor House National Nature Reserve, northern England. Here, deep peats have developed at high altitudes where high rainfall and low evapotranspiration combine to cause excessive soil wetness that retards decomposition. These peatlands are of special significance because they represent a significant (ca. 30%) global store of terrestrial C. Image by Richard Bardgett.

altering the nutritional quality of leaf litter that is produced annually, as well as the biomass and turnover of roots. As will be discussed in Chapter 4, such changes in vegetation, both above-ground and below-ground, can have profound effects on the functioning of terrestrial ecosystems.

Topography

Variations in topography influence soil development, largely through effects on soil drainage and erosion. Soil drainage is primarily affected by the position of a soil on a slope; soils at, or near, the top of a slope tend to be freely drained with a water-table at some depth, whereas those at, or near, the valley bottom tend to be poorly drained with a water-table close to the soil surface. These differences in drainage strongly influence soil development, leading to the development of a hydrological sequence (Figure 1.7): well-drained soils on hilltops have deep orange-brown subsurface horizons indicative of oxidation processes (Fe in ferric state); as drainage deteriorates towards the valley bottom, the soil profile becomes increasingly anaerobic and blue-grey in colour, indicative of a dominance of reduction processes (Fe in ferrous state) and the process of gleying (Box 1.1). In extremely wet valley-bottom soils, deep O horizons develop on the soil surface because of retardation of decomposition processes under anaerobic conditions.

Slope characteristics also greatly influence soil erosion processes, which in turn impact on soil formation. In general, soils on ridges and steeper parts of slopes are shallower than those on lower slopes in valley bottoms, due to the movement of soil particles down-slope by wash and soil creep. Because erosion preferentially moves finer particles down-slope, the soils of lower slopes and valley

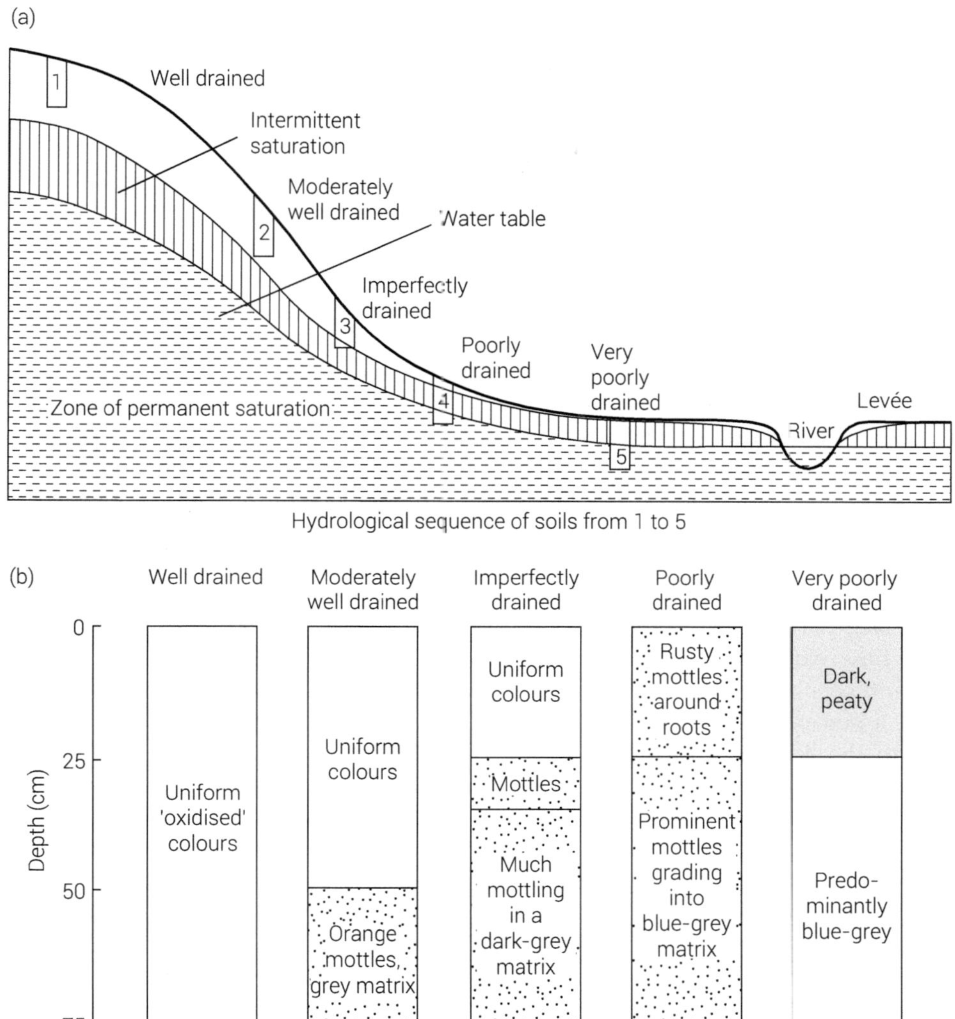

Figure 1.7 Section of a slope and valley bottom showing a hydrological soil sequence (a) and changes in soil profile morphology (b). From White (2005) by permission of John Wiley and Sons.

bottoms also differ in their mineralogical composition, being more fine-textured. In humid regions, soils of lower slopes and valley bottoms also tend to be enriched in base cations and salts, due to seepage of solutes from higher slopes and hilltops. Soil movement down-slope also leads to the formation of distinct morphological features on slopes, such as terraces. These features are especially common in alpine regions where steep slopes and freeze-thaw motion leads to instability of surface soil and consequent down-slope creeping, a process termed solifluction. What happens here is that soils become saturated with water, freeze, and then melt; the expansion associated with freezing makes the surface soil very unstable when it thaws. This leads to downward movement of soil even on gentle slopes, and especially if the subsurface soil is frozen.

Time

Soil age is a major factor underlying variation in soils and ecosystem properties. Soils become increasingly weathered with age, and, therefore, soil profiles generally become more differentiated and deeper, with more abundant and thicker horizons. Rates of soil formation vary considerably depending on the other soil-forming factors, such as the type of parent material and climate, and they tend to be faster in humid than in arid parts of the world. As a rule, however, the average rate of soil formation, over the entire surface of Earth, has been estimated to be around 10 cm for every thousand years, or just 0.1 mm $year^{-1}$ (Stockmann et al. 2014).

As soils age, they go through stages of development from youth to maturity. A key process involved in this ageing is the leaching downwards of elements and minerals in percolating water. During the process of podzolisation, for example, progressive downward movement of Fe and Al leads to the development of a spodic subsurface horizon that is enriched in these minerals, and an upper E eluvial horizon that is bleached in appearance, from where these elements have been removed (Box 1.1). Over time, clay minerals are also leached down profile, a process called lessivage (Box 1.1), which leads to the development of a subsurface argillic horizon. Other important changes in soils that occur as soils age include increases in organic matter and nitrogen (N) content (Crocker and Major 1955, Walker et al. 2010). Over very long timescales (hundreds of thousands or millions of years), a progressive reduction in the availability of soil phosphorus (P) occurs due to its loss from the system and fixation in mineral forms that are not available to plants, as discussed below.

The relation between time and soil development is best illustrated by examining soil chronosequences, which are places where, for various reasons, a sequence of differently aged, but otherwise similar, geologic substrates exist. Glacier Bay, on the coast of southeast Alaska, is one of the best places for studying soil and ecosystem development because of the continuous retreat of the glaciers since 1794. Also, records of the retreat, over some 100 km, have been kept since this time, so the age of the glacial moraines is known. This has resulted in a site chronology upwards of 200 years, and from this, a pattern of soil development can be tracked along with the succession in vegetation from initial pioneer plant communities on recent moraine through to the climax spruce forest on the oldest moraines (Box 1.2). This chronosequence represents stages in the development of podzol: as time progresses, the soil profile becomes progressively deeper and differentiated in its horizons, the oldest soils being acidic in nature with a thick organic surface horizon, a bleached E horizon, and subsurface iron-rich spodic Bs horizon (Crocker and Major 1955). As organic matter steadily accumulates in the surface organic horizon, so does the amount of N in soil; the organic carbon (C) and N of underlying mineral soil also builds up (Figure 1.8). Of particular significance for organic matter and N accumulation is the early stage of vigorous alder growth on terrain that has been ice-free for some 75 years; here, N accumulates at a rate of about 4.9 g N m^{-2} $year^{-1}$, reaching values of some 250 g N m^{-2} within around 50 years of soil development (Crocker and Major 1955). Although there is no single explanation of the soil development sequence at Glacier Bay, it appears that establishment of plants, especially N-fixing plants, and intensive leaching play important roles (Matthews 1992, Chapin et al. 1994).

Studies at Glacier Bay demonstrate progressive soil development over relatively short timescales towards the climax, or maximal biomass phase, during which SOM, nutrient content, and plant biomass increase over time. Other sites, however, demonstrate soil change over hundreds of thousands or even millions of years, far beyond the maximal biomass phase. The Hawaiian Island archipelago, for example, presents a chronology of soil development of over 4 million years: Kilauea volcano on the island of Hawai'i is the youngest, being still active, while the island of Kaua'i on the northwest end of the high islands is the oldest, being an estimated > 4 million years old (Figure 1.9). A range of intermediate-aged islands occurs in between, all being derived from volcanic lava of similar mineralogy and with vegetation dominated by the same tree, *Metrosideros polymorha* (Figure 1.10). A key feature of the Hawaiian chronology is that soils of the older sites have lower P availability, which causes a decline in forest stature and productivity (Crews et al. 1995). This fall in P availability follows the theoretical model of soil development proposed by Walker and Syers (1976): as soils age, P becomes surrounded,

Box 1.2 Glacial moraine succession at Glacier Bay, southeast Alaska

Many glaciers in the Northern Hemisphere have been retreating during the past 200 years. Glacier Bay, on the coast of south-east Alaska, is one of the best-known places for research on primary succession because of the continuous retreat of the glaciers since 1794. Furthermore, records of the retreat, over some 100 kilometers, have been kept since this time, so the age of the glacial moraines is known. This produces what is known as a site chronology over a period of 200 years. The map of Glacier Bay fjord complex shows the rate of ice retreat since 1760. The dashed lines show the approximate edge of the ice in 1760 and in 1860.

At Glacier Bay, there are four stages of succession:

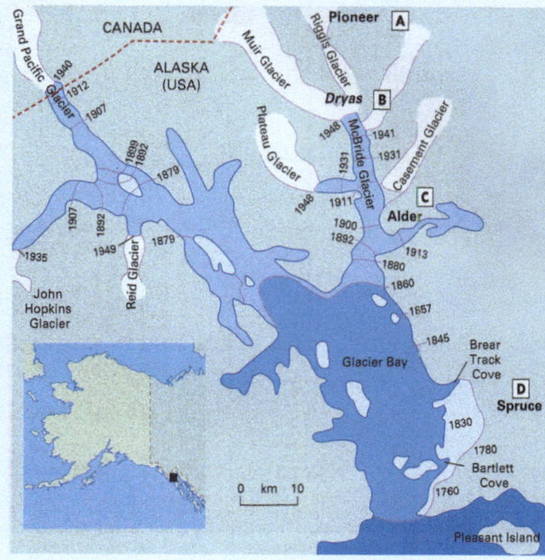

The **pioneer stage** (a) has been ice-free for 10-20 years. The ground is mainly bare glacial till with patches that have been colonized by algal crusts, lichens, mosses, *Dryas* and scattered willows.

The **Dryas stage** (b) is on ground that has been ice free for some 30 yeas. Here, the ground surface is an almost continuous mat of *Dryas* with scattered willow, cottonwood, and alder.

The **alder stage** (c) forms after about 50-70 years. Here the *Dryas* is taken over by dense thickets of the nitrogen-fixing plant alder.

The **spruce stage** (d) forms after 100-150 years of succession. Here, spruce trees that have replaced the alder dominate the canopy.

a.

b.

c.

d.

Images by Richard Bardgett

or occluded, by Fe and Al oxides, rendering the P largely unavailable to plants and the soil biota (Figure 1.11). This process of occlusion is especially likely to occur in very old soils because prolonged weathering of minerals leads to the formation of Fe and Al oxides that have a strong affinity for P. Moreover, P limitation in older soils becomes intensified because low soil P availability sets in

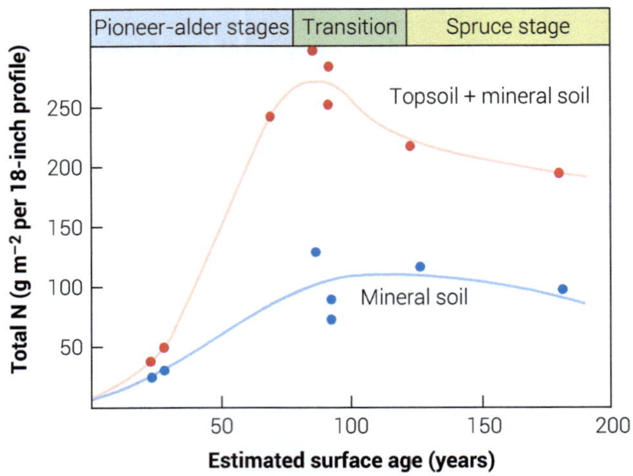

Figure 1.8 Total N content of soils recently uncovered by glacial retreat at Glacier Bay, Alaska. Plant succession is shown along the top. From Crocker and Major (1955) by permission of John Wiley and Sons.

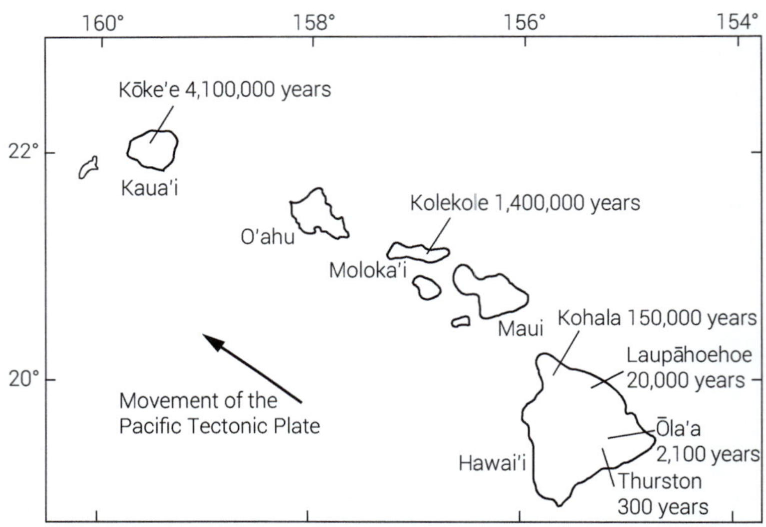

Figure 1.9 The Hawaiian Island archipelago presents a chronosequence of soil development of more than 4 million years. Kilauea volcano on the island of Hawaii is still active, while Kaua'i, the oldest site, is estimated to be over million years old. All sites receive similar amounts of precipitation (~ 2,500 mm year^{-1}) and are dominated by the same tree species, *Metrosideros polymorpha*. From Crews et al. (1995) by permission of John Wiley and Sons.

motion a feedback, whereby low P concentration in plant litter slows its decomposition and the microbial recycling of P, leading to declines in P supply and standing plant biomass, a phenomenon termed ecosystem retrogression (Crews et al. 1995, Wardle et al. 2004b, Vitousek 2006, Peltzer et al. 2010).

Ecosystem retrogression appears to be a general phenomenon associated with older soils that have not been subject to major disturbances (Wardle et al. 2004b, Peltzer et al. 2010). For example, upward of 400,000 years of progressive weathering and leaching of free-draining sand dunes on the subtropical coast of Queensland, Australia, has led to the development of so-called giant podzols (Thompson 1981) with soil profiles of greater than 20 m depth, which are characterised

Figure 1.10 *Metrosideros polymorpha* ('ohi'a lehua) forest on the Hawaiian island of Molokai, with low stature and biomass. This ecosystem developed in the absence of catastrophic disturbances during the past 1.4 million years. Depletion of K at this site has led to a decline in forest productivity. Image by Richard Bardgett.

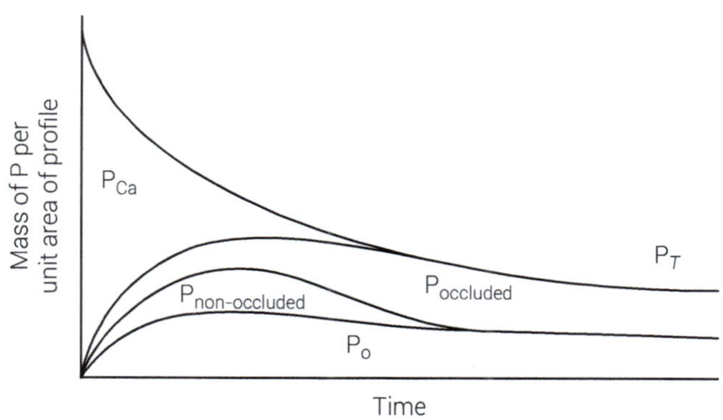

Figure 1.11 Generalised model of Walker and Syers (1976) of the effects of long-term weathering and soil development on the distribution and availability of P in soil. P_{Ca} = calcium phosphates; P_T = total phosphorus; P_o = P bound to organic matter. From Crews et al. (1995) by permission of John Wiley and Sons.

by extreme P limitation and stunted tree growth (Figure 1.12). Also, at Glacier Bay, ecosystem retrogression occurs in boreal forests that have been ice-free for 12,000–14,000 years, and these forests are likewise characterised by declines in standing biomass, decomposition, and rates of nutrient cycling as they age (Wardle et al. 2004b). Further to this, studies of the 2-million-year-old

Jurien Bay chronosequence in southwestern Australia reveal that plant productivity peaks within a few thousand years, but then declines steadily during the retrogression due to increasing P limitation (Laliberté et al. 2012). For more information on ecosystem retrogression the reader is referred to the review of Peltzer et al. (2010), which includes a summary of the most intensively studied long-term soil chronosequences around the world.

Figure 1.12 Stunted tree growth resulting from P limitation in old-growth (400,000 years) eucalyptus forest on the subtropical coast of Queensland, Australia. Image by Richard Bardgett.

Organisms

As already mentioned, the role of organisms, including vegetation, microorganisms, and animals, as a soil-forming factor is not so clear-cut as for other factors, in that biota are both affected by and influence the properties of soil. Despite this, many examples exist in the literature of organisms, especially vegetation, acting as a true soil-forming factor. The expansion of the Devonian forests some 400 million years ago, for example, triggered the initial formation of soils: roots and their mycorrhizal associations penetrated underlying rock, stimulating its weathering and the liberation of minerals to support plant growth. The expanding forest produced large amounts of detritus, providing a substrate for soil organisms, which stimulated the recycling of the organic matter. A more recent example used by Jenny (1941) is the prairie–forest transition, which once extended across the centre of North America from Illinois in the north to Texas in the south. The reasons for this abrupt boundary in vegetation are still debated and much of it has been obliterated by agriculture. But its impact on soil formation remains striking: prairie soils (Mollisols) have greater SOM content and are darker in colour than forest soils, and are deep and nutrient rich, being fed by the extensive root systems and litter of productive grassland plants, whereas forest soils have less organic matter and are more strongly leached, and often have a bleached subsurface spodic horizon of a podzol.

Other notable examples, which will be considered in detail later in this book, include the classic distinction of mull and mor soils, first noted by the Danish forester P. E. Müller (Müller 1884) to describe the contrasting soils associated with deciduous and coniferous vegetation, respectively. Müller noted that soils of coniferous forest and heathland had features typical of podzolic soils, most notably a deep, acidic surface organic horizon, or 'mor' humus, underlain by a subsurface spodic horizon, whereas those of deciduous forest had a less distinctive organic horizon, or 'mull' humus layer, which was intermixed with underlying mineral soil because of high biological activity and rates of decomposition. Also, as will be discussed in Chapter 4, soils can be transformed when new plant species invade ecosystems. A dramatic example of this comes from the Hawaiian Islands where the invasion of the N-fixing shrub *Myrica faya*—which was first introduced to the Hawaiian Islands from the Azores and Madeira in the 1800s as an ornamental or medical plant—into native *Metrosideros polymorpha* forests, which are strongly N-limited, has completely transformed the forest soils, quadrupling rates of N cycling and creating a nutrient-rich soil (Vitousek and Walker 1989).

Anthropogenic impacts on soil development are also substantial and growing, with upward of one-third of the Earth's land surface suffering from moderate to severe soil degradation caused by human activities (Montanarella et al. 2016). Human activities are not only transforming the characteristics of soils, principally via agriculture and urbanisation, but also are increasing the extent of rare and endangered soils. A study of the diversity of soils in China, for example, revealed that 88 types of soil are endangered across the nation and a further 17 have become extinct due to pressures of farming and urbanisation (Shangguan et al. 2014). Likewise, a study of soil diversity across the USA revealed a total of 500 'endangered' soils at risk from farming and urbanisation, and several that have gone extinct (Amundson et al. 2003). Impacts of human activities on soil development can also be surprisingly long lasting. For example, ancient irrigation practices of Hohokam prehistoric populations living in Arizona, in the US Southwest, substantially altered the texture of soils, and the legacy of this on soil and vegetation can still be detected today, seven hundred years after the land was abandoned (Hall et al. 2013). Also, in Western Europe, the impact of farming during Roman times on soil nutrient availability and forest vegetation can still be detected today, almost 2,000 years on (Dupouey et al. 2002), and medieval cultivation in England has left its legacy in the form of ridge and furrow, which gives pastures an undulating appearance with distinctive patterns of drainage. Effects of human activities on soil biota and functions will be discussed further in Chapter 2.

Soil properties

Variations in soil-forming factors determine the physical and chemical nature of soils, which in turn strongly influences the structure and activity of soil biota, and hence ecosystem properties of decomposition and nutrient cycling. Variation in soil properties, especially the physical matrix of the soil, also strongly influences the movement of water and associated materials both within and between ecosystems. This section examines some of the key soil physical and chemical properties that most strongly influence soil organisms and their activities, and the structure and functioning of terrestrial ecosystems.

Soil texture and structure

The term soil texture refers to the relative proportions of various-sized particles—sand (0.05–2.0 mm), silt (0.002–0.05 mm), and clay (< 0.002 mm)—within the soil matrix (Figure 1.13). It is primarily dependent on the parent material from which the soil is formed and the rate at which it is weathered, as discussed above. Soil texture is of importance largely because it determines the ability of the soil to retain water and nutrients: clay minerals have a higher surface area to volume ratio than sand and silt, and hence soils with a high clay content are better able to hold water by adsorption and to retain cations on their charged surfaces. The ability of a soil to retain cations (e.g., Ca^{2+}, Mg^{2+}, NH_4^+) is referred to as its cation exchange capacity, which reflects the capacity of clay minerals to hold cations on negatively charged surfaces. This retention of cations on clay minerals represents a major short-term store of nutrients for plant and microbial uptake.

Soil structure reflects the binding of the above various-sized mineral particles into larger aggregates termed peds. The formation of stable aggregates requires the action of a range of physical, chemical, and biological factors. These include: freeze-thaw and soil shrinkage and swelling, which help to mould the soil into aggregates; mechanical impact of rain and physical disturbances such as ploughing, which reorganises soil materials; activities of burrowing animals, such as earthworms, which mix mineral and organic materials together, aiding the formation of stable organo-mineral complexes, and faeces of soil animals, which act as nuclei for aggregate formation; and, as will be discussed in Chapter 3, soil microbes produce a range of polysaccharide glues that bind soil particles together, and fungal hyphae hold together, or enmesh, mineral particles and organic matter. Together, these factors combine to produce stable aggregates in soil. As will be discussed in Chapter 3, large soil animals, such as earthworms and termites, can substantially affect soil structure by creating macropores and channels because of their feeding and burrowing activities, which in turn increases water infiltration rates and the drainage of water through soil.

Soil aggregation and structure is of concern to the soil ecologist not only because the activity of the soil biota strongly affects it, but also because the structure of soil determines the physical nature of the living space of biota. Aggregation determines the pore distribution of soil, which affects the distribution of water in soil (specifically the degree to which pores are filled with water) and the extent that biota can enter and occupy pore space, which is controlled by pore neck diameter and the size of the organism. For example, nematodes are approximately 30 μm in diameter, so their migration in soil is restricted by pores of diameter < 30 μm and is optimal in soils with particle sizes in the region of 150–250 μm. Fungi and bacteria are much smaller in size, so they can occupy much smaller pores (1–6 μm) than can nematodes. Because the primary food source of many nematodes is microbes, these smaller pores can therefore provide a refuge for microbes from their predators, thereby affecting the nature and intensity of trophic interactions in soil.

In sum, good soil structure is a key attribute of fertile and biologically active soil, because it increases the flow of water through soil, reducing the possibility of anaerobic conditions developing, which would be detrimental to soil biota and their activities and harmful to plant growth. Good soil structure promotes free movement of biota, increasing opportunities for trophic interactions, allows roots to proliferate, and enables aerobic microbial processes to dominate.

Soil organic matter

SOM content varies considerably across and within ecosystems in terms of both its chemical

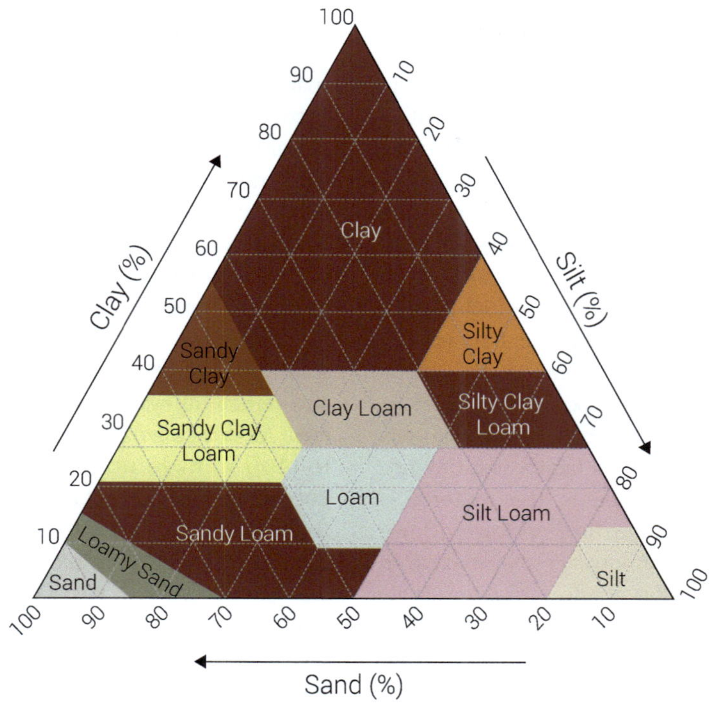

Figure 1.13 Composition of the textural classes of soils, based on percentages of sand, slit, and clay. For example, a soil with 60% sand, 10% silt, and 30% clay is a sandy clay loam.

composition and quantity. As already discussed, this depends on a variety of interacting factors such as vegetation type, climate, parent material, soil drainage, and the activity of soil biota; a particular combination of these factors generally leads to the formation of either mull or mor humus forms. SOM is of particular importance because it promotes soil structural stability (see above), thereby preventing soil erosion, and it is extremely effective at retaining water within the soil matrix. It is also of importance for soil biota, being the primary source of nutrients and C. Through a range of activities, soil biota decomposes organic matter, converting it into simple organic molecules that biota can assimilate and use for energy and growth. Most components of SOM occur as large molecules (e.g., cellulose, amino sugars, proteins, nucleic acid, lignin), which, over time, are depolymerised by extracellular microbial enzymes to yield simpler units which can then be assimilated by the microbes for energy and C. For example, the enzyme cellulase breaks down cellulose—the principal component of plant tissue—into glucose units which are readily assimilated by microbes and used for energy and C.

Microbial enzymes also produce soluble organic nutrients (e.g., amino acids), which are either absorbed by microbes to meet their nutrient requirement (e.g., for C and N) or, in some circumstances, taken-up directly by plants and mycorrhizal fungi, as will be discussed in Chapter 3. Soluble organic nutrients that are absorbed by microbes are retained to meet their C and N needs, or, when they only require C to support their energy needs, the excess N is secreted into the soil environment as inorganic N (ammonium), which is then available for plant uptake. This process is called mineralisation; it is of high importance at the ecosystem scale because this process determines the availability of inorganic nutrients for plant uptake and hence plant productivity. Soil fauna also contribute to the process of nutrient mineralisation, both directly, by mixing and fragmenting it into smaller units that are more accessible to microbial attack, and indirectly, by feeding on microbes, affecting their growth and activity, and excreting nutrients that are in excess of the animal's requirements into the soil environment.

The rate that organic inputs to soil are decomposed depends primarily on their quality, which is dependent on the type of compounds that are present within it. Rapidly decomposing material, such as litter of deciduous trees and animal faeces, generally contains high amounts of labile substances such as amino acids and sugars, and low concentrations of recalcitrant compounds such as lignin. In contrast, litter of coniferous trees is decomposed slowly, being rich in large, complex structural compounds such as lignin, and defence compounds such as polyphenols; this material is also unpalatable to soil fauna, further slowing down its decomposition. The importance of variation in the rate of decomposition at the ecosystem scale relates to the production of carbon dioxide (CO_2) from heterotrophic microbial activity and its evolution into the atmosphere, as well as the conversion of organic nutrient forms to simple inorganic nutrients (e.g., ammonium and phosphate) that are available for plant uptake, a strong determinant of plant productivity. These factors will be discussed in detail in later chapters.

Soil water

It is water that renders the soil environment habitable. Together with dissolved nutrients, water makes up the soil solution, an important medium for supplying nutrients and water to growing plants. It also provides a medium for many soil organisms to live and move around in. Nematodes and protozoa, for example, live in water films and in free water, along with the bacteria that they feed upon. Larger fauna, such as microarthropods (mites and Collembola), live in open pore spaces, but are very sensitive to desiccation and migrate out of soil when it becomes dry.

There are certain water-holding characteristics of soils that determine the amount of water that is available to plants and soil biota. Under normal conditions, soil pores will contain air as well as water, and most water is held in pore spaces as films or as water absorbed to soil particles. Under these conditions the soil is said to be unsaturated. After a period of heavy rain or irrigation, however, pore spaces in soil become filled with water and the soil becomes saturated. After a period of drainage, when the amount of water held in soil is in equilibrium with gravitational suction (5–10 kPa), field capacity is reached and no more water drains from the soil. The pores that are drained at field capacity are the macropores, which are those that are created by the burrowing activities of animals and by plant roots. As plant roots continue to absorb water from the soil, the amount of water held in films is reduced and remaining water adheres tightly to soil particles. At this stage, water moves through thin films to plant roots in response to a gradient of water potential; if water is not replenished and plants continue to take up water, a situation is reached when plants can no longer remove water that is strongly adhered to particle surfaces. This is the permanent wilting point, and the difference between this measure and field capacity is referred to as the water-holding capacity of the soil. In general, the water-holding capacity of soil is higher in soils that contain large amounts of clay and/or organic matter, because these components have high surface areas that readily retain water. Also, clay soils have more small pores, which readily retain water under gravitational suction, than do sandy soils, which have larger pores and are more easily drained.

Soil pH

A large proportion of the Earth's soils are acid, especially in the tropics where ecosystems persist at soil pH values of 4 or less (pH is a measure of the concentration of H^+ ions in soil water). Many northern ecosystems also have very acid soils: pH values of soils of boreal forests and heathlands are often of 4 or less. In many parts of the world, soil acidity is further exacerbated through the use of inorganic fertilisers and acid rain. Soils become acid if base cations (e.g., Ca^{2+}, Mg^{2+}, K^+) are leached from the soil profile, to be replaced by H^+ and Al^{3+} ions on cation exchange sites. Acidity in soils can come from various sources: carbonic acid, which is formed by the dissolution of CO_2 in water, dissociates to yield H^+ ions; microbial oxidation of ammonium ions (NH_4^+)

to nitrate (NO_3^-), the former being derived from mineralisation and fertiliser inputs, also yields H^+ ions; atmospheric pollution (acid rain) and natural sources of acids, including volcanic eruptions and thunderstorms that yield sulphur dioxide and oxides of N, respectively, producing sulphuric and nitric acids that acidify soils (it has been proposed that the widespread occurrence of acidic soils in tropical and subtropical regions is in part a result of high thunderstorm activity in these regions); and decomposition of organic matter that has high concentrations of phenolic and carboxyl groups, which liberates H^+ ions.

Soil pH is of concern to the soil ecologist because it controls nutrient availability and because it directly impacts on soil biota. Acid soils, for example, are characteristically high in soluble aluminium (Al^{3+}), which can be toxic to plants and microbes. This is because Al occurs largely as insoluble forms, for example, as part of clay minerals whose structure becomes unstable at low pHs (4–5), releasing Al ions into soil solution. P availability in soil is typically low under acid conditions, due to the formation of Fe and Al phosphates. These phosphates dissolve to release P into soil solution as pH rises, making it available for plant uptake; the availability of P is typically greatest between pH 6 and 7.

Effects of pH on soil organisms are well documented, and approximate tolerances of major groups of soil organisms are known. Most microbes grow within the pH range of 4–9, although acidophiles are known to survive at pHs as low as 1: for example, sulphate-oxidising bacteria (species of the genus *Thiobacillus*) which occur in hot springs and mine wastes. Growth in extremely alkaline conditions is restricted to a few fungi and bacteria, but such conditions only tend to occur in soda lakes and deserts: for example, those in Egypt which have pH values between 9 and 11. Soil animals are also very sensitive to acidic conditions. For example, earthworms occur in very low numbers, with few species, in acidic soils, and they become progressively more abundant as soil pH increases to neutrality. Acid soils tend to be dominated by enchytraeid worms (potworms), which replace earthworms as the functionally dominant soil animal.

Conclusions

Within most landscapes, there are a tremendous variety of soil types, reflecting spatial variations in the operation of a range of interacting soil-forming factors and pedogenic processes, as well as the influences of man. The need for the soil ecologist to recognise this soil variation, at both spatial and temporal scales, cannot be stressed enough because it means that the role of soil biota, relative to other abiotic factors, in controlling ecosystem functions (e.g., decomposition, nutrient cycling, and plant productivity) and soil formation itself, is context dependent. In other words, the role of soil biota and the factors that regulate the structure and activities of below-ground communities varies from soil to soil, depending on the dominant physical and chemical characteristics of that soil. In view of this, it is essential that studies on the soil biota and their activities are accompanied by detailed characterisation of the physical and chemical nature of their habitat, and an appreciation of the soil-forming factors that led to the development of that soil in the first place.

CHAPTER 2

The diversity of life in soil

Key messages

- A vast diversity of microorganisms and animals live in soil and constitute the soil food web, whose primary role is to recycle dead organic matter from the plant-based food web.
- Microbes are the most diverse members of the soil food web, but many animals also live alongside them, which are broadly categorised as microfauna, mesofauna, and macrofauna.
- A hierarchy of factors shape soil biodiversity across spatial scales, including abiotic factors (e.g., soil pH, climate), biotic factors (e.g., vegetation), and multiple global change stressors.
- Soil communities change constantly over time, with seasonal and interannual shifts in the abundance of different organisms, and changes in soil biodiversity over successional time.

Ecologists have long been fascinated by the vast diversity of organisms that live on Earth. Indeed, making sense of the complex patterns of this diversity, and the dominant forces that control them, has long been a major theme of community ecology. This area of ecology, however, has traditionally had an above-ground and aquatic focus. Much less effort has been put into characterising and understanding the distribution and functional importance of below-ground biodiversity (Bardgett and van der Putten 2014). As a result, information on the diversity and distribution of soil organisms is lagging what is known about above-ground organisms, especially at the species level. Also, despite mounting efforts to characterise the distribution of soil organisms at regional and global scales (e.g., Fierer and Jackson 2006, Wu et al. 2011, Tedersoo et al. 2014, Delgado-Baquerizo et al. 2018b, Caruso et al. 2019b, Phillips et al. 2019, Van den Hoogen et al. 2019, Guerra et al. 2022, Lavelle et al. 2022, Potapov et al. 2023, Graham et al. 2024), available information is restricted to relatively few ecosystem types and taxonomic groups, with a strong bias to the northern hemisphere (Guerra et al. 2020). This historic lack of attention to below-ground biodiversity is surprising given a recent estimate that soil is home to 59% of all species on Earth, making it the most biodiverse habitat on Earth (Anthony et al. 2023).

Despite the historic lack of interest in soil biota, the last few decades have witnessed a surge of interest in the biology of soil. This has been prompted by a growing awareness that soil organisms play crucial roles in regulating major ecosystem functions, such as organic matter turnover, nutrient cycling, and plant production. But evidence is mounting that soil biota act as important drivers of plant community dynamics and terrestrial ecosystem responses to global change. Technological developments have also contributed to this surge of interest in the biology of soil, especially the rapid development of molecular tools that enable in-depth characterisation of the diversity of soil life. This chapter introduces the major groups of organisms that live in soil and discusses some of what is known about their abundance, diversity, and distribution. The chapter also considers our current understanding of factors that act as determinants of soil biodiversity across a range of spatial and temporal scales.

The Ecology of Soil. Richard D. Bardgett, Oxford University Press. © Richard D. Bardgett (2025). DOI: 10.1093/9780191858710.003.0002

The soil biota

The vast array of microorganisms and animals that live in soil constitute the soil food web, whose primary role in ecosystems is the recycling of litter and dead organic matter from the plant-based food web (Figure 2.1). The most numerous and diverse members of the soil food web are microbes (bacteria, fungi, archaea, viruses, and protists), which constitute the soil microbiome. But many animal species of varying size also live in soil, including microfauna (body width less than 0.1 mm, e.g., nematodes), mesofauna (body width 0.1–2.0 mm, e.g., microarthropods and enchytraeids), and macrofauna (body width > 2 mm, e.g., earthworms, termites, and millipedes) (Figure 2.2). These animals cross a range of trophic levels and are typically allocated to functional groups based on their feeding habit (Figure 2.3) (Hunt et al. 1987, de Ruiter et al. 1995, Moore et al. 2003). Some animals feed primarily on microbes (microbivores) or litter (detritivores), and others feed principally on plant roots (herbivores) or other animals (carnivores). Omnivores are also common, with many soil animals feeding across trophic levels and/or multiple types of resources (Potapov et al. 2022). Detailed descriptions of the major components of the soil food web are beyond the scope of this book; here, the main players are briefly introduced. The reader is referred to *Soil Fauna Assemblages* (Nielsen 2019) for more detailed accounts of soil fauna and approaches used to study them, and to *Soil Microbiology, Ecology and Biochemistry*, 5th Edition (Paul and Frey 2023) for information on soil microbes. The *Global Soil Biodiversity Atlas* (Orgiazzi et al. 2016) also provides much information on the diversity of organisms that live in soil.

The soil microbiome

The primary consumers of the soil food web are the microbes, including bacteria, fungi, and archaea, which are primarily responsible for breaking down and mineralising complex organic substances, transforming them into various inorganic and gaseous forms. The soil microbiome also

Figure 2.1 A selection of organisms that make up the soil food web, including ectomycorrhizal (a) and decomposer fungi (b), bacteria (c), nematodes (d), testate amoebae (e), collembolans (f) and mites (g), enchytraeid worms (h), millipedes (i), centipedes (j), earthworms (k), ants (l), woodlice (k), flatworms (n), and moles (o). All photographs are from the *European Atlas of Soil Biodiversity*, Joint Research Centre (JRC), courtesy of A. Jones; individual photo credits are K. Ritz (b, c), H. van Wijnen (d), E Mitchell (e), P. Henning Krog (f), D. Walter (g), J. Rombke (h), J. Mourek (i, j), D. Cluzeau (k), S Taiti (m), and H. Atter (o). From Bardgett and Van der Putten (2014).

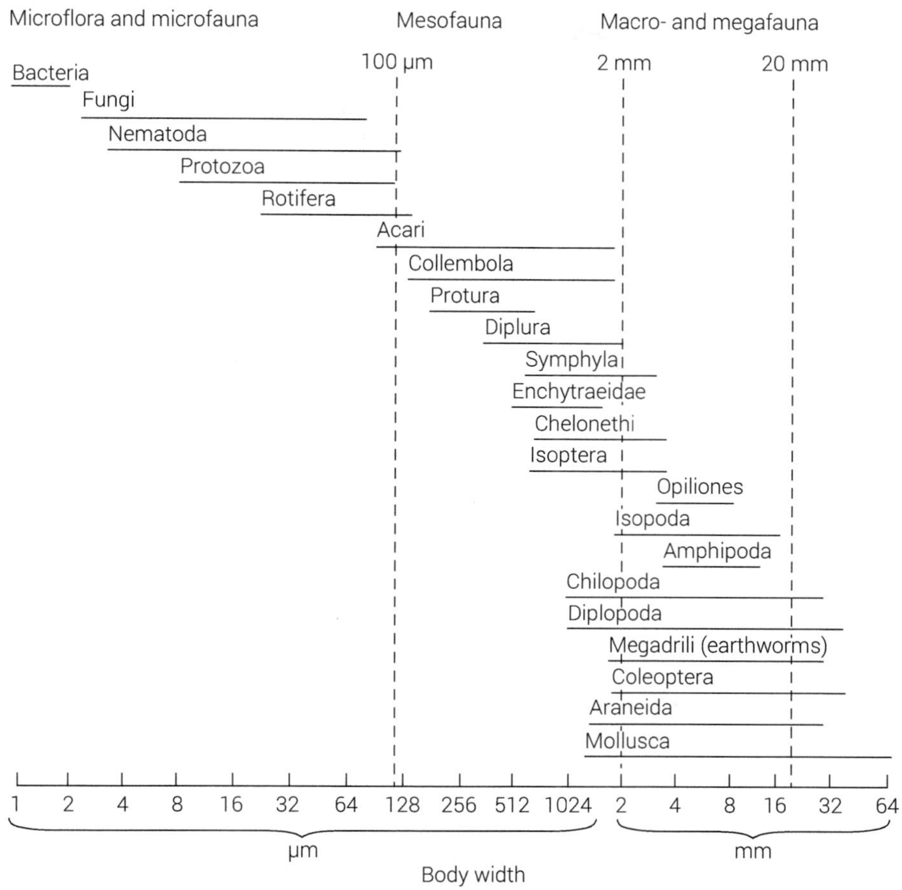

Figure 2.2 Classification of soil organism groups based on their body size. From Swift et al. (1979) by permission of John Wiley and Sons.

includes protists, which are unicellular eukaryotes that occupy many roles in soil food webs, acting as consumers of bacteria and fungi, photoautotrophs, and parasites of plants, animals, and even larger protists (Geisen et al. 2018). Microbes are by far the most numerically abundant and diverse members of the soil food web, with upward of 1,000 kg microbial biomass C ha^{-1} and tens of thousands of individual species (Roesch et al. 2007, Fierer et al. 2009, Fierer 2017). While extremely diverse, microbial communities are typically dominated by a relatively small number of phyla, with most individuals being relatively rare (Lynch and Neufeld 2015, Mahé et al. 2017, Delgado-Baquerizo et al. 2018b, Egidi et al. 2019).

Bacteria and fungi

The two most abundant groups of microbes in soil are the fungi and bacteria. Fungi are distinct from bacteria in that they are eukaryotic and produce filamentous hyphae that can effectively penetrate and explore the labyrinth of channels and pores that exist in soil (Ritz and Young 2004). In contrast, bacteria are prokaryotic and unicellular. For mobility, bacteria rely on the presence of flagella that enable them to move through water films, or if flagella are absent, on passive transport through soil via roots, fauna, or the general movement of water through soil. While fungi and bacteria dominate the soil microbial biomass, archaea can also make up a sizable portion, with

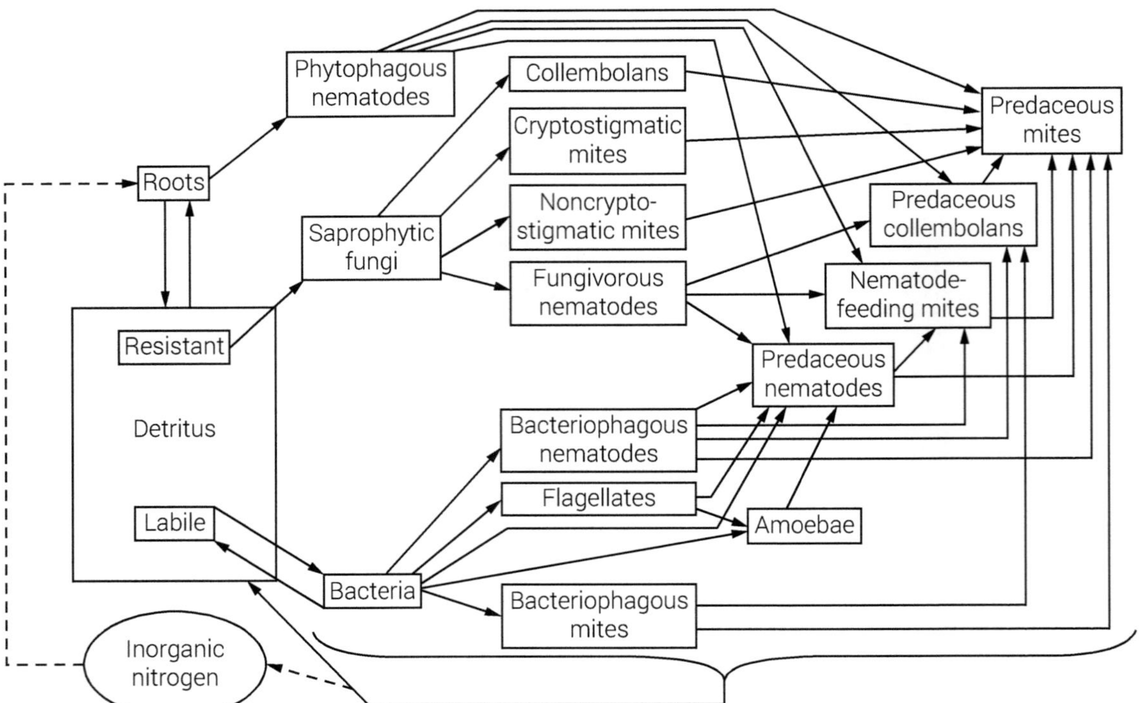

Figure 2.3 Schematic of the below-ground food web with species aggregated into functional groups based on food choice and life-history parameters. Material flows (C and N) are represented by solid arrows and net N flows are represented by dashed arrows. Plant roots, through the rhizodeposition of labile C exudates and the turnover of root cells and hairs rich in resistant cell walls, initiate the dominant material flows in the rhizosphere. Material flows to the detritus pools and inorganic N pool, for example through death and the excretion of waste products, are represented as a single flow at the base of the diagram. From Moore et al. (2003) based on Hunt et al. (1987) by permission of John Wiley and Sons.

archaeal abundance typically ranging from 0% to 10% of the total prokaryotes (Bates et al. 2011, Fierer 2017). The biomass of protists is more uncertain, although a single gram of soil can contain tens of thousands or more individuals (Adl and Gupta 2006, Geisen et al. 2018). Until relatively recently, it was impossible to detect most microbes in soil since most are unculturable using traditional culturing techniques. However, the development of culture-independent molecular methods over recent decades has revolutionised the study of the soil microbiome, enabling a more complete characterisation of the taxonomic and functional diversity of microbial life in soil. For more information on the molecular and biochemical approaches for studying soil microbial communities and their functioning the reader is referred to Paul and Frey (2023).

The primary role of fungi is to decompose organic matter, such as wood and plant litter, via the production of a vast array of extracellular enzymes; as such, fungi represent a key driver of decomposition and hence the functioning of terrestrial ecosystems. Fungi vary greatly in their decomposition ability, forming a spectrum from slow-growing fungi that are generally poor decomposers, to fast-growing, highly competitive fungi with fast rates of decomposition (Lustenhouwer et al. 2020). The biomass of fungi also varies greatly across biomes, being greatest in tundra soils, lowest in deserts, and higher in boreal forests than tropical and temperate forests (Table 2.1) (He et al. 2020). By virtue of their large biomass, they also represent a significant portion of the ecosystem C and nutrient pool. It has been estimated, for example, that fungal biomass C makes up some 1.8% of the global

Table 2.1 Biome-level fungal biomass carbon (FBC), bacterial biomass carbon (BBC), and fungal to bacterial (F:B) ratio (He et al. 2020)

Home	FBC (mg kg^{-1} soil)	BBC (mg kg^{-1} soil)	F:B ratio
Unvegetated ground	192.744^{de} (54.79–677.99)	24.604^{d} (7.68–78.72)	3.90^{bc} (2.20–6.91)
Desert	16.92^{f} (14.40–19.89)	6.83^{e} (6.10–7.65)	3.14^{bcd} (2.20–4.49)
Grassland	215.19^{de} (168.98–274.03)	62.60^{cd} (50.44–77.92)	4.03^{b} (3.52–4.62)
Pasture	632.15^{bc} (288.99–1382.81)	270.65^{a} (129.07–567.53)	2.48^{cde} (1.62–3.80)
Cropland	212.69^{de} (150.35–300.88)	65.77^{cd} (46.30–93.42)	3.28^{bcd} (2.89–3.73)
Shrub	218.14^{cde} (106.01–448.9)	45.42^{cd} (23.48–87.85)	4.82^{b} (3.72–6.25)
Savanna	103.36^{e} (60.62–176.23)	44.37^{cd} (25.94–75.92)	1.82^{e} (1.57–2.11)
Tropical/subtropical forest	451.40^{bc} (362.32–562.99)	209.96^{ab} (179.03–246.24)	2.22^{de} (1.87–2.63)
Temperate forest	258.39^{de} (189.16–352.95)	53.05^{cd} (38.71–72.70)	4.92^{b} (4.39–5.51)
Boreal forest	1234.08^{b} (870.72–1749.06)	226.37^{ab} (172.79–296.58)	5.03^{b} (4.23–5.98)
Tundra	3683.59^{a} (1678.49–8083.94)	428.37^{a} (236.98–774.31)	8.60^{a} (6.71–11.01)
Natural wetland	329.81^{cde} (194.80–558.4)	92.58^{bc} (50.99–168.10)	4.13^{b} (3.50–4.86)

Values are presented as means, with 95% confidence boundary in parentheses. Different superscript letters in one column mean significant difference at the significance level of $P = 0.05$, while the same letters indicate no significant difference.

soil organic carbon (SOC) pool, with bacteria making up around 0.6% (He et al. 2020), and in grassland, upward of 250 kg dry fungal hyphae ha^{-1} can be found in the surface soil (equivalent to over 3 km hyphae g^{-1} dry soil), containing as much as 8 kg of N and 2 kg of P (Bardgett et al. 1993a). Even greater fungal biomass is found in deciduous woodland soils, where the soil surface can be literally packed with fungal mycelium, representing a major ecosystem nutrient pool (Frankland 1982). Current data also suggest that fungal diversity is much greater in soil than other environments such as aquatic ecosystems. Even within a single handful of soil, hundreds of fungal species can be found (Peay et al. 2016) and a global survey of soil fungal diversity identified > 80,000 species (Tedersoo et al. 2014). Further, upward of a thousand fungal taxa were found in an exhaustive survey of boreal forest soil in Alaska, which equates to a fungal to plant ratio of 17:1; a global extrapolation of this ratio would suggest that there are some 6 million species of fungi found in soil (Taylor et al. 2014).

As well as degrading relatively permanent pools of organic matter in soil, extensive mycelial networks can exploit vacant resource patches that become available, as well as relocate nutrients gained from such exploitation to other parts of the mycelial network (Boddy 1999). Fungi have other important roles: not only do they act as plant pathogens, but also they bind soil particles together, thereby enhancing soil structural stability, and provide a food source to microbial-feeding fauna. Some fungi, the mycorrhizae, form mutualistic associations with plant roots, supplying growth-limiting nutrients such as N and P to the

Box 2.1 Mycorrhizal fungi

The term mycorrhiza literally means *fungus-root*, derived from the Greek *mykes* and *rhiza*, which respectively mean fungus and root. Some 95% of vascular plants belong to genera that are associated with mycorrhizal fungi, which colonise the cortial tissue of roots during active periods of plant growth. The mycorhizal association is characterised by the movement of plant-derived C to the fungus and fungal-acquired nutrients to the plant. Benefits to the plant include enhanced nutrient acquisition, especially of P, and consequently increased growth and improved fitness. Plants that have mycorrhizas also tend to be more competitive and better able to tolerate environmental stress than non-mycorrhizal ones. In return, the fungus gains a habitat that is comparatively free of competition and a supply of photosynthetic C from the plant. Three main types of mycorrhizal association are detailed below, whereas their significance for plant nutrient supply and community dynamics is considered in Chapters 3 and 4, respectively.

Arbuscular mycorrhizal (AM) fungi are most widespread and infect a wide variety of plants, including grasses, herbs, agricultural crops, and some legumes. The diagnostic feature of AM fungal associations is a highly branched arbuscule within the root cortical cells. The arbuscule is thought to be a site of exchange between the fungus and plant, an assumption that is based on the large surface area of the arbuscule interface. Some AM fungi also have vesicles, which are thin-walled, lipid-filled structures in intercellular spaces that are used for storage. External structures of AM fungi are the hyphae that extend into soil, but also spores that function as resting stages and propagules. These are produced asexually from swellings on hyphae in soil. AM fungi are classified in the order Glomales.

Ectomycorrhizal (ECM) fungi are most abundant in temperate and boreal forest ecosystems, where they infect roots of trees and shrubs, especially coniferous species that are often obligatory ECM. The diagnostic feature of ECM associations is the presence of fungal hyphae between root cortical cells to form a complex, intercellular system called the Hartig net. Some fungi also form a sheath or mantle of fungal tissue that completely covers the root. Contiguous with the mantle are hyphal strands that extend into the soil. There are over 4,000 species of ECM fungi, primarily Basidiomycetes, and more than one species can infect an individual plant.

Ericoid mycorrhizal (ERM) fungi infect ericaceous plants, for example heathers and dwarf shrubs, that commonly occur in alpine and tundra ecosystems. These are typically associated with acidic, nutrient-poor, highly organic soils. A distinguishing feature of these associations is that the fungus penetrates the root cortex cells but does not form arbuscules. The fungi also produce a loose welt of hyphae on the root surface, but no mantle is formed. The fungi involved are Ascomycetes.

plant, and in reward they gain a supply of C and space free of competition (Figure 2.4). There are three main types of mycorrhizal fungi: arbuscular mycorrhiza (AM), which are the most common type that form associations with most (~80%) vascular plant species, including herbaceous plants and trees; ectomycorrhiza (ECM), which form associations with woody plants; and ericoid mycorrhiza (ERM), which only form associations with ericaceous plants (Box 2.1). Additionally, plants are categorised based on their mycorrhizal status, being either consistently colonised (i.e., obligatory mycorrhizal), sometimes colonised (i.e., facultatively mycorrhizal), or non-mycorrhizal.

In general, the distribution of these different mycorrhizal types follows a general pattern with increasing latitude in the northern hemisphere, ranging from vegetation dominated by AM plant species at lower latitudes (grasslands), to temperate and boreal forests dominated by ECM trees, to high latitude and altitude heathlands dominated by ERM plants (Read 1991, Read and Perez-Moreno 2003) (Figure 2.5). Moreover, studies show that the abundance of facultatively mycorrhizal plant species increases with increasing latitude, whereas obligately mycorrhizal plants decrease, and non-mycorrhizal plants peak at high latitudes where the harsh climate inhibits fungal growth (Bueno et al. 2017). It should be noted, however, that other soil-borne fungi (and bacteria) also inhabit plant tissues without apparently causing harm to the host: the so-called endophytes. While the diversity and distribution of fungal and bacterial endophytes is poorly understood, they play important roles in ecosystems, for instance through producing metabolites that promote host resistance to disease (Harrison and Griffin 2020, Xia et al. 2022) and enhancing plant uptake of N (Upson et al. 2009).

Figure 2.4 Extensive ECM hyphal networks forage for nutrients and excrete a range of extracellular enzymes that degrade complex organic compounds. Image by Damian P. Donnelly and Jonathan R. Leake.

Soil bacteria are striking in that they are extremely abundant and diverse, and they are ubiquitous, inhabiting the most extreme environments and soils. As for fungi, bacterial biomass is greatest in tundra and lowest in deserts, and higher in tropical than temperate forest soils (He et al. 2020) (Table 2.1). But most startling is their diversity, which is immense: although most bacterial taxa remain uncultured (Fierer 2017, Steen et al. 2019), it has been estimated that upward of 4×10^6 taxa occur soil, and around 10,000 genetically distinct prokaryotic types can be found within a single handful of soil (Curtis et al. 2002, Torsvik et al. 2002). Their abundance is equally astounding, with as many as 10^{10}–10^{11} individual bacterial cells g^{-1} soil (Curtis and Sloan 2005). Like fungi, heterotrophic bacteria have a general role in soil of degrading a range of organic compounds via the production of extracellular enzymes. Reflecting their vast taxonomic diversity, the metabolic diversity of bacteria is equally immense, with them playing significant roles in all major biogeochemical cycles and the degradation of xenobiotic compounds that are toxic to other organisms in soil (Escalas et al. 2019).

Certain groups of bacteria also perform specific functions of major ecological importance. For example, chemoautotrophic bacteria such as nitrifiers (*Nitrosomonas* and *Nitrobacter*) are of importance for nutrient cycling since they oxidise ammonium (NH_4^+) to nitrate (NO_3^-), the process of nitrification. Symbiotic N-fixing bacteria of the genus *Rhizobium* infect roots of legumes, greatly increasing N supply to the system with important consequences for plant productivity, as do free-living, non-symbiotic heterotrophic bacteria of

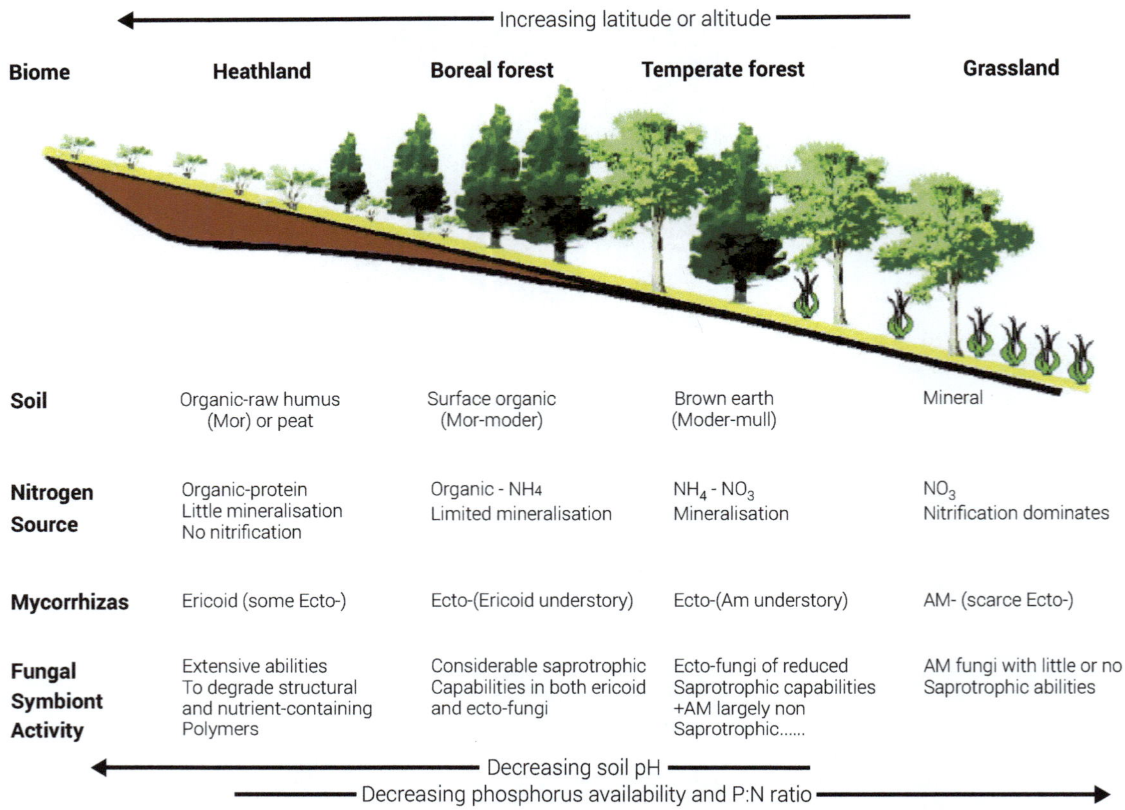

Figure 2.5 Relationship between the distribution of biomes in the northern hemisphere along environmental gradients, prevailing mycorrhizal associations, and their roles in nutrient capture. From Read and Perez-Moreno (2003) by permission of John Wiley and Sons. *New Phytologist*, Volume: 157, Issue: 3, Pages: 475–492. First published: 03 March 2003, DOI: 10.1046/j.1469-8137.2003.00704.x.

the genus *Azotobacter* and *Clostridium* that also perform the process of N fixation. Bacteria that prey on other bacteria are also common in soils and play an important, but underappreciated, functional role in regulating the structure and function of the soil microbial food web (Hungate et al. 2021). For example, the myxobacteria have a 'wolf-pack hunting' strategy, whereby they congregate as a swarm and secrete a cocktail of lysing enzymes, antibiotics, and toxins that kill their prey (Petters et al. 2021, Groß et al. 2023). These predatory bacteria also secrete fungicides that inhibit the growth of fungi and germination of fungal spores (Norén and Odham 1973). Other predatory bacteria that are widespread in soil include the *Bdellovibrio*, which penetrate the cell membrane of their prey and grow in their periplasmic space, eventually lysing and killing them, and the *Vampirovibrio*, which insert 'fang-like' cytoskeletal protrusions into their prey and extract the cytoplasm from the attacked cell (Hungate et al. 2021, Petters et al. 2021). Given the importance of microbial mortality for biogeochemical cycles (Sokol et al. 2022), it is likely that predatory bacteria play a significant role in regulating the functioning of soils.

Protists and viruses

Protists also represent a key, but often neglected, component of the soil microbiome. Not only are they ubiquitous and abundant in soil, but also they are highly diverse and perform a wide range of roles in the soil food web. The abundance of protists varies greatly across soils, but they typically reach 10^4–10^7 individuals g^{-1} soil in surface soil (Adl and Gupta 2006), declining to as low as 100 individuals g^{-1} soil in deeper soil horizons where organic matter and prey availability are low (Schnürer et al. 1986). Their true diversity is largely unknown and difficult to assess due to methodological shortcomings; unbiased methods capable of targeting the entire diversity of soil protists do not yet exist (Geisen et al. 2018, Oliverio et al. 2020). Nevertheless, studies using DNA sequence-based methods suggest that protist diversity in soil is immense (Bates et al. 2013, Mahé et al. 2017), with a single handful of soil containing hundreds of species of protists, spanning several orders of magnitude in length from a few micrometres to cells of > 1 mm (Geisen et al. 2018).

Traditionally, protists were classified based on their modes of nutrition and morphology, distinguishing photoautotrophs (microalgae) from heterotrophs (protozoa and lower fungi, *pro parte*) that ingest and digest other microorganisms. Furthermore, heterotrophic protists were classified as part of the animal kingdom. However, the classification of protists has been revised substantially in recent years to reflect their tremendous ecological and functional versatility (Figure 2.6). Most protists are consumers feeding on bacteria and fungi (Oliverio et al. 2020, Mazel et al. 2022), which play an important role in regulating nutrient fluxes and plant nutrient uptake: a large proportion (~ 60%) of nutrients ingested are excreted by protists into the soil, which is available for plant uptake—termed the microbial loop (Clarholm 1981). However, protists that feed on arthropods and other animals (i.e., parasites) also make up a significant portion of the protist community, especially in tropical forests (Mahé et al. 2017), along with protists that contain chlorophyll and serve as primary producers (traditionally called algae), saprotrophs that degrade organic matter, and protists that live in symbiosis with fungi, plants, and animals. As such, protists perform multiple roles within soil food webs through which they impact ecosystem functioning (Geisen et al. 2018).

Another neglected group of soil microganisms are the viruses, which can reach upward of 10^{10} g^{-1} soil (Williamson et al. 2017) and display an astonishing level of taxonomic and functional diversity, which has only recently become apparent using molecular tools (Fierer et al. 2007b, Starr et al. 2019, Liao et al. 2022, Graham et al. 2024). For example, a recently published *Global Soil Virus Atlas* uncovered 616,935 uncultivated viral genomes from global soil samples, although most soil viral diversity remained unexplored (Graham et al. 2024). Viruses cannot replicate independently of a living host, and, in soil, most are thought to be bacteriophages that infect bacteria and introduce their genetic material into them, essentially hijacking their metabolism to serve the purpose of viral replication. However, eukaryotes, including protists, fungi, and various soil animals, also serve as hosts for soil viruses, which similarly cause host mortality (Williamson et al. 2017, Starr et al. 2019).

While remarkably little is known about their abundance and distribution of viruses in soil, they likely play a profound, but underappreciated, role in regulating microbial life and biogeochemical cycling in soil. As in marine ecosystems (Suttle 2007, Jover et al. 2014), viruses are a major cause of microbial mortality, resulting in the release of significant amounts of microbially sourced organic C into the soil environment (Kuzyakov and Mason-Jones 2018, Nicolas et al. 2023). It was recently estimated that viruses are responsible for upward of 46% of microbial death following the wet-up of seasonally dry soils, thereby contributing significantly to microbial turnover and organic C dynamics (Nicolas et al. 2023). Moreover, these organic cell constituents released following microbial death via viral lysis will either be rapidly used by surviving microbes or become bound to soil minerals and protected from decomposition, thereby contributing to SOC sequestration and stabilisation (Kuzyakov and Mason-Jones 2018), a topic that will be considered in more detail in Chapter 7. Viruses also have many other unknown roles in soil, such as the movement of genetic material between different microbial species (i.e., horizontal gene transfer). But as noted by Williamson et al. (2017), '*we simply don't know enough*' about their impacts in soil ecosystems and more work is needed to understand their ecology and functional importance in soil.

Soil fauna

Soil fauna exists alongside microbes, feeding on them, each other, SOM, and roots. Soil is home to a tremendous variety of animals, with much variation in their feeding specialisation,

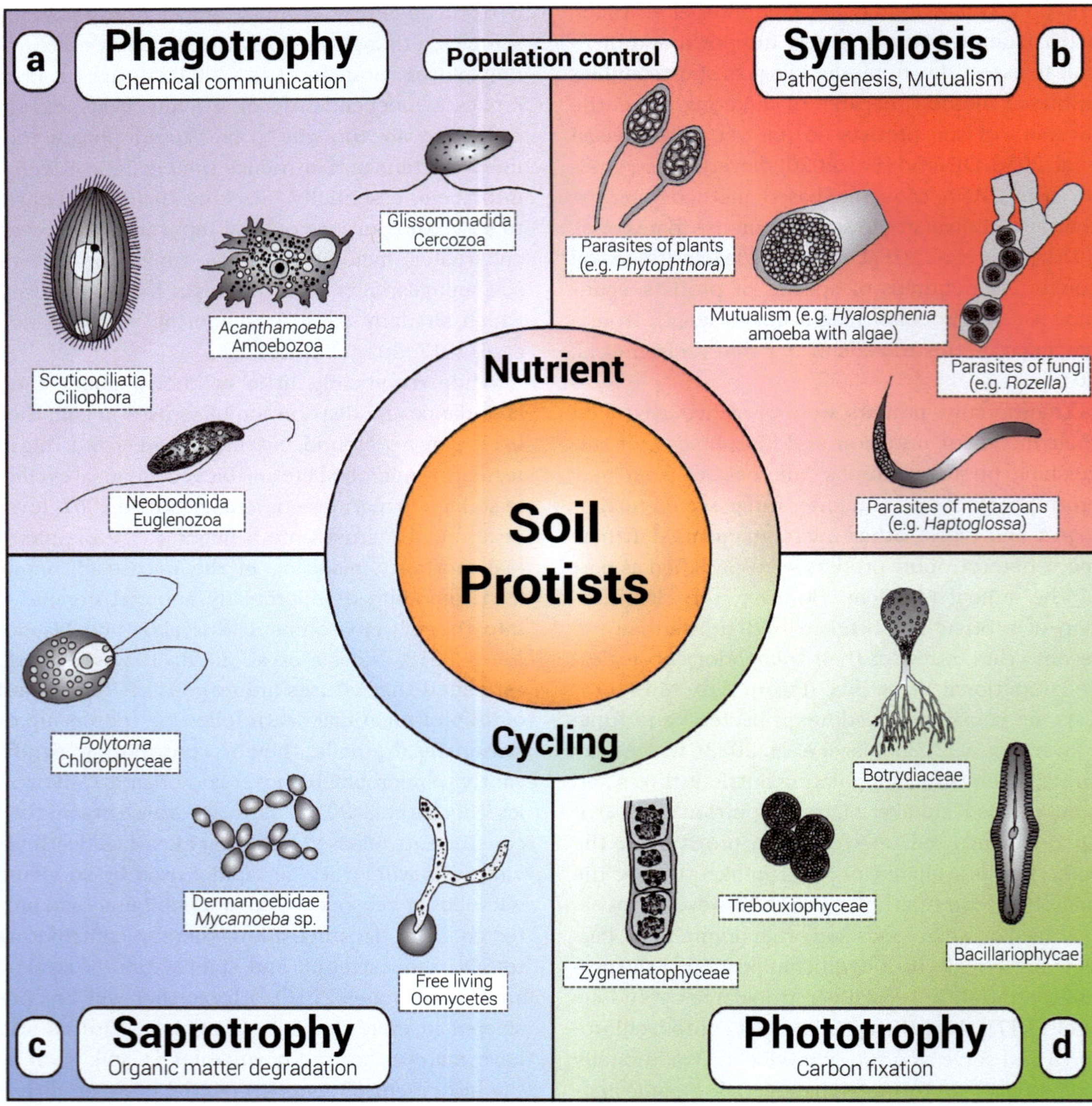

Figure 2.6 The functional and ecological versatility of protists. (a) Many soil protists are unicellular phagotrophs feeding on bacteria, whereas some feed on fungi, other protists, and animals; (b) some protists live in symbiosis, including parasitism, commensalism, and mutualism, with fungi, other protists, plants, and animals; (c) some soil protists, such as oomycetes, can participate as saprotrophs in organic matter degradation; and (d) some soil protists contain chlorophyll (usually called algae) and can be phototrophic or mixotrophic. From Geisen et al. (2018) by permission of Oxford University Press. *FEMS Microbiol Rev*, Volume 42, Issue 3, May 2018, Pages 293–323, https://doi.org/10.1093/femsre/fuy006.

life history strategies, and spatial distribution. As with microbes, species of soil animals also differ markedly in their response to abiotic factors and show much interspecific variation in their tolerance to soil moisture, food availability, and drought (Bardgett and van der Putten 2014), which may determine the species-specific spatial and temporal distribution of animals in soil. Furthermore, many soil animals can undergo periods of inactivity when conditions are unfavourable, thereby allowing them to tolerate periods of harsh soil conditions. All these factors contribute to making the study of soil animals very challenging, in terms of both their sampling and their characterisation into functional groups. As already mentioned, the simplest way of classifying soil animals is based on their body size, ranging from the smallest nematodes, which typically vary from 5 to 80 μm in body diameter (Andriuzzi and Wall 2018a), to large macrofauna, such as earthworms, termites, ants, and moles. This section considers some of the important groups of organisms within each of the body size categories, namely microfauna (< 0.1 mm), mesofauna (0.1–2.0 mm), and macrofauna (> 2 mm).

Microfauna

The most abundant group of microfauna are the nematodes, or roundworms. Soil nematodes live in water films around soil particles, in which they move, feed, and reproduce, although they have evolved mechanisms to survive unfavourable conditions, such as during extremely dry conditions (cryptobiosis, a complete cessation of metabolism, and dormant juvenile stages). Nematodes are generally grouped according to the type of food that they consume, based on the morphology of their mouthparts (Figure 2.7). The most common feeding groups of nematodes are plant feeders, fungal feeders, bacterial feeders, omnivores, and predatory nematodes (Yeates et al. 1993). As with protists, the role of nematodes in soil systems relates mainly to their feeding activities, and can be direct, for instance via the excretion of nutrients into soil (i.e., the microbial loop), or indirect, through altering the size, composition, and activity of microbial communities. They also play an important role in the inoculation of new substrates in soil through the transport of microbes adhering to them and the excretion of viable microbes. Through feeding on plant roots, plant-parasitic nematodes also directly impact plant production and community structure, as will be discussed in later chapters.

Nematodes are the most numerous multicellular animals in soil: an estimated $4.4 \pm 0.64 \times 10^{20}$ nematodes inhabit the upper layer of soils across the globe (van den Hoogen et al. 2019) and densities can reach 30 million m^{-2} in deciduous forest (Volz 1951) and 10 million m^{-2} in productive grasslands (Yeates et al. 1997). Even in soils severely polluted with heavy metals, densities can reach 260,000 m^{-2} (Bardgett et al. 1994). Regarding their diversity, a review of global data revealed that nematode species richness was greatest in temperate broadleaf forest at 61.7 nematode species per sample (Boag and Yeates 1998). Also, a maximum of 89 nematode species were identified from a single soil core taken from a tropical forest in Cameroon, and a total of 431 species and 194 genera were identified from an intensive sampling of 24 sites (Bloemers et al. 1997). Likewise, 112 nematode species were identified in a single soil core taken from a beech forest in Romania (Popovici 1989) and 154 species were found in a chalk grassland in the UK (Hodda and Wanless 1994). Many species of nematodes also inhabit soils of infertile ecosystems: a study of a New Zealand sand dunes yielded 44 nematode species under a single coloniser plant, *Ammophila arenaria* (Yeates 1968), and 27 nematode species were found in unvegetated sand dunes in Scotland

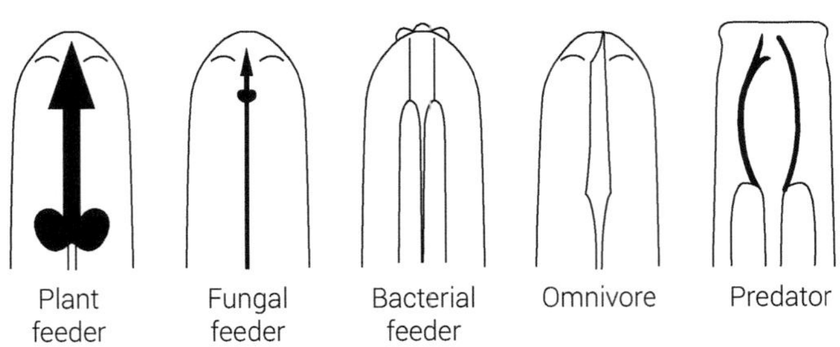

Figure 2.7 Schematic of mouthparts of nematode feeding groups commonly found in soil. From Bardgett (2005) redrawn from Bardgett and Griffiths (1997) by permission of Oxford University Press.

(a) (b)

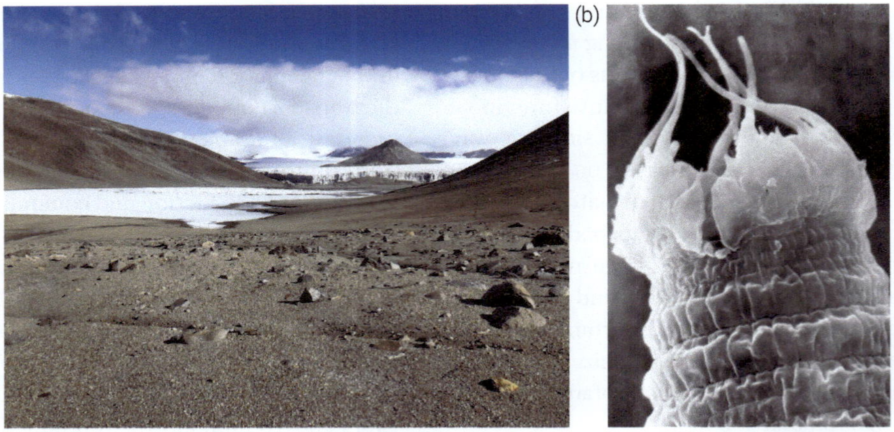

Figure 2.8 Antarctic Dry Valleys (a) and the dominant species of nematode, *Scottnema lindsayae* (b), which plays a significant role in soil C cycling in these harsh ecosystems. Images respectively by Diana Wall and Manuel Mundo.

(Wall et al. 2002). Even in soils of Antarctic Dry Valleys, the coldest and driest place on Earth, five genera of nematodes exist, including the dominant species *Scottnema lindsayae*, which has a unique tolerance to low soil moisture and high salinity and plays a key role in soil C cycling in Antarctica (Virginia and Wall 1999, Adams et al. 2014) (Figure 2.8).

Two less-studied groups of soil microfauna that similarly have a remarkable ability to survive harsh environmental conditions are rotifers and tardigrades. Rotifers are microscopic (typically 0.1–0.5 mm body length, although they range in lengthfrom 0.05 to 2 mm), transparent-bodied animals that live in water films in the soil and in mosses and litter, reaching abundances upward of 100,000 m^{-2} in wet, organic soils (Wallace 2002). Like nematodes, they feed on bacteria and fungi, as well as protists and other rotifers, and while they live and move in water films, they can survive extended periods in a desiccated state, a phenomenon called anhydrobiosis (Orgiazzi et al. 2016). Tardigrades, known as water bears, are also microscopic (typically 0.05–1.2 mm in body length), live in water films, and can reach abundances upward of 75,000 individuals m^{-2} in forest soils (He et al. 2024). They are ubiquitous in their distribution, inhabiting virtually all terrestrial, freshwater, and marine ecosystems, and, along with rotifers and nematodes, constitute a major component of the soil food web of extreme environments, such as Antarctica (Adams et al. 2006, Smykla et al. 2012). As with rotifers, little is known about their ecological roles. But they occupy a wide range of trophic levels, feeding on microbes, plants, protists, rotifers, and even nematodes (Potapov et al. 2022). As such, they likely perform a wide range of functions in soils and terrestrial ecosystems.

Mesofauna

The microarthropods are the best known of the mesofauna. The two most abundant microarthropod groups in soil are the Collembola, which are small (typically 1–5 mm in length), wingless hexapoda, with six abdominal segments and biting mouthparts, and the Acari (mites), which have sac-like bodies that may be divided into anterior and posterior sections (Figure 2.9).

The taxonomy of Acari is more controversial, but the major groups that dominate in the soil belong to the orders Crytostigmata, Mesostigmata, Prostigmata, and Astigmata. Microarthropods are the most abundant non-aquatic faunal groups in soils of most ecosystems. Densities of soil microarthropods can be as high as 300,000 m^{-2} in permanent grassland with dense root systems and high organic matter content (Bardgett and Griffiths 1997). In temperate grasslands, the biomass of mites and collembolans is often reported to be similar, whereas in tropical grasslands the biomass of mites can be 2–5 times that of Collembola (Petersen and Luxton

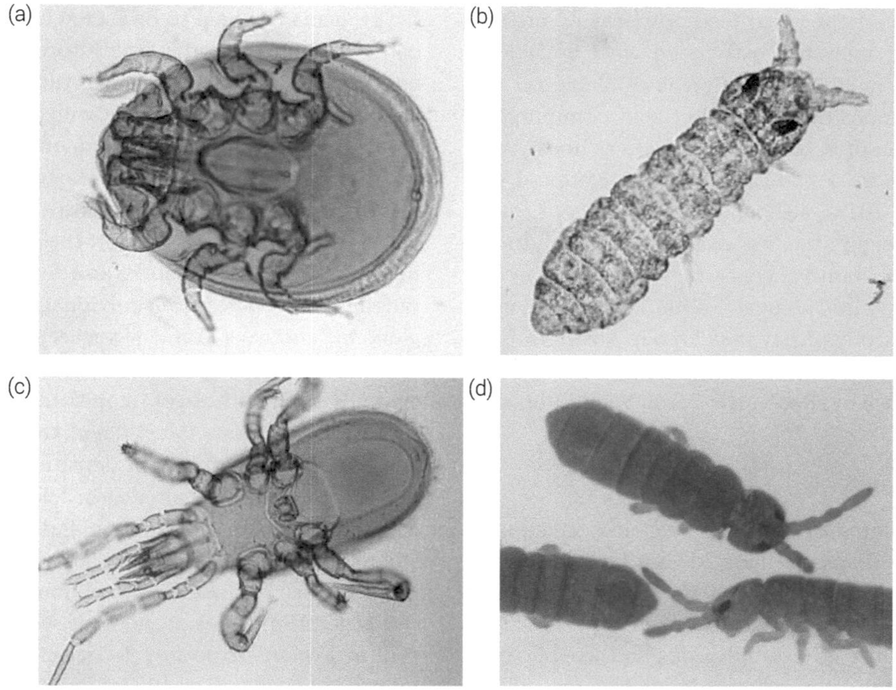

Figure 2.9 A selection of soil microarthropods extracted from a temperate grassland soil in the UK. (a) The mite *Uropoda minima*; (b) the collembolan *Friesea mirabilis*; (c) the mite *Lysigamasus truncus*; and (d) the collembolan *Isotoma anglicana*. Images by Lisa Cole.

1982). The biomass of Collembola varies considerably across latitudes, with global datasets showing that maximum densities occur in tundra (median = 131,422 individuals m^{-2}) and minimum densities occur in tropical forests (median = 5,831 m^{-2}) and agricultural ecosystems (median = 3,438 m^{-2}) (Potapov et al. 2023). Many records of microarthropod diversity also exist. For example, a total of 396 mite species were recorded across 35 sites in the North American arctic and subarctic (Behan-Pelletier 1978), and 250 species of mites within 1 m^2 of litter and soil, to just 5 cm depth, in mixed deciduous/coniferous forest in Quebec, Canada (Behan-Pelletier, personal communication). Also, a total of 108 species of microarthropods were found in 500 cm^2 of soil of an unmanaged Dutch grassland (Siepel and Van de Bund 1988), and 58 prostigmatid and 59 oribatid species of mites were found in a Danish pasture (Weis-Fogh 1948). Lower diversities of Collembola have been reported for other grassland systems: 27 species of Collembola were found in tallgrass prairie in Illinois (Brand and Dunn 1998), whereas only 12 species were found in low productivity, acid grassland in Scotland (Cole et al. 2006). Moreover, a recent analysis of global data on Collembola showed that local species richness was highest in tropical forest (mean = 36.6 species per site) and lowest in temperate agricultural (19.5 species per site) and grassland ecosystems (22.8 species per site) (Potapov et al. 2023).

Another notable, but less widely studied group of mesofauna are the enchytraeids (Oligochaeta), commonly known as potworms (Figure 2.1h). These animals are considered keystone species in acid mor soils, such as those of boreal forest, tundra, and heathland, where they can reach densities of up to 400,000 individuals m^{-2}, making up some 75% of the total faunal biomass (Didden 1993). Much lower densities of 4,000 m^{-2} are found in agricultural and mull forest soils, where earthworms typically dominate the faunal biomass (Didden 1993). They are less conspicuous because of their smaller size than other soil annelids like earthworms, being only 1–5 mm long. Their main food source is fungal mycelium, but they also feed on decaying organic matter and microbes associated with it (Didden 1993). Not much information is available on the species composition of enchytraeid communities across different habitats, but species richness typically ranges from 1 to 15 species (Bardgett and Van der Putten 2014), although higher levels of diversity have been found. For example, 23 enchytraeid species were found across a range of Czech mountain forest soils (Chalupský 1995), 28 species were found in a study of hay meadows in northern England (Standen 1984), and 35 species were recorded across a range of sites in the Cerna Valley, Romania (Ştefan 1977).

Macrofauna

The macrofauna are the most conspicuous fauna of soil, and perhaps the most documented in terms of their biology and impact on soil fertility. They include a wide array of taxonomic groups that cross several trophic levels: millipedes and woodlice are major consumers of organic debris in forests; dung beetles are especially abundant in tropical forests and savannas, where they transfer large quantities of vertebrate faeces into soil, with major effects on nutrient cycling and plant growth; various insect larvae, such as those of the cranefly (Tipulidae), are important consumers of root material; whereas centipedes, spiders, scorpions, and beetles tend to be the dominant predators in soil and litter. Ants are another group of soil macrofauna that are amongst the most ubiquitous insects on the planet. They have a variety of feeding habits, acting as predators, microbial and plant feeders, and even opportunistic omnivores. In the biomes where ants are abundant, they affect many soil processes through their nesting activities. These nests can be > 2 m in height and at densities of between 200 and 1,000 mounds ha^{-1}, and they contribute greatly to the creation of patch mosaics that characterise the soils and vegetation of many landscapes (Figure 2.10). They also greatly affect critical ecosystem processes such as nutrient cycling and water redistribution.

Earthworms are probably the best known of the macrofauna and there is a huge literature on their biology and importance for soil fertility in natural and agricultural systems (Lee 1985, Edwards and Bohlen 1996, Edwards 2004, Edwards et al. 2020). Indeed, Cleopatra (69–30 BC), the last pharaoh of Egypt, declared them to be sacred because of their contribution to Egyptian agriculture, and literature on their importance for soil fertility dates as far back as Aristotle (384–322 BC), who called them the '*intestines of the earth*'. The importance of earthworms for soil fertility was also the topic of Darwin's last book, published in 1881 (Darwin 1881). Earthworms usually make up most of the animal biomass in soil, with total abundances and diversity typically ranging between 5 and 50 individuals m^{-2} across the globe with between 0 and 4 species per site (Phillips et al. 2019) (Figure 2.11). However, in highly fertile soils, such as those of temperate grasslands and pastures, deciduous forests, and tropical pastures and rainforests, earthworm densities can reach as high as 400–500 m^{-2}, whereas in acid mor soils typical of boreal forest and heathland and in deserts where water is limiting they are virtually absent. A review of earthworm species found under various vegetation types around the world reported a typical diversity to be only 2–5 species, with a maximum of 11 species (Lee 1985). However, a worldwide survey of earthworms in tropical regions reported that a total of 51 exotic and 151 native species are commonly found in tropical agroecosystems (Fragoso et al. 1999).

Earthworms are typically divided into three groupings based on their behaviour: epigeic earthworms live in and feed on leaf litter on the soil surface, and play a key role in the breakdown and decomposition of these organic materials; endogeic earthworms live in and feed on soil, mixing mineral and organic material and creating networks of burrows that improve soil structure; and anecic earthworms burrow deep into soil and feed on leaf litter on the soil surface, which they mix with mineral particles in their gut and egest as surface or subsurface casts. They also create large vertical burrows that increase water infiltration and macropores that improve soil porosity.

Another group of macrofauna that are considered keystone species in the tropics and subtropics are the termites. As with earthworms, much has been written about their biology and important roles in these ecosystems (Abe et al. 2000, Bignell et al. 2011). Termites feed mainly on deadwood, decaying organic matter, including the dung of large mammals, and roots, and are best known for their dramatic effects on soil nutrient

Figure 2.10 Large ant nest formed by the mound-forming ant *Formica aquilonia*, a red wood ant species of the *Formica rufa* group, in boreal forest, northern Sweden, overlooked by Finnish soil ecologist Heikki Setälä. Image by Richard Bardgett.

cycling and the physical architecture and hydrology of soils, through their mound building and gallery excavating activities (Figure 2.12). Moreover, in tropical rainforests, termites are responsible for most of the deadwood decomposition, and hence represent a large, but often overlooked, contributor to the global C cycle (Griffiths et al. 2019).

Most termites depend on symbiotic gut bacteria for energy metabolism, whereas others, the so-called fungus growers, lack gut flora and rely on externally cultivated basidiomycete fungi for their nutrition. Some termites also contain populations of protists within their guts that can degrade cellulose and other plant polysaccharides. Others have the capacity to fix atmospheric N through the action of their gut bacteria, and as a result they are often the primary providers of N in many soils and food chains, with termite mounds serving as foci for nutrient redistribution in some landscapes. The effects of termites on vegetation are most pronounced in termite savanna ecosystems, a term coined for vegetation formations in Africa and South America that consist of grasslands with islands of woodland based on large termite mounds (Harris 1964).

Termite abundance varies from less than 50 individuals m^{-2} in arid savannas to more than 7,000 m^{-2} in some African forests, where they make up some 95% of the insect biomass in soil (Aber et al. 2000), and upward of 10,000 individuals m^{-2} can be found in some tropical forests (Eggleton et al. 1996). More than 2,300 species have been described in 270 genera and 6 families, although new species are being recorded all the time (Aber et al. 2000). However, spatial patterns of termite diversity are variable, ranging from around 10 species in semi-arid savanna to upward of 100 species in tropical rainforest (Casalla and Korb 2019), and they are strongly influenced by factors such as latitude, being maximal at the equator, and climate, with both termite abundance and diversity being greater in regions of

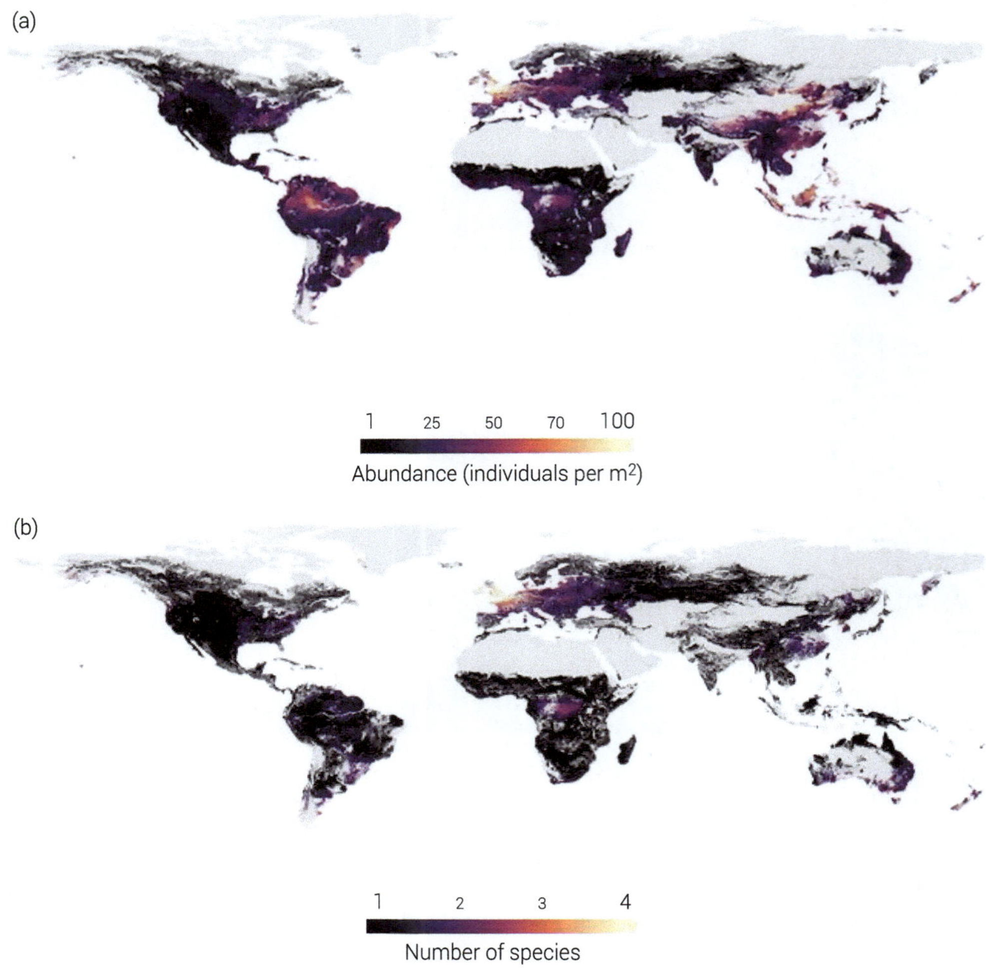

Figure 2.11 Predicted global distribution of earthworm abundance (a) and species richness (b). Yellow indicates high abundance and richness; dark purple, low abundance and richness; grey areas lack samples. From Phillips et al. (2019) by permission of the American Association for the Advancement of Science.

higher rainfall (Casalla and Korb 2019, Aber et al. 2000).

Patterns of soil biodiversity

One of the fascinations of biological diversity is that it is not the same everywhere. There is almost infinite variation in patterns of diversity across and within the Earth's ecosystems. Furthermore, these patterns change with time because of processes of succession, ecosystem development, and evolutionary change. This tremendous spatial and temporal complexity in diversity, which occurs both above-ground and below-ground, makes it difficult to evaluate and explain patterns of species richness. However, carefully framed questions about species diversity, which recognise the existence of such complexity, can yield insights into the wide range of processes that control diversity on a range of spatial and temporal scales. This section examines some of what is known about patterns of diversity in soils, at global, landscape, and local

(a)  (b) 

Figure 2.12 Termite mound in savanna, Lolldaiga Hills, Kenya (a) formed by termites of the genus *Macrotermes* (b), which are widely distributed throughout Africa and South East Asia. Images respectively by Richard Bardgett and Kate Parr.

scales, and discusses how soil biological diversity changes with time.

Global patterns of soil biodiversity

A host of global patterns of spatial variation in biodiversity have been explored, but perhaps the most prominent and intensively studied are latitudinal gradients of species richness. A long-standing view in ecology is that species richness is maximal in the tropics and gradually declines towards the poles (Huston 1994). However, information on latitudinal gradients of taxonomic groups is mostly based on above-ground organisms, with much less being known about patterns of species diversity in soil; indeed, major taxonomic and geographic '*blind spots*' exist in the availability of soil biodiversity data across the globe (Cameron et al. 2018, Guerra et al. 2020).

The few studies that have explored global patterns of soil biodiversity indicate that the composition of soil communities varies markedly across biomes (e.g., Petersen and Luxton 1982, Fierer and Jackson 2006, Fierer et al. 2009, Wu et al. 2011, Bates et al. 2013, Delgado-Baquerizo et al. 2018b, Guerra et al. 2022). However, available data suggest that relationships between latitude and species richness for most soil organisms differ from those that occur above-ground. Studies of soil nematode diversity, for example, reveal that species richness is similarly high along most of the latitudinal gradient, although it peaks in temperate regions where soil resources (e.g., nutrients and moisture) are relatively high and drops markedly towards the poles due to low resource availability and harsh environmental conditions (Boag and Yeates 1998, Nielsen et al. 2014, Song et al. 2017). The diversity of oribatid mites at the country level has also been shown to be greatest in temperate and boreal zones (Maraun et al. 2007), and the local diversity and abundance of earthworms typically peaks at mid latitudes (Figure 2.11), which contrasts with most above-ground organisms (Phillips et al. 2019). Collembolan diversity similarly deviates from latitudinal gradients reported for above-ground diversity, being similarly high in tropics, temperate regions, and even tundra (Potapov et al. 2023). Soil microbial diversity also displays dramatic departures from patterns found above-ground. For example, while the overall richness of soil fungi increases towards the equator, ECM fungi and several fungal classes run counter to this, being most diverse in temperate and boreal ecosystems (Tedersoo et al. 2014), and soil bacterial diversity peaks in temperate soils (Bahram et al. 2018, Guerra et al. 2022). In fact, the only group of soil organisms that does seem to follow a clear latitudinal gradient of increasing in diversity towards the tropics are termites (Eggleton and Bignell 1995) (Figure 2.13); for other soil organism groups, available data point to a lack of coupling with above-ground diversity at global scales, with soil biodiversity generally peaking in temperate to boreal regions of the world.

Several studies have also attempted to identify the dominant drivers of soil biodiversity patterns at global scales, showing that soil organism

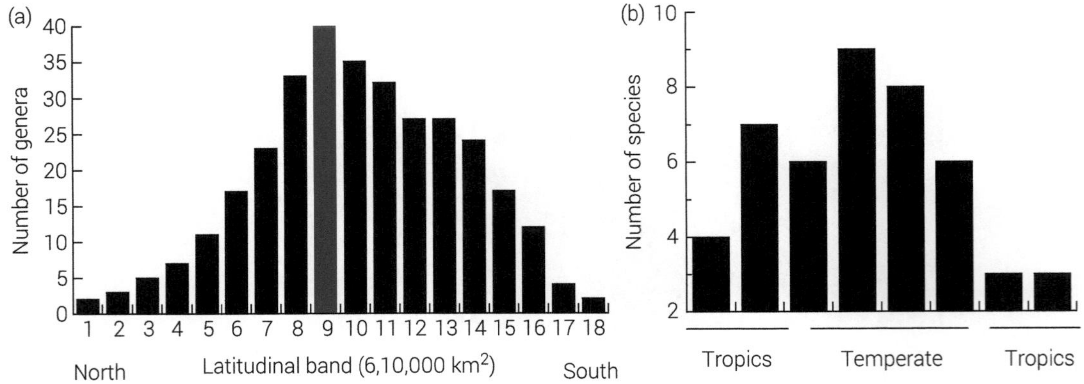

Figure 2.13 Latitudinal patterns of (a) termites (data from Eggleton and Bignell 1995) and (b) earthworms (data from Lavelle et al. 1995). From Bardgett (2005) by permission of Oxford University Press.

groups differ in their environmental preferences. For example, global- and continental-scale patterns of bacterial diversity are mainly attributed to variation in soil abiotic properties, being greatest in ecosystems with neutral pH soils and/or high resource availability and lowest in ecosystems with acidic soils (Fierer and Jackson 2006, Delgado-Baquerizo et al. 2018b). Soil water content also acts as an important driver of global patterns of bacterial diversity, being greatest at intermediate soil water contents, where isolated aqueous habitats are numerous (Bickel and Or 2020). This is especially important within drylands, which cover almost half of the global land surface, wherein increases in aridity can result in significant declines in bacterial diversity (Maestre et al. 2015). Climatic factors, especially temperature, followed by soil abiotic properties, have been shown to be the best predictors of soil fungal richness and community composition at the global scale (Tedersoo et al. 2014). Climate also appears to be the predominant factor explaining global-scale patterns of protist community composition (Oliverio et al. 2020), Collembola diversity (Potapov et al. 2023), and earthworm diversity and abundance (Phillips et al. 2019), suggesting that these soil organisms are highly vulnerable to climate change. In contrast, soil abiotic properties (e.g., texture, SOC, pH, and cation exchange capacity) have been shown to be the predominant factors driving nematode abundance at a global scale, with effects that largely overwhelmed those of climate (van der Hoogen et al. 2019). Vegetation also plays an important role in shaping soil invertebrate communities at global scales, with studies indicating that reductions in forest cover and plant diversity and declines in plant production associated with increases in aridity could negatively impact the diversity of soil invertebrates globally (Bastida et al. 2020).

Historically, soil microbial ecologists held the view of Baas Becking (1934) that '*everything is everywhere, and the environment selects*'. The rationale for this is that many soil organisms, especially free-living microbes, are cosmopolitan, being able to migrate extensively across the globe (Roberts and Cohen 1995, Finlay et al. 1999). Moreover, it has been argued that the abundance of individuals of free-living microbial species is so great that their dispersal is rarely (if ever) restricted by geographical barriers (Finlay 2002). Evidence for the cosmopolitan nature of soil organisms, for example, comes from the discovery that approximately one-third of the global diversity of soil protozoa was present within a single patch of grassland in Scotland, and that those species that were locally rare were similarly rare on a global scale (Figure 2.14) (Finlay et al. 2001, Finlay 2002). However, a growing number of studies point to a high degree of endemism among soil organisms, thereby challenging this view. For instance, a global study of the distribution of soil animals found very little overlap in species composition across different biomes

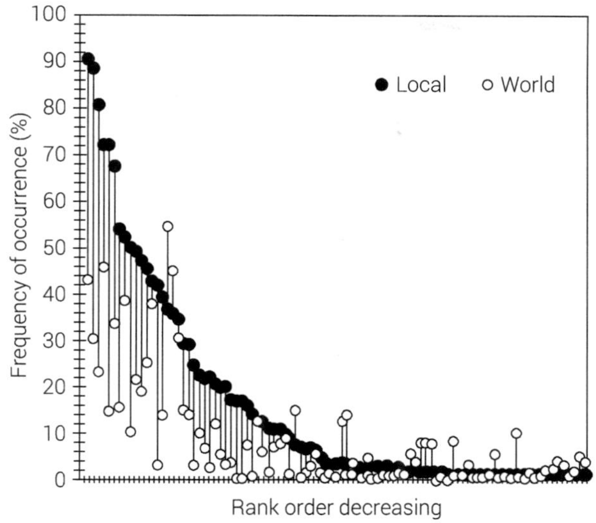

Figure 2.14 Frequency of detection of 95 ciliate species in 150 soil samples from 1 ha of grassland in Scotland and of the same species in 606 samples taken worldwide, showing a significant agreement between local and global diversity; i.e., that those species that were locally rare were similarly rare on a global scale. From Finlay et al. (2001) by permission of Elsevier.

of the world, suggesting that there is a high degree of endemism among soil animals (Wu et al. 2011). Further, while some soil microbial taxa are ubiquitous in their distribution (Delgado-Baquerizo et al. 2018b), studies show that others have restricted distributions and a high degree of endemicity (Barberán et al. 2014), and that dispersal limitation can contribute to regional endemism of certain soil bacteria (Cho and Tiedje 2000, Fulthorpe et al. 2008). Moreover, a large-scale soil sampling effort across France revealed considerable spatial turnover of soil bacterial diversity related to soil habitat heterogeneity and, to some extent, dispersal limitation (Ranjard et al. 2013), again calling into question the argument that '*everything is everywhere, and the environment selects*'.

Soil fungal communities have also been shown to display a high degree of endemism (Talbot et al. 2014), with endemicity peaking in tropical habitats, including Amazonia, Yucatan, West Central Africa, Sri Lanka, and New Caledonia (Figure 2.15) (Tedersoo et al. 2022). Consistent with this, it has been suggested that most fungal species do not have cosmopolitan distributions, and patterns of fungal diversity at the species level are strongly influenced by biogeographical factors, such as climate and geographic isolation (Peay et al. 2016). Nevertheless, in drawing such conclusions it should be noted that estimates of the global distribution of soil organisms, and hence the degree of endemism, remain highly speculative because only a small fraction of habitats across the globe have been sampled. There is also a strong sampling bias to sites in the northern hemisphere, and in many parts of the world, such as Africa, South America, and India, soil biodiversity data are scarce, and in many cases nonexistent (Guerra et al. 2020). As such, it has been suggested that greater sampling depth could reveal that endemism in soil communities is stronger than suspected, but also it might reveal more sharing of soil microbial species across continents, meaning that endemism is less strong than data suggest (Peay et al. 2016).

Field- and ecosystem-scale patterns of soil biodiversity

The patterning of soil biodiversity at the field and ecosystem scale, which cover distances of tens of metres to kilometres, is mostly related to spatial patterns of soil abiotic properties and topography,

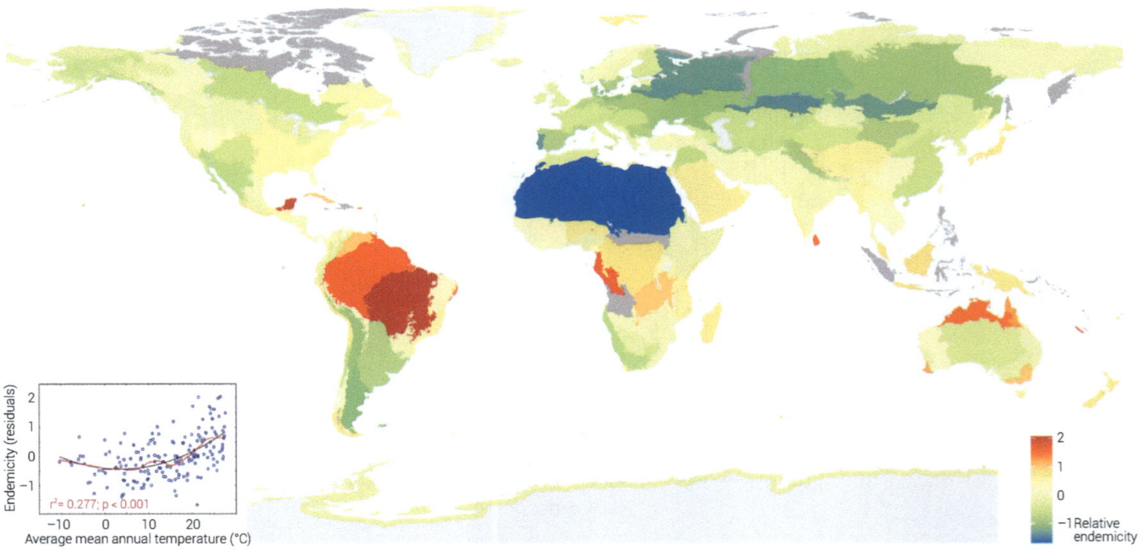

Figure 2.15 Survey of endemicity of soil fungi across ecoregions or the world based on available fungal sequence data. The inset graph indicates that endemicity has a positive, U-shaped relationship with mean annual air temperature, the best predictor of endemicity, and grey ecoregions were excluded because of insufficient data availability. From Tedersoo et al. (2022). *Global Change Biology*, Volume: 28, Issue: 22, Pages: 6696–6710. First published: 19 August 2022, DOI: (10.1111/gcb.16398).

as well as vegetation and anthropogenic disturbance. At the broadest level, the characterisation of soil into mull and mor humus types provides a general framework for understanding how and why soil biodiversity varies at the ecosystem scale (Ponge 2003). Specifically, fertile mull soils associated with productive ecosystems, such as deciduous forest and temperate grassland, tend to support a high level of diversity due to ample provision of resources, neutral pH, and a high level of microscale heterogeneity caused largely by the activities of the soil organisms themselves. In contrast, mor type soils, typical of unproductive ecosystems such as coniferous forests and heathlands, are acidic in nature and rich in recalcitrant organic matter, and, as such, they support fewer animal and microbial species, with key groups of fauna, such as earthworms, being absent (Ponge 2003). This kind of patterning is also evident within ecosystems, for instance in forests, where differences in litter quality beneath coexisting tree species produce zones of influence that shape the distribution of soil organisms and ecosystem processes (Saetre and Bååth 2000, Weber and Bardgett 2011). Similarly, in semiarid ecosystems, patterns of soil biodiversity are strongly related to the patchiness of vegetation, with plants serving as resource islands that support a greater diversity of soil organisms than in adjacent vegetation-free zones (Herman et al. 1995). In ecosystems where vegetation is sparse, however, abiotic factors play a prominent role in the patterning of soil biodiversity. For example, in the McMurdo Dry Valleys of Antarctica, spatial patterns of soil nematode diversity are explained by abiotic factors such as organic matter, salts, and soil moisture (Courtright et al. 2001), although biotic interactions, such as competition for resources among nematode species, also play a role (Caruso et al. 2019a).

A factor that is of increasing importance for shaping soil biodiversity patterns at the field and ecosystem scale is disturbance from human activities, especially the conversion of natural habitats to agriculture and agricultural intensification, which generally have negative impacts on soil biodiversity (Phillips et al. 2024). There are numerous examples of this. Conventional tillage is one such example, which causes physical disturbance to soil, can severely reduce the abundance and richness of earthworms (Springett 1992,

Briones and Schmidt 2017), with epigeic and larger, deep-burrowing worms (e.g., *Lumbricus terrestris*) being especially vulnerable (Briones and Schmidt 2017). Intensification of agricultural management has also been shown to consistently reduce the diversity and complexity of soil food webs, with larger-bodied soil animals (i.e., earthworms and microarthropods) and higher trophic levels (i.e., predaceous mites) being especially vulnerable (Tsiafouli et al. 2014). The conversion of primary tropical forest to agriculture has also been widely shown to reduce the abundance and diversity of macrofauna (Lavelle and Pashanasi 1989, Franco et al. 2019b), and the diversity of termites (Eggleton et al. 2002), microathropods (Tian et al. 1992), and nematodes (Bloemers et al. 1997) (Figure 2.16). Similarly, tropical rainforest conversion to rubber and oil palm plantations has been shown to reduce the diversity of soil invertebrates but increases their total biomass due to an increase in the biomass of earthworms (Potapov et al. 2019). The conversion of native prairie grassland to agriculture similarly reduces the diversity of soil nematodes (Yeates and King 1997) and Collembola (Brand and Dunn 1998), and earthworm communities of agroecosystems are less species rich, with fewer native species than those of undisturbed ecosystems (Fragoso et al. 1997, Dlamini and Haynes 2004).

The response of soil microbial diversity to land use change and intensification is more variable and often diverges from that of soil animals. An extensive survey of different ecosystem types across Wales, UK, for example, revealed that local richness of soil microbes (i.e., bacteria, fungi, and protists, but not archaea) was greatest in the most intensively managed systems, whereas the converse was true for soil animal richness, which was reduced by intensive land use (George et al. 2019). A large-scale survey of sites across Europe also revealed that bacterial and fungal diversity were greater in more disturbed agricultural soils than in less disturbed woodland soils (Labouyrie et al. 2023), and the local richness of soil bacterial communities was found to increase following tropical forest deforestation (Rodrigues et al. 2013), as does the diversity of soil microbial decomposers in response to land use intensification in grasslands (Gossner et al. 2016). However, some microbial groups are vulnerable to land use change and intensification, which also causes major shifts in soil microbial communities. For example, a recent global synthesis indicated that while the conversion of native forests to plantations, grasslands, and croplands generally increases bacterial diversity, it results in less diverse and more homogenous fungal communities dominated by pathogens and with lower abundance of symbionts (Qu et al. 2024)

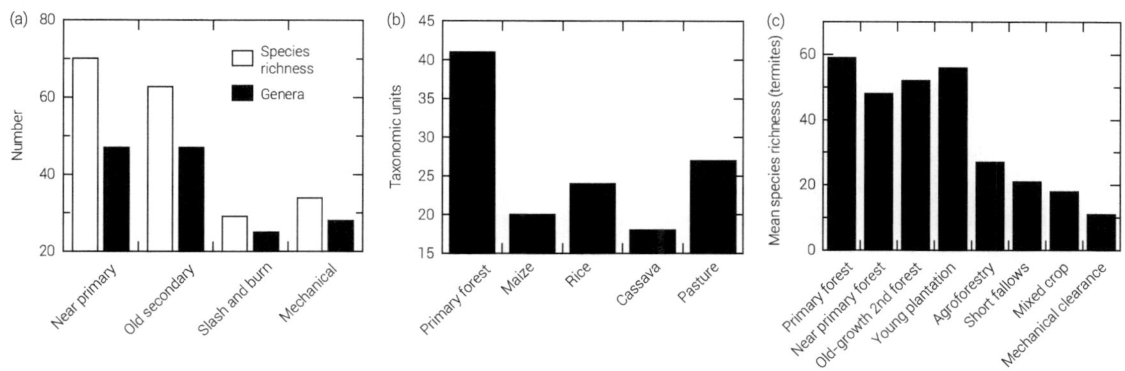

Figure 2.16 Anthropogenic disturbances and the diversity of soil fauna: (a) effects of forest disturbances on the number of species and genera of nematodes in soil of tropical forests, Cameroon (data from Bloemers et al. 1997); (b) effects of land management on the number of taxonomic units of macrofauna in Peruvian Amazonia (data from Lavelle and Pashanasi 1989); and (c) termite diversity across a gradient of anthropogenic disturbance in the humid forest zone of West Africa (data from Eggleton et al. 2002). From Bardgett (2005) by permission of Oxford University Press.

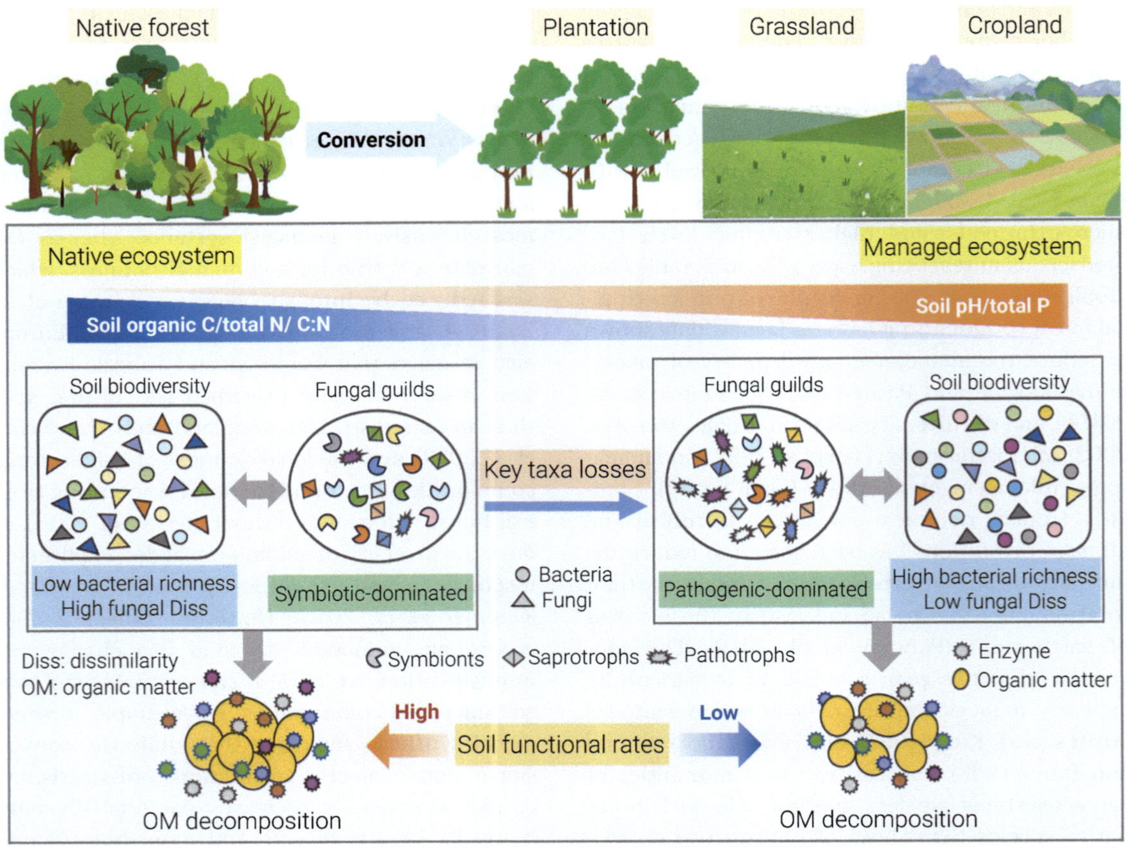

Figure 2.17 The conversion of native forests to plantations, grasslands, and croplands results in higher soil bacterial diversity and more homogeneous soil fungal communities dominated by pathogens and with a lower abundance of beneficial symbionts. Such conversions also result in significant reductions in soil C storage, nutrient cycling, and soil functional rates related to organic matter decomposition. From Qu et al. (2024).

(Figure 2.17). Similarly, in the above-mentioned European survey, more disturbed agricultural soils were found to harbour fewer plant beneficial fungal symbionts and a greater proportion of fungal plant pathogens than soils of less disturbed ecosystems, such as woodlands and extensively managed grasslands (Labouyrie et al. 2023). The abundance of viruses was also found to be lower in agricultural soils subject to high levels of disturbance than in forests (Liao et al. 2022), as is also the case with AM fungal diversity (Daniell et al. 2001), and species diversity of ECM fungi has been shown to decline with clear-cutting of forests (Hagerman et al. 2001). A study of 60 farmlands across Switzerland, covering a wide range of management regimes and soil conditions, also revealed that agricultural intensification markedly reduces network complexity and the abundance of key mycorrhizal taxa of the root microbiome (Banerjee et al. 2019). Soil fungal communities have also shown to become less complex and homogenous along a gradient of degradation in alpine grasslands; moreover, this degradation-induced simplification of fungal communities did not recover following 10 years of grassland restoration (Luo et al. 2023).

The above studies focussed on how land conversion and intensification impact patterns of local diversity (i.e., α diversity) and reveal divergent responses among soil organism groups, but overall declines in soil food web complexity. As well as affecting local diversity, intensive land use can negatively impact soil biodiversity at larger, landscape scales (i.e., β diversity, compositional dissimilarity between sites), making soil communities more similar or homogenised. This homogenisation of soil biodiversity at larger spatial scales has been shown to occur for soil bacteria in response to tropical forest deforestation (Rodrigues et al. 2013), for soil fungi in response to arable farming (Banerjee et al. 2024), and for mycorrhizal fungi, decomposer microbes, herbivores, and secondary consumers in response to land use intensification of grassland (Gossner et al. 2016). Furthermore, agricultural practices were found to effectively erase latitudinal gradients in soil nematode diversity driven by climate, which again is indicative of homogenisation at larger scales (Li et al. 2020c). Such biotic homogenisation—the process by which the similarity of communities increases over space and/or time—in response to land use conversion and intensification, but also urbanisation (Delgado-Baquerizo et al. 2021), threatens soil β diversity by homogenising communities within and across landscapes. In addition, it could counterbalance any increases in α diversity due to ecosystem conversion or land use intensification, thereby resulting in a net decline in soil biodiversity (Rodrigues et al. 2013, Gossner et al. 2016).

There are numerous, interacting reasons for changes in soil biodiversity resulting from ecosystem conversion to agricultural land. Physical disturbance associated with site preparation (e.g., deforestation, burning, and cultivation) and declines in the amount and complexity of organic residues returned to the soil are important factors, as are declines in SOM caused by reduced organic matter return to soil and regular tillage, which enhances organic matter mineralisation and soil C loss. Also, organic material returned to soil from plants in agricultural systems, where it is often derived from a single crop, is of much lower diversity than in natural ecosystems. The diversity of decomposers is also typically greater in more complex litter mixtures than in simple soil substrates (Hansen 2000, Szanser et al. 2011), and declines in the diversity of certain soil animals (e.g., termites) due to the conversion of primary forest to cropland has been linked to reduced diversity and the heterogeneity of plant litter inputs to soil (Gillison et al. 2003, Jones et al. 2008). Moreover, the conversion of tropical forest to oil palm and rubber plantation has been shown to result in substantial changes in energy flux below-ground, reducing energy flux to higher trophic levels and promoting a shift from slower fungal-based to faster bacterial-based energy channels with greater consumption of SOM and, hence, depletion of soil C stocks (Potapov et al. 2024).

As noted earlier, disturbance can increase the diversity of soil communities, especially when it enhances the spatial heterogeneity of soil, thereby providing opportunities for more species to coexist, or when it increases soil pH, for instance via liming acid soils. An example of this is in soils of the riparian zone (the zone between terrestrial and aquatic ecosystems) that possess an unusually high diversity of fauna and microflora (Ettema et al. 2000). This diversity is maintained by a variety of disturbances (e.g., periodic flooding, drought, freezing, abrasion, erosion, and occasionally toxic concentrations of nutrients) that create a spatial and temporal mosaic with few parallels in other systems (Ettema et al. 2000). Also, increases in local diversity (i.e., α diversity) of soil bacterial communities following deforestation and conversion of tropical forest to agriculture has been linked to a neutralisation of soil pH through lime application (Rodrigues et al. 2013, Franco et al. 2019b). Similarly, higher α diversity of soil microbial communities in more intensively managed land use types across Wales, UK, was attributed to consistent nutrient inputs and greater disturbance (e.g., from tillage), which induce non-equilibrium dynamics that prevent competitive exclusion by dominant taxa (George et al. 2019).

A note on above-ground theory

A frequently asked question concerning patterns of soil biodiversity at the field and ecosystem scale is whether the same determinants of above-ground diversity operate below-ground. Two factors have traditionally been emphasised as determinants of above-ground diversity: (1) productivity or resource supply, i.e., the productivity–diversity hypothesis; and (2) consumption or physical disturbance, i.e., the intermediate disturbance hypothesis (Figure 2.18). The general idea of the

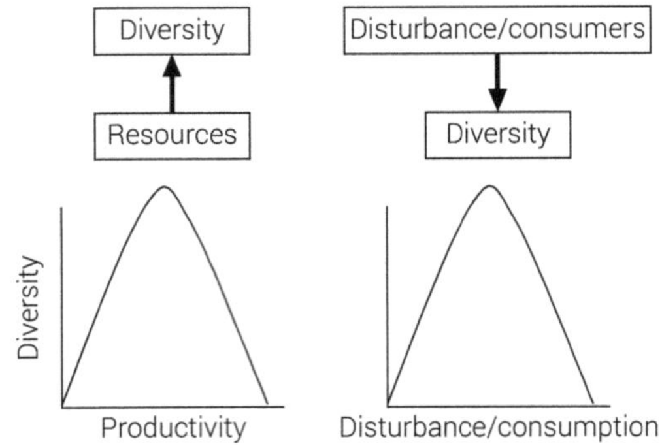

Figure 2.18 Disturbance/consumer versus resource control of species diversity. Both models predict that diversity will be maximal at intermediate levels of disturbance and productivity. From Bardgett (2005) by permission of Oxford University Press.

productivity–diversity hypothesis, or humped-back model, is that species richness first increases and then decreases with increasing productivity (Grime 1973, Al Mufti et al. 1977), with the decrease in diversity at high productivity being due to competitive exclusion. However, competitive exclusion can be prevented by periodic mortality events caused by consumption (e.g., herbivory) or physical disturbance, which releases subordinate species from competition (Connell 1978, Huston 1979); these factors likewise display unimodal relationships with diversity in line with the intermediate disturbance hypothesis (Connell 1978). It has also been proposed, and shown in marine food webs, that these factors interact with each other, i.e., disturbance-diversity relationships depend on productivity and vice-versa (Huston 1979, Worm et al. 2002).

While these diversity relationships are common in above-ground communities, the general lack of data on the diversity of soil organisms prevents rigorous testing of whether they apply below-ground. But data that are available provide mixed support for the productivity–diversity or intermediate disturbance hypothesis operating below-ground. Regarding the productivity–diversity relationship, for example, it is known that food webs become longer and more diverse in more productive situations, with more energy being diverted to higher trophic levels (Moore and de Ruiter 2000). However, an extensive literature synthesis of available data on soil organisms found no support for the humped-back model operating below-ground, i.e., the diversity of soil organisms does not peak at intermediate levels of productivity (Wardle 2002). Similar conclusions have emerged from field-based studies of the productivity–diversity hypothesis for soil microarthropods, where declines in microarthropod abundance and diversity were not detected at the highest end of a productivity gradient based on fertiliser N addition to grassland (Cole et al. 2008). There is also little support in the literature for optimisation of soil diversity at intermediate levels of disturbance, although, as previously mentioned, disturbance can increase local diversity of soil communities, especially when it enhances spatial heterogeneity of soil (Bardgett and Wardle 2010).

The above studies lend support to the idea that competitive exclusion is a more important driver for above-ground than below-ground communities, largely on the basis that many soil organisms are spatially separated in soil (Bardgett et al. 2005b). However, recent studies have challenged this view, at least for soil microbes: significant unimodal productivity–diversity relationships were detected for soil bacterial communities in simple laboratory systems (Geyer and Barrett 2019), and for both soil bacterial and fungal communities across global biomes (Bastida et al. 2021). Together, these studies suggest that resource availability and competitive exclusion are important drivers of soil microbial diversity, and therefore shared forces structure diversity above-ground and below-ground. Nevertheless, caution is needed in drawing conclusions on soil productivity–diversity and diversity–disturbance relationships, given a paucity of data, and because drivers of soil diversity vary both with spatial scale and for different organism groups (Thakur et al. 2020).

Another important determinant of above-ground diversity at an ecosystem scale is island or habitat size, which is part of the island biogeography theory (MacArthur and Wilson 1967). Island size determines, amongst other things, the availability of resources available for organisms, and, in general, larger islands have a greater variety of resources and hence more species are found—the so-called species-area relationship. This idea was tested in old-growth temperate forest in Northland, New Zealand, by examining the diversity of groups of organisms in organic matter of suspended soils produced by spatially separated epiphytes or tree-top islands in the crowns of canopy trees (Wardle et al. 2003a). Larger islands were found to support a greater diversity of macrofauna and microarthropods, consistent with the island size theory, but physical proximity of islands to sources of colonising organisms did not influence the diversity of these faunal groups, which contrasts with expectations of the island biogeography theory (Wardle et al. 2003a). Studies on microbial communities also lend support to the species-area relationship operating in soil. For example, across 29 islands of Thousand Island Lake (a reservoir in subtropical China created by damming, which transformed former hilltops into islands) both bacterial and fungal diversity increased with island area but were unaffected by island isolation (Li et al. 2020a). Further, the diversity of root-associated fungal communities was found to increase with forest area across a range of lava-fragmented forests in Hawaii (Vannette et al. 2016), and soil bacterial richness was found to be positively related to habitat size and resource availability in an experimental microcosm study (Delgado-Baquerizo et al. 2018a). While these studies lend support to the species-area relationship operating in soil, future research is needed to fully test the applicability of these and other biodiversity theories for explaining patterns of biodiversity below-ground (Thakur et al. 2020).

Local-scale patterns of soil biodiversity

Local-scale patterns of soil biodiversity, from the micrometre-scale to metres, are predominately related to soil habitat heterogeneity, in terms of the micro-scale structural complexity of the soil physical environment, the chemical complexity and availability of organic resources, and influence of plant roots. As discussed in Chapter 1, the soil is extremely variable in its structure and composition, and all soil systems are spatially heterogeneous across many orders-of-magnitudes of scale, thereby providing unrivalled opportunities for organisms to establish niches and to coexist (Figure 2.19). This notion is supported by several studies that show that spatial heterogeneity in soil promotes soil biodiversity, and that the greatest diversity is found in microhabitats, such as those created by soil aggregates and the pores within and around them. For example, bacterial diversity has been shown to be greatest in microaggregates due to access to more chemically complex organic matter in micropores, which increases niche space and favours coexistence of many taxa (Bach et al. 2018), and it has long been known that animal diversity is positively related to soil habitat complexity (Gill 1969, Anderson 1978, Stanton 1979, Hansen 2000). Recent studies also indicate that pore space morphology is a key determinant of microbial growth and functioning in soil. In particular, the use of Soil Chips, which allow manipulation of micrometre-scale structural heterogeneity, has revealed profound differences in the response of different fungal decomposer species to micrometre-scale constrictions, sharp turns, and obstacles, with implications for their foraging behaviour and capacity to colonise new territories (Aleklett et al. 2021). Soil Chips have been used to show that fungal and bacterial growth and interactions are affected by habitat geometry, with implications for substrate degradation (Arellano-Caicedo et al. 2021).

Roots also create spatial heterogeneity in soil, not just through effects on soil physical structure, but also via their death and decomposition (i.e., root turnover) and the exudation of easily degraded compounds from roots into the zone of soil immediately adjacent to the root: the rhizosphere (Figure 2.20) (Hiltner 1904). Those compounds lost from plant roots by exudation, which include low molecular mass exudates and secretions, polysaccharide mucilages, and lysate, represent a high quality nutrient source for rhizosphere microorganisms. As such, they create a hot spot of microbial growth

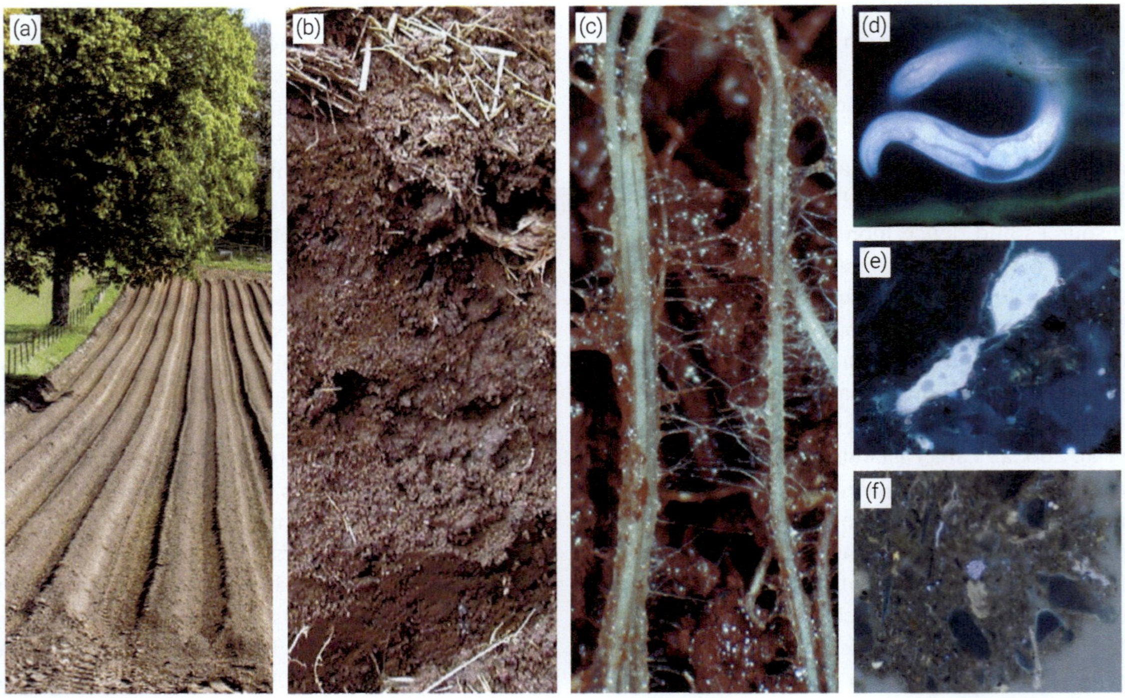

Figure 2.19 Views of arable soil across several orders of spatial magnitude: (a) field scale; (b) profile scale (image width = 25 cm); (c) root scale (image width = 10 mm); (d) nematode in pore (image width = 200 μm); (e) naked amoeba (image width = 100 μm); (f) bacterial colony in soil aggregate (image width = 100 μm). Images by Karl Ritz.

and diversity in the rhizosphere, where microbial communities differ substantially in their composition from those of bulk soil just a few centimetres away (Philippot et al. 2013). This in turn creates a hot spot of prey availability for microbial-feeding protists and nematodes, which are often at their most abundant in the rhizosphere (Bardgett and Griffiths 1997, Li et al. 2024b).

The amounts of plant C allocated to the rhizosphere via root exudation can be considerable (i.e., 4–18% of photosynthetically fixed C released as root exudates) (Farrar et al. 2003, Dror and Klein 2022), with a large portion also being directed to mycorrhizal fungi (Farrar et al. 2003, el Zahar Haichar et al. 2014, Hawkins et al. 2023) (Figure 2.21). Indeed, the global root exudation flux was recently estimated to be as much as 13.4 (10.1–20.2) pentagrams (Pg) C year^{-1}, or about 9% (7–14%) of global annual gross primary productivity (GPP) (Chari et al. 2024). It has also been estimated that as much as 13.12 Pg of CO_2 equivalents fixed annually by terrestrial plants is allocated to mycorrhizal fungi, which equates to around 36% of current annual CO_2 emissions from fossil fuels (Hawkins et al. 2023). As such, below-ground plant C allocation to root exudates and mycorrhizal fungi is recognised as a major driver of the rhizosphere microbial and faunal communities and SOC dynamics, often to a greater extent than plant litter inputs from above-ground (Sokol and Bradford 2019, Zhou et al. 2023). Root exudates also play a key role in regulating plant health, via their role in mediating plant–microbial interactions related to plant nutrient acquisition, defence, and tolerance to environmental stress, such as drought (el Zahar Haichar et al. 2014, Williams and De Vries 2020). However, many factors can alter the quantity and chemical composition of C compounds released from roots into the rhizosphere, such as plant identity, physiological state, age, and environmental conditions, which in turn has differential effects on the structure and functioning of

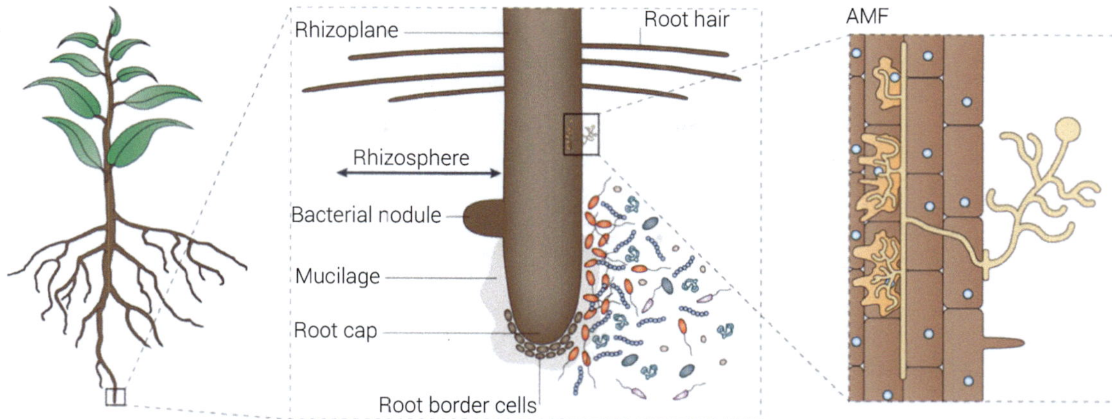

Figure 2.20 Schematic of the rhizosphere, a narrow zone of soil of a few millimetres width that surrounds and is influenced by plant roots. The schematic shows magnified pictures of the rhizosphere, containing saprophytic and symbiotic bacteria and fungi, including arbuscular mycorrhizal fungi (AMF). From Philippot et al. (2013) by permission of Springer Nature.

the rhizosphere microbiome, with feedback consequences for plant growth and ecosystem functioning (Philippot et al. 2013), as will be discussed in Chapter 4.

It is important to note that the complexity of the soil environment occurs at a range of spatial scales. Therefore, heterogeneity will potentially influence the diversity of different groups of organisms in different ways, depending on their body size and the size of their habitat unit, or domain. Microscale heterogeneity, for example, at the scale of the soil particle can be extremely high, providing steep environmental gradients in oxygen concentration over a few millimetres. Such microscale heterogeneity could therefore act as an important determinant of microbial diversity in soil. In contrast, heterogeneity over larger spatial scales, such as that caused by resource hot spots in the form of buried litter, dead animals, roots, and other organic materials, or patterns of root C flow, and variations in soil structure that affect the physical dimensions of habitable space in soil, will impact on both microbes and larger organisms. In sum, variations in spatial and temporal heterogeneity of the soil environment, at the scale of the soil particle to the soil horizon, coupled with the incredible specialisation of the soil biota, act as the major determinants of patterns of soil biodiversity at the local scale.

Temporal patterns of soil biodiversity

Soil communities are not static. They change constantly over time, ranging from daily to seasonal shifts in the abundance of different groups of soil organisms and species, to changes in soil biodiversity over successional time of hundreds or thousands of years, to the evolution of trophic diversity in soil food webs over geological time, dating back to the Early Paleozoic (Schaefer and Caruso 2019). For instance, significant diurnal shifts in the composition and functioning of rhizosphere bacterial communities can occur between night and day (Baraniya et al. 2018, Fickling et al. 2024), and abrupt seasonal transitions in the composition and functioning of soil microbial communities have been detected in alpine ecosystems triggered by snowmelt (Broadbent et al. 2021). Seasonal variation in soil fungal communities has also been detected that is equivalent to the level of spatial variation observed over thousands of kilometres (Averill et al. 2019), and it has long been known that the abundance and composition of soil animal communities displays significant seasonal and interannual variation (Wardle et al. 1999b, Yeates et al. 2000, Doblas-Miranda et al. 2007). However, the temporal dynamics of soil microbial and animal communities vary considerably across ecosystems depending on factors such as soil fertility and plant productivity (Wardle et al. 2012), as well as soil depth, being typically greater in

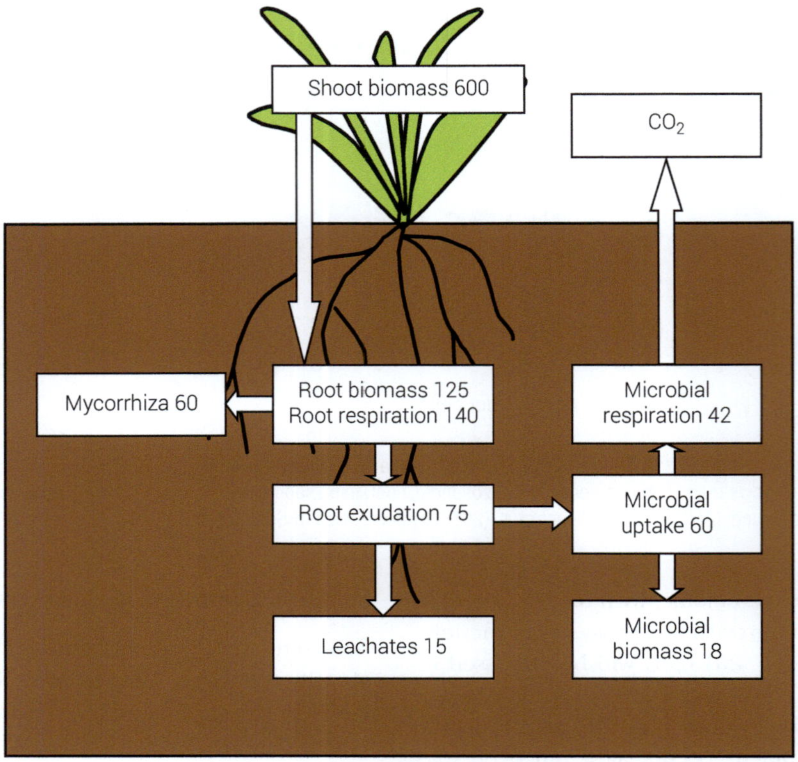

Figure 2.21 Schematic of the partitioning of C and the plant distribution of root exudates within the soil after net fixation of 1,000 units of C. From Farrar et al. (2003) by permission of John Wiley and Sons.

surface soil than in deeper soil (Berg and Bengtsson 2007). Temporal dynamics of below-ground communities can also be complicated by the fact that many soil organisms undergo long periods of inactivity when conditions are unfavourable, allowing them to tolerate harsh soil conditions (Bardgett 2005).

Soil biodiversity changes over timescales of tens or hundreds of years, through processes of succession and ecosystem development, for instance on newly exposed terrain following the retreat of glaciers (Figure 2.22). Several studies give insights into the kinds of patterns that occur in soil communities over these longer timescales. For example, for soil microbes it is evident that during the initial stages of succession—the build-up phase—and towards maximal plant biomass (i.e., community climax), soil microbial communities typically become increasingly abundant and physiologically active (Insam and Haselwandter 1989, Ohtonen et al. 1999, Tscherko et al. 2003) and more taxonomically and functionally diverse (Ortiz-Álvarez et al. 2018, Delgado-Baquerizo et al. 2019, Farrer et al. 2019, Hu et al. 2021c) (Figure 2.23). Also, microbial communities often become more fungal dominated as succession proceeds, reflecting an increase in the abundance of fungi relative to bacteria (Ohtonen et al. 1999, Bardgett et al. 2007) (Figure 2.23), and they display marked shifts in microbial community composition. For example, in a study of soil microbial community succession in the foreland of the Damma glacier, central Switzerland, soil fungi shifted from Ascomycota-dominated communities in young soils to more Basidiomycota-dominated communities in older soils (Zumsteg et al. 2012).

Root-associated mycorrhizal communities also change as primary succession proceeds. In early succession, ruderal plants are generally non-mycorrhizal, whereas in mid succession, the dominant herbaceous plants tend to have a

Figure 2.22 Forelands of retreating glaciers, such as the Damma Glacier, Switzerland (a), the Svínafellsjökull Glacier, Iceland (b), the Odenwinkelkees Glacier, Austria (c), and the Rotmoos Glacier, Austria (d), provide a natural laboratory for studying primary succession both above- and below-ground. The chronosequences resulting from glacial retreat represent ecological gradients of increasing terrain age with a sequence of different communities. Images by Richard Bardgett.

facultative requirement for AM fungi; in climax communities the trees and shrubs, which dominate the vegetation, often have an obligate need for ECM fungi (Read 1994) (Figure 2.24). Similar shifts in microbial community composition occur during secondary succession; for example, the abandonment of agricultural management typically leads to a shift in the soil microbial community over time from one that is dominated by bacteria to one dominated by fungi (Bardgett et al. 2001b, Zeller et al. 2001, Van der Wal 2006, Hannula et al. 2017), which is associated with a shift in fungal community composition from one composed of fast-growing and pathogenic fungal species to one consisting of beneficial and slower-growing fungal species (Hannula et al. 2017). Also, below-ground community networks become more connected during secondary succession, which corresponds with increased efficiency of nutrient cycling and C uptake (Morriën et al. 2017). It is not entirely clear why these patterns in microbial communities occur in response to primary and secondary succession, but they most likely relate to a build-up in the amount and complexity of SOM, as well as changes in the quantity and quality of resource inputs to soil resulting from vegetation change. They are also likely related to changes in soil abiotic properties, for example a drop in pH that commonly occurs in late succession (Bardgett and Wardle 2010). However, it should be noted that the rate and trajectory of soil microbial community change during primary succession varies tremendously across biomes and with differences in climate, parent material, and vegetation type (Delgado-Baquerizo et al. 2020a).

Much less is known about how soil faunal communities change during primary and secondary succession, but it is reasonable to expect that fungal-feeding fauna become increasingly abundant, and that predators become more abundant as more energy becomes available for higher trophic levels. This was found to be broadly the case in a chronosequence study of restoration sites on ex-arable lands in the Netherlands, where the number of taxa of fungivorous and predaceous mites, root-feeding nematodes, and bacterivorous nematodes increased during secondary succession (Morriën et al. 2017). However, other studies show that soil invertebrate groups responded very differently to secondary

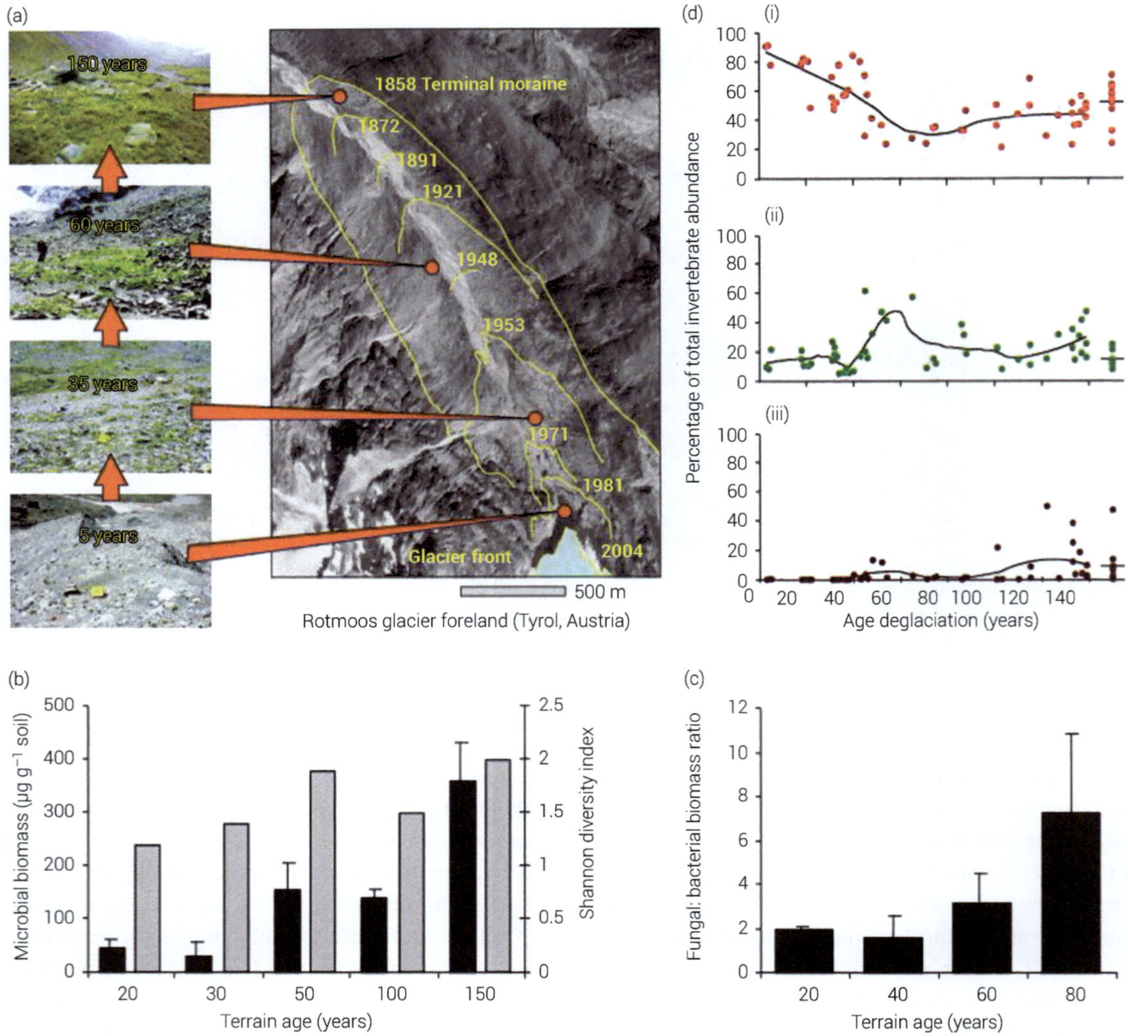

Figure 2.23 Patterns of below-ground community development with increasing terrain age on the (a) Rotmoos glacier foreland, Austria, showing: (b) increases in soil microbial biomass (black bars) and the functional diversity of the microbial community (grey bars), calculated as the Shannon diversity index of a range of enzyme activities (data from Tscherko et al. 2003); (c) increase in the ratio of fungal to bacterial biomass with increasing terrain age in front of the Lyman Glacier, Washington, USA, indicative of a progressive shift towards a fungal-dominated food web in older sites, reflecting a greater abundance and complexity of resources (data from Ohtonen et al. 1999); and (d) changes in trophic composition of invertebrates, with the first colonisers being (i) predators (ground beetles and spiders), followed by (ii) herbivorous groups after 40–50 years when plant cover has increased, and finally (iii) detritivores (mostly millipedes) which become abundant after 100 years when an organic surface soil horizon has developed (data from Kaufmann 2001). From Bardgett et al. (2005) by permission of Elsevier.

succession (from a wheat field to a beechwood), suggesting that responses of soil animals to succession are difficult to predict (Scheu and Schulz 1996). Further, a detailed study of surface-active micro- and macro-arthropods in the Central Alpine glacier foreland of the Rotmoostal, Tyrol, Austria, likewise showed that patterns of invertebrate succession are not always as might be expected: although faunal colonisation and succession in Alpine glacier forelands generally followed a predictable pattern,

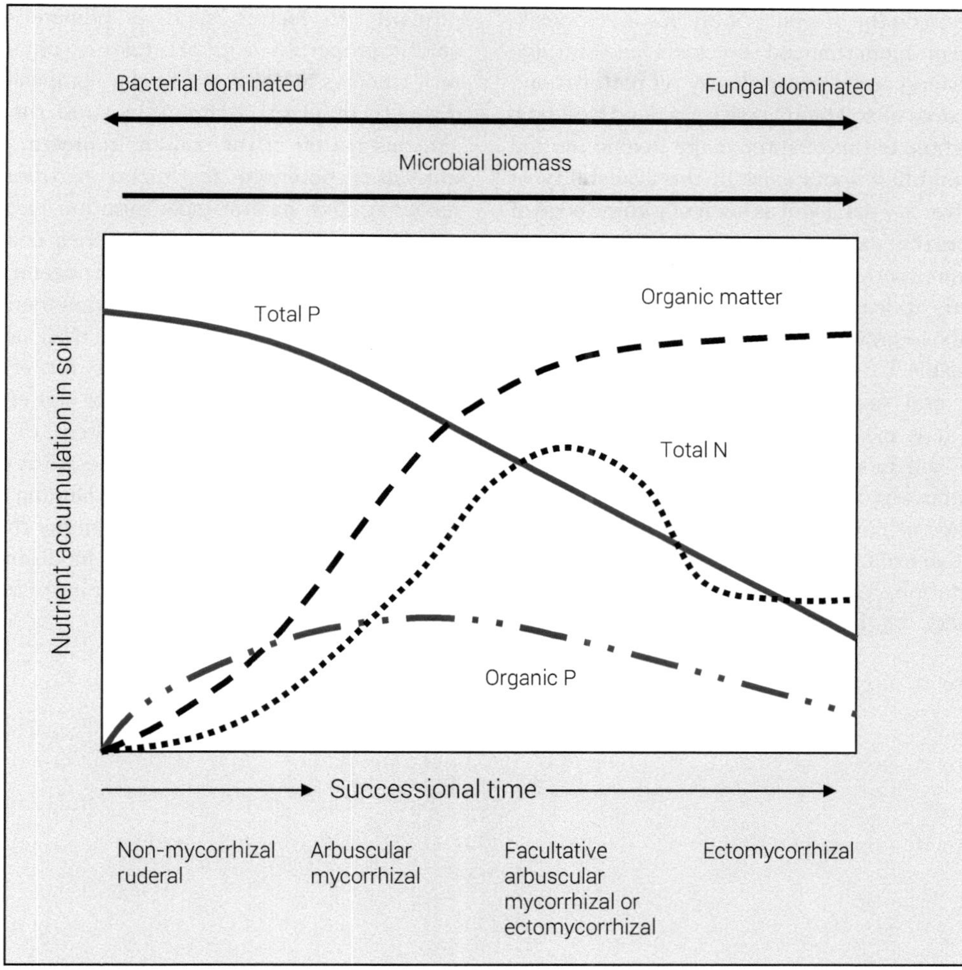

Figure 2.24 Schematic of broad-scale changes in soil microbial community structure, root-associated mycorrhizal communities, and soil nutrient pools during primary succession. From Bardgett (2005) by permission of Oxford University Press.

the first colonisers were almost exclusively predators, and herbivores and decomposers appeared later (Figure 2.23d) (Kaufmann 2001). Similarly, counter-intuitive patterns of invertebrate community assembly have been found during primary succession on glacier forelands in Ny-Ålesund, Spitsbergen, Svalbard (78° N): spiders are the earliest colonisers of newly exposed moraine substrates, where their densities are highly correlated with allochthonous inputs of potential prey items, predominantly chironomid midges (Hodkinson et al. 2001). In these systems, large allochthonous inputs of insects potentially provide significant quantities of N and P to the developing ecosystem from the earliest stages of succession, even before a conspicuous cyanobacterial crust has formed. That spiders entrap nutrients in such a way could be of high importance in early ecosystem development in these extreme environments.

Conclusions

The soil is home to a vast diversity of organisms and understanding the factors that regulate this diversity, and its consequences for ecosystem function, represents a major challenge for soil ecology.

Information on the diversity of life in soil is increasing, but a problem that soil ecologists face in trying to understand and develop theory on patterns and determinants of soil biodiversity is a dearth of data; as highlighted in this chapter, major taxonomic and geographic blind spots exist in the availability of soil biodiversity data, and as such our knowledge of the factors that explain patterns of soil biodiversity at different spatial and especially temporal scales is limited, at least compared to what is known about above-ground biodiversity. Despite this, it is possible to make certain judgements about patterns and determinants of soil biodiversity. Notably, it is evident that patterns and determinants of soil biodiversity differ from those that operate above-ground, especially in relation to the importance of competition and the prevention of competitive exclusion, which are important drivers of vegetation change. In soil, patterning of biodiversity at larger spatial scales is related primarily to factors such as climate and soil abiotic properties (e.g., pH, texture, organic matter), whereas at local scales the composition and diversity of plant communities, and the heterogeneous nature of the soil environment, provides unrivalled potential for niche partitioning, or resource and habitat specialisation, leading to avoidance of competition and hence coexistence of species. Furthermore, anthropogenic disturbances are becoming increasingly dominant factors in shaping soil food webs and their activities, often with major consequences for ecosystem functioning. A major challenge for soil ecologists, therefore, is to identify the hierarchy of controls on soil biological diversity that operate at different spatial and temporal scales, including anthropogenic disturbances, and to determine the role of spatiotemporal patterning of soil biodiversity as a driver of the structure and functioning of terrestrial ecosystems.

CHAPTER 3

Soil biota and ecosystem functioning

Key messages

- Soil microorganisms play crucial roles in regulating organic matter decomposition and nutrient transformations that are key to the supply of nutrients to plants.
- Soil microbial communities are regulated by trophic interactions with soil-dwelling animals, which ultimately control the flux of energy and nutrients through the soil food web.
- Soil animals also modify the physical structure of soil, often acting as ecosystem engineers, thereby affecting the habitat and activities of other fauna and microbes.
- Soil food web complexity and diversity regulates ecosystem functioning, although responses vary with species traits and abiotic factors, especially climate and soil fertility.

In the previous chapter (Chapter 2), the main performers of the soil food web and the factors that shape spatial and temporal patterns of soil biodiversity were introduced. This chapter extends on this by considering why these soil organisms, and their interactions with one another, matter for the functioning of terrestrial ecosystems, especially through their roles in the decomposition of organic matter and recycling of key nutrients. Decomposer organisms are essential for the functioning of terrestrial ecosystems because they decompose dead organic material in soil, converting it into CO_2 and soluble nutrients that provide resources for other biota and the growth of plants. In natural ecosystems, most N and a portion of the P required for plant growth are supplied through decomposition of organic matter, which relies almost entirely on the activities of soil organisms. Even in agricultural systems that typically receive large inputs of inorganic fertilisers and manures, the supply of plant nutrients via the decomposition of organic matter is central to the maintenance of sustainable crop and livestock production.

This chapter explores how the activities of soil biota, including their trophic interactions, influence these processes of decomposition and nutrient cycling, and examines the significance of this for material flow and plant production in terrestrial ecosystems. The focus is on N and P cycling, as these are the two nutrients that most limit primary productivity in natural and managed terrestrial ecosystems. First, the issue of how soil microbes regulate the internal cycling of nutrients in terrestrial ecosystems will be discussed. This will be followed by a discussion of how soil animals influence nutrient cycling and plant growth through their feeding activities on microbes and other fauna. The issue of how soil organisms and their interactions impact on ecosystem properties by modifying the physical environment in which they live will then be discussed. Finally, a related, but very topical issue will be considered: that is, how do changes in the diversity and structure of soil food webs impact ecosystem functioning?

Microbial control of soil nutrient availability

Upward of 90% of primary production enters the soil system as dead plant litter, and the primary decomposers of this material are bacteria

The Ecology of Soil. Richard D. Bardgett, Oxford University Press. © Richard D. Bardgett (2025). DOI: 10.1093/9780191858710.003.0003

and fungi. Initial fragmentation and ingestion of fresh organic material by detritivores increases the material's surface area for microbial colonisation, but the microbes, via the production of extracellular enzymes, do most of the chemical alteration of this dead material, catalysing the transformation of nutrients into plant-usable forms. As noted in Chapter 2, soil microbes have the capacity to produce a vast array of enzymes that can degrade almost all plant-derived compounds, and which catalyse nutrient and C transformation in soil.

N mineralisation

The mineralisation of nutrients—the process by which soil microorganisms break down soluble and insoluble organic matter and convert it into inorganic forms—is of critical importance for ecosystem function since, in many ecosystems, it directly determines the availability of nutrients to plants. For example, in productive ecosystems, such as temperate deciduous forests, plant N supply is strongly correlated with rates of N mineralisation (Nadelhoffer et al. 1985) (Figure 3.1), and worldwide, many agricultural systems rely on SOM mineralisation as the main source of N for crops (Lu and Tian 2017). A summary of the soil N cycle is given in Figure 3.2.

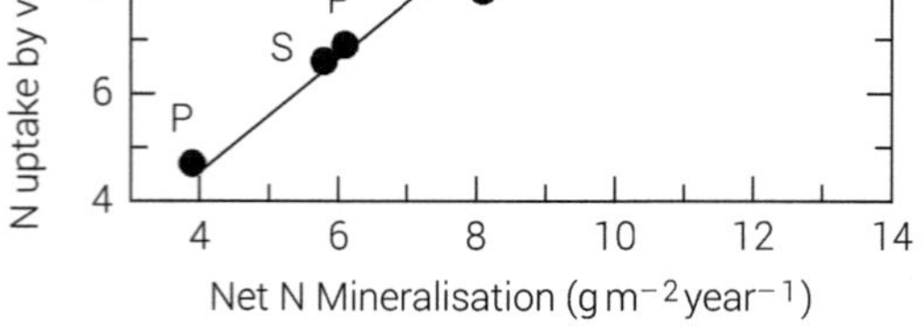

Figure 3.1 Plant N uptake in relation to net N mineralisation in the surface 0–20 cm of soil across a range of vegetation types. Symbols designate dominant genera on sites: O = oak, B = birch, M = maple, P = pine, S = spruce. From Bardgett (2005) adapted from Nadelhoffer et al. (1985) by permission of Oxford University Press.

Most N in soil (some 96–98%) is contained in dead organic matter as complex insoluble polymers such as proteins, nucleic acids, and chitin. These polymers are too large to pass through microbial membranes, so microbes produce extracellular enzymes (e.g., proteinases, ribonucleases, and chitinases) which break them down into smaller, water-soluble subunits (e.g., amino acids) that can be absorbed by microbial cells. This material is called dissolved organic nitrogen (DON), and it can constitute a large proportion of the total soluble N pool, especially in ecosystems with acidic mor soils, such as boreal forest and arctic tundra (Northup et al. 1995, Jones and Kielland 2002). Even in agricultural soils with regular applications of inorganic N fertiliser, soil concentrations of DON can be high relative to inorganic N, and the leaching of DON can represent a significant pathway of N loss from soil (Van Kessel et al. 2009). Indeed, in some ecosystems, such as undisturbed forests, DON losses by leaching can exceed those of inorganic N (Hedin et al. 1995, Perakis and Hedin 2002), thereby representing a significant N 'leak' from the soil system (Neff et al. 2003).

The majority of DON in soil solution is absorbed by free-living soil microbes, which use the C and N contained within it for their growth. How microbes use DON, however, depends mainly on whether they are C or N limited. Under conditions when microbial growth is C limited, microbes use the C from DON to support their energy needs for growth and they excrete plant-available NH_4^+ as a waste product into soil, i.e., N is mineralised by the microbial biomass. When DON is not sufficient to meet microbial N demand, however, microbes absorb additional inorganic N from soil solution, i.e., N is immobilised by the microbial biomass, thereby reducing the availability of inorganic N for plant uptake. Aside from plant uptake, NH_4^+ produced by microbial mineralisation has several potential fates: it can be absorbed onto negatively charged surfaces of clay minerals and organic matter and, when oxygen is plentiful, it can be oxidised to NO_3^- via the process of nitrification (Box 3.1). Nitrification is generally viewed as a two-step process: ammonia (NH_3) is first oxidised to nitrite (NO_2^-) by ammonia-oxidising bacteria (AOB)

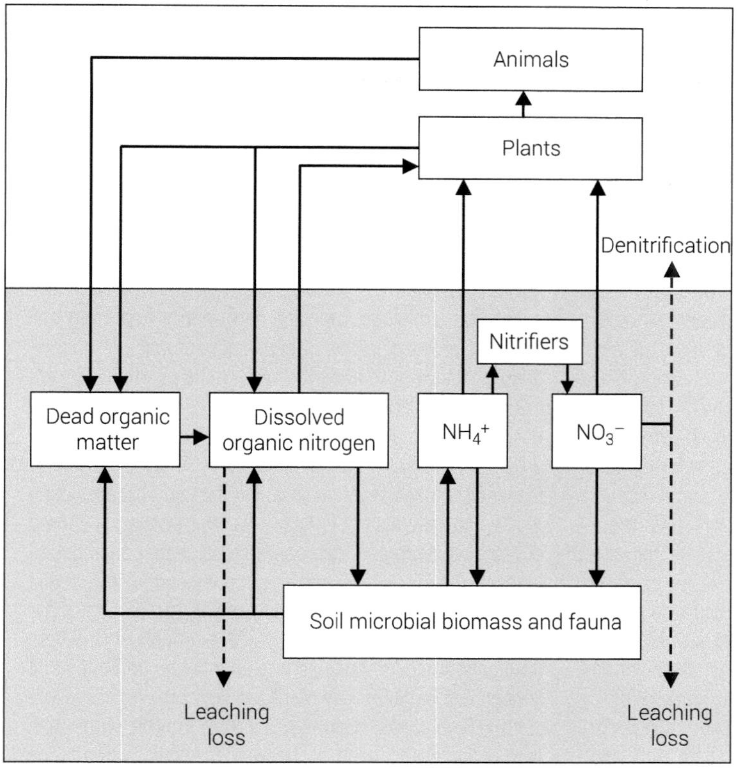

Figure 3.2 Simplified schematic of the terrestrial N cycle. From Bardgett (2005) by permission of Oxford University Press.

and ammonia-oxidising archaea (AOA), and then to NO_3^- by nitrite-oxidising bacteria (NOB) (Prosser and Nicol 2008, Lu et al. 2020). However, the recent discovery that some *Nitrospira* bacteria can catalyse the complete oxidation of NH_3 to NO_3^- has challenged this two-step view of nitrification, pointing to completely nitrifying bacteria as a key component of N-cycling microbial communities (Daims et al. 2015, Van Kessel et al. 2015).

Both immobilisation and mineralisation of N occur simultaneously in soil, but *net* N mineralisation occurs when microbes are predominately C limited, whereas *net* immobilisation occurs when N limits their growth. The balance between mineralisation–immobilisation is determined by a range of factors such as the availability of DON and predation by soil animals, but one factor that is of high importance is the relative demand by microbes for N and C, which is determined by the C:N ratio of the organic substrate they are utilising. Heterotrophic bacteria and fungi have C:N ratios ranging from 4:1 to 12:1, but as they break down organic matter, they respire some 50% of the C contained within it and use the remaining 50% for biomass production (Kaye and Hart 1997). Therefore, there is a widely cited critical substrate C:N ratio of around 30:1 which is required to meet microbial needs for N (Kaye and Hart 1997). If substrate C:N ratios are higher, which is often the case for plant litters, microbes become N limited and hence use (or immobilise) exogenous sources of inorganic N while decomposing the substrate (Kaye and Hart 1997). In such situations, plants and microbes should theoretically compete for N in soil solution. Several factors influence substrate quality in terrestrial ecosystems, such as inherent soil fertility, plant community composition, and global change factors that can modify the quality of plant litter inputs to soil, as will be discussed in Chapter 6.

Not all DON in soil solution is utilised by microbes: significant quantities of DON are leached

Box 3.1 Nitrification

Nitrification is the process by which ammonium (NH_4^+) is oxidised to nitrite (NO_2^-) and subsequently to nitrate (NO_3^-). This process is traditionally viewed as a two-step process: ammonia (NH_3) is first oxidised to nitrite (NO_2^-) by ammonia-oxidising bacteria (*Nitrosomonas* and other *Nitros-* genera) and archaea, and then to NO_3^- by nitrite-oxidising bacteria (*Nitrobacter* and other *Nitro-* genera). However, some *Nitrospira* bacteria have been identified that can catalyse the complete oxidation of NH_3 to NO_3^-, thereby challenging the traditional two-step view of nitrification. The process of nitrification is influenced by several soil factors, most notably the availability of NH_4^+, which is the sole energy source for autotrophic nitrifiers. Hence, rates of nitrification tend to be elevated in agricultural soils that receive large inputs of fertiliser N and in natural ecosystems where atmospheric N deposition is high. Oxygen is crucial for nitrification to occur, so factors that influence the diffusion of O_2 through soil, such as moisture and soil structure, will alter rates of nitrification in soil. Nitrification varies with soil temperature, being optimal at 30–35°C, but it is also known to occur in cold, but unfrozen soils under snow. Rates of nitrification are affected by soil moisture and tend to be very low in dry soils, where thin water films restrict the diffusion of NH_4^+ to nitrifiers. Nitrification is an acidifying process in that the oxidation of NH_4^+ to NO_3^- yields H^+ ions, although it is also sensitive to soil pH, being negligible below pH 4.5. Whilst rates of nitrification tend to be relatively low in temperate acid forest soils, it is elevated after disturbances, such as clear-cutting and fire, that result in elevated rates of NH_4^+ production by microbial mineralisation.

Plants and microbes take up NO_3^- produced by nitrification, but significant amounts of NO_3^- may be lost from soil via leaching and denitrification. Compared to positively charged cations (e.g., NH_4^+) that are retained in soil by absorption to negatively charged soil colloids (e.g., clay minerals), NO_3^- is highly mobile in soil solution. Hence, when not used by plants or microbes, or when soils have a limited anion exchange capacity, it is readily lost from soil to drainage waters. This is especially the case in coarse-textured sandy soils and after rain events or when soils are irrigated. This leaching of NO_3^- is a major problem in fertilised agricultural soils, where it often leads to pollution of groundwater, lakes, and streams. NO_3^- may also be lost from soil by denitrification, which is the microbial reduction of NO_3^- to gaseous NO, N_2O, and N_2. This process is carried out by a wide range of microorganisms under conditions of high NO_3^- and C supply and low oxygen, and is prevalent in wetlands either that receive inputs of NO_3^- from outside the systems or when they have a surface aerobic zone where nitrification occurs, such as in paddy fields. Denitrification is prevalent in intensively managed agricultural soils, especially in high rainfall regions where soils may be wetter than field capacity for much of the year. In these systems, the combination of high rainfall and fine-textured, compacted soils, coupled with high NO_3^- from fertiliser application, creates conditions that are especially conducive to denitrification. High rates of denitrification are of particular concern since it is a major source of nitrous oxide (N_2O), a potent greenhouse gas that contributes to the depletion of ozone in the stratosphere.

from soil, as mentioned above, but also plants can uptake DON directly from soil in the form of amino acids, thereby bypassing the microbial mineralisation step. The traditional view of plant N nutrition is that plants uptake inorganic forms of N, namely NH_4^+ and NO_3^-. However, it has long been known that plants also have the capacity to absorb organic N compounds through their roots, principally as amino acids; whether this represents a significant pathway of plant N nutrition, however, remains a matter of debate (Näsholm et al. 2009, Moreau et al. 2019). Studies show, for example, that direct uptake of organic N by plants in the form of amino acids is prevalent in low productivity, N-limited ecosystems, such as alpine (Raab et al. 1996, Miller and Bowman 2003, Liu et al. 2025), arctic tundra (Chapin et al. 1993, Kielland 1994, Schimel and Chapin 1996, Henry and Jefferies 2003), boreal forest (Näsholm et al. 1998, Nordin et al. 2001, Persson and Näsholm 2001), and low productivity temperate grassland (Bardgett et al. 2003, Wilkinson et al. 2015, Liu et al. 2025), and as more complex organic N forms (i.e., peptides) in maritime Antarctic soils (Hill et al. 2011a). Agricultural plants, including cereals, have the capacity to directly uptake amino acids and other more complex organic N forms, such as peptides (Näsholm et al. 2000, 2001, Hill et al. 2011b). However, organic N is unlikely to be of importance for crop N nutrition due to very rapid microbial mineralisation of organic N to inorganic N forms in agricultural soils, resulting in inorganic N being the dominant N form (Owen and Jones 2001, Wilkinson et al. 2014) and the selection of high yielding crop varieties with high N requirements (Salvagiotti et al. 2008). An exception to this, however, could be where amino acid availability is high, such as in patches or hot spots of nutrients in soil. Many plants have the capacity to exploit such resource patches by root proliferation and via root-associated mycorrhizal fungi; hence, in these situations, the acquisition of organic N could be a significant pathway of plant nutrition (Hodge et al. 1998, Cheng et al. 2016).

N fixation

Another way that plants gain N is via their association with soil microorganisms that have the capacity to use the enzyme nitrogenase to catalyse the reduction atmospheric N_2 to NH_3: the process of N fixation. This process of biological N fixation is especially important in natural ecosystems, where it can often be the main source of N entering the system (Cleveland et al. 1999), but it is also of major importance for agriculture, where it provides an annual global input of 120 Tg N (Herridge et al. 2008). Biological N fixation is performed by both free-living and symbiotic microbes, the latter forming nodules in roots where they receive sugars derived from photosynthesis to meet their energy needs, and in exchange they provide fixed N to the host plant. Details on the establishment of symbiosis, the biochemistry of N fixation, and the factors that influence rates of N fixation are given in Paul and Frey (2023), and their significance for plant community development is discussed in Chapter 4.

The most widely known N fixers are the legumes that are associated with bacteria of the genus *Rhizobium* and non-legumes such as alder that have the actinomycete *Frankia* as their endophyte. These symbiotic associations can supply large amounts of N to soil systems in both natural and agricultural ecosystems, and their benefits for soil fertility have been recognised since pre-Roman times. In productive pastures, for example, legumes, such as white clover (*Trifolium repens*), can contribute upward of 250 kg N ha^{-1} $year^{-1}$ to the system (Peoples et al. 2012) and as much as 80% of the total N used by neighbouring grasses (Boller and Nösberger 1987, Ledgard 1991). Also, the dominant crop legume, soybean (*Glycine max*), contributes around 16.4 million tonnes N $year^{-1}$, representing 77% of the N fixed by crop legumes globally (Herridge et al. 2008). In natural ecosystems, such as tropical forests where N-fixing trees are abundant, N-fixing bacterial symbionts can contribute upward of 20% of all N acquired by plants annually (Van der Heijden et al. 2008) and play a key role in supplying N to support rapid post-disturbance forest regrowth on nutrient-limited tropical soils (Batterman et al. 2013). Recent evidence also indicates that fast-growing N-fixing trees in tropical forests can promote specialised groups of microbes involved in mineral weathering, thereby unlocking mineral nutrients to sustain forest growth (Epihov et al. 2021). Specifically, the acidification of soil in the immediate vicinity of N-fixing trees, caused by enhanced C mineralisation and carbonic acid production, was found to promote acidophilic Fe-reducing bacteria, which enhance the dissolution of Fe-bearing minerals and release of inorganic P occluded within them to support forest growth (Epihov et al. 2021). N that is fixed by N-fixing symbionts can also be transferred to neighbouring non-N-fixing plants, thereby benefitting their growth and overall ecosystem productivity. This happens via several mechanisms, including: (i) decomposition of legume root tissue enriched in N and its subsequent uptake by neighbouring plants of mineralised N; (ii) exudation of N-rich compounds from legume roots and uptake of this N by neighbouring plants; and (iii) direct transfer of legume N to neighbouring plants via root-associated mycorrhizal fungal networks (Thilakarathna et al. 2016).

Another important pathway of N fixation is through free-living and associative N-fixing bacteria, which are ubiquitous in terrestrial ecosystems and can fix relatively small, but nevertheless significant amounts of N (< 3 kg N ha^{-1} $year^{-1}$) (Cleveland et al. 1999). For instance, there is evidence that plants can recruit N-fixing bacteria that do not form root nodules, but live in the rhizosphere, where they can supply significant quantities of N to the plant

(Bouffaud et al. 2016). Also, N-fixing cyanobacteria can contribute significant amounts of N to terrestrial ecosystems, especially dry and/or cold regions, where the growth of vascular plants is limited. For example, in arid ecosystems they occur in biological soil crusts (Figure 3.3) (Box 3.2) composed of assemblages of cyanobacteria, lichens, and mosses (Belnap 2003) and contribute around 1 kg N ha^{-1} $year^{-1}$, representing the dominant source of N to the system (Belnap 2003, Belnap and Lange 2003, Bowker et al. 2010).

Cyanobacteria are abundant as biological crusts in recently exposed glacial terrain (Nemergut et al. 2007, Schmidt et al. 2008), where they play a vital role in soil development, fixing appreciable amounts of N and stabilising surface soil particles. They also have an important influence on the N economy of boreal forest, where N fixation has traditionally been thought to be extremely limited. For instance, an N-fixing symbiosis between a cyanobacterium (*Nostoc* sp.) and the ubiquitous feather moss *Pleurozium schreberi* has been shown to fix significant quantities of N (1.5–2.0 kg N ha^{-1} $year^{-1}$) and contribute as much as 50% of the total N input, especially in late successional and old growth boreal forests (DeLuca et al. 2002, 2008, Lagerström et al. 2007, Gundale et al. 2010) (Figure 3.4). This finding is especially significant since *P. schreberi* is the most common moss on Earth, being found throughout North and South America, Greenland, Asia, Europe, and Africa, and in boreal forests it often forms a continuous carpet, accounting for up to 80% of the ground cover (DeLuca et al. 2002).

Microbial P mineralisation

In contrast to N, most soil P is in a range of insoluble inorganic forms that are unavailable for plant uptake. As a result, many of the reactions that govern plant availability of P are geochemical, rather than biological, and involve the removal of available phosphate from soil solution into the solid phase

Figure 3.3 Biological soil crust, Canyonlands National Park, Utah, USA. Image by Richard Bardgett.

Box 3.2 Biological soil crusts

Extreme environments such as deserts are often thought to be lifeless places. However, despite their barren appearance, they are teaming with life; rocks and sand on the soil surface are often covered with a thin film of cyanobacteria, microfungi, lichens, and/or mosses, collectively known as biological soil crusts. Most of the cohesion in biological crusts is due to large filamentous cyanobacteria, whose filaments link otherwise loose particles together to form larger soil aggregates. Key to their survival in extreme environments is their ability to tolerate prolonged periods of dehydration, resulting from lengthy periods of high heat, strong light, and no water. Many species have also evolved ways of prolonging their activity under harsh conditions. For example, many lichens can photosynthesise over a wide range of temperatures, whilst others require little water to begin metabolism. Pigments are also used by many crust species to reflect and/or absorb excessive radiation (Belnap 2003).

Biological crusts have many roles in the formation, stabilisation, and fertility of soils (Belnap 2003, Belnap and Lange 2003). They capture nutrient-rich dusts, thereby increasing both the fertility and water-holding capacity of soils. Crusts add nutrients to soil via fixation of atmospheric C and N; in deserts, C inputs from cyanobacterial crusts range from 0.4 to 2.3 g m^{-2} and N inputs from N fixation are typically around 1 kg ha^{-1} $year^{-1}$ (Belnap 2003), representing the dominant source of N in arid and semi-arid ecosystems. They enhance soil fertility by exuding negatively charged polysaccharides that increase soil aggregation and retain positively charged nutrients (e.g., NH_4 and Ca^{2+}), thereby preventing their leaching from soil. Soil food webs beneath crusts are often more abundant and diverse, thereby further promoting rates of decomposition and nutrient cycling. All these factors contribute to increased soil fertility, often facilitating the growth and nutritional status of vascular plants. These functionally important soil organisms are extremely vulnerable to climate change and disturbances such as off-road vehicles and trampling by grazing livestock, and their recovery from such disturbances is slow, taking decades to centuries.

(Box 3.3). For this reason, most agricultural soils are supplemented with inorganic P fertilisers to sustain crop production, which often has adverse environmental impacts due to the leakage of excess P into groundwater and adjacent waterways, with consequences for water quality (Sharpley et al. 2013, Haygarth et al. 2014). Nevertheless, soil microbes play a central role in regulating soil P availability through the solubilisation of sparingly available inorganic P forms and the mineralisation of organic P and the immobilisation of P into microbial biomass (Richardson et al. 2009, Richardson and Simpson 2011) (Figure 3.5).

Box 3.3 Phosphate availability

Phosphate is the main inorganic form of P that is available to plants. This phosphate is often divided into two forms: (1) the *labile* P pool, which constitutes the P on mineral surfaces that can be easily desorbed by plants and phosphate ions in soil solution. This form of P is immediately available to plants; and (2) P that precipitates with Fe and Al compounds, becoming insoluble and unavailable to plants. This latter form of P is referred to as *occluded* P.

Amounts of labile P in soil are very low relative to the total P pool, because any P released by mineralisation is rapidly adsorbed by soil particles, a process called *fixation*, rendering it unavailable to plants. The rate of fixation depends on a range of soil factors, but mostly soil pH. The most serious fixation occurs in very acid soils (pH < 5) where there are large amounts of Fe and Al hydroxides in soil solution. These combine with phosphate to produce Fe and Al phosphates that are insoluble and hence unavailable to plants. This is especially prevalent in highly weathered soils, such as those of the tropics (i.e., Oxisols and Ultisols), where the growth of plants is often strongly P limited. In calcareous soils, where pH values exceed 7, a different type of fixation occurs. Here, phosphate combines with Ca to form insoluble phosphates such as tricalcium phosphate ($Ca_3(PO_4)_2$). Fixation of P is at its lowest in soils of pH 5.5–6.5, since fixation by Fe, Al, and Ca is at minimal. Hence, availability of P to plants is maximal at this pH range.

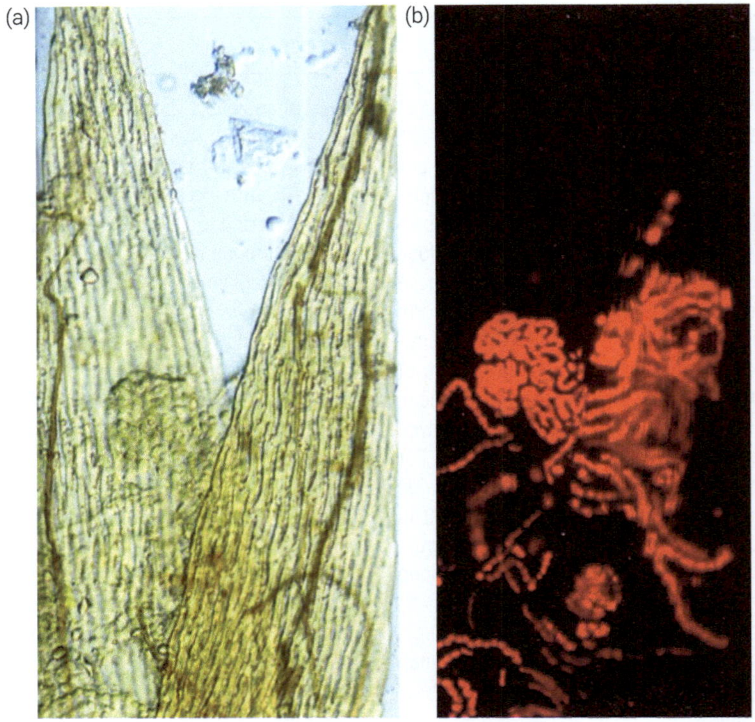

Figure 3.4 N-fixing cyanobacteria which inhabit the incurves of leaves of the feather moss *Pleurozium schreberi* fix significant quantities of N and act as a major contributor to N accumulation and cycling in boreal forests, especially in late successional systems and old-growth forests. (a) Under light microscope and (b) under ultraviolet-fluorescence micrograph with a green filter. Coiled chains of *Nostoc* are hidden in the leaf under light microscopy but are readily observed as red cells under ultraviolet-fluorescence microscopy with a green filter. From De Luca et al (2002) with permission from Springer Nature. Images by P. Lundgren and U. Rasmussen, respectively.

The mineralisation of organic P is a key pathway of making P available to plants, which is governed by the production of plant and microbial phosphatases, which are especially active in P-deficient soils and in the rhizosphere. These enzymes act by cleaving ester binds in organic matter to liberate phosphate (PO_4^{3-}) that can be taken up by the plant or microbial biomass. As with N, the mineralisation of organic P is partly regulated by the C:P ratio of organic substrates under microbial attack. In general, when the C:P ratio rises above ~ 100, P is immobilised by microbes that have a relatively high P requirement (1.5–2.5% P by dry weight compared to 0.05–0.5% for plants), and, as a result, microbes compete aggressively with plants for P. The importance of microbial P immobilisation is illustrated by the fact that the microbial biomass contains as much as 20–30% of the total soil organic P pool (Jonasson et al. 1999), which is much larger than the proportion of C (~ 1–2%) or N (~ 2–10%) that is typically contained in microbes. Microbial P is therefore an important potential source of plant-available P, which is released and made available to plants via a variety of mechanisms. These include recurring dry–wet cycles, which lead to microbial cell death and the liberation of previously locked-up P, and predation of microorganisms by protists and microfauna and subsequent excretion of consumed microbial P into the soil environment, i.e., the 'microbial loop'. Mycorrhizal fungi also play an important role in the transfer of P in soil systems, as will be discussed in the next section.

As noted above, a large amount of inorganic P in soil is in insoluble forms that are unavailable to plants due to precipitation reactions with aluminium (Al^{3+}), iron (Fe^{3+}), and calcium (Ca^{3+}).

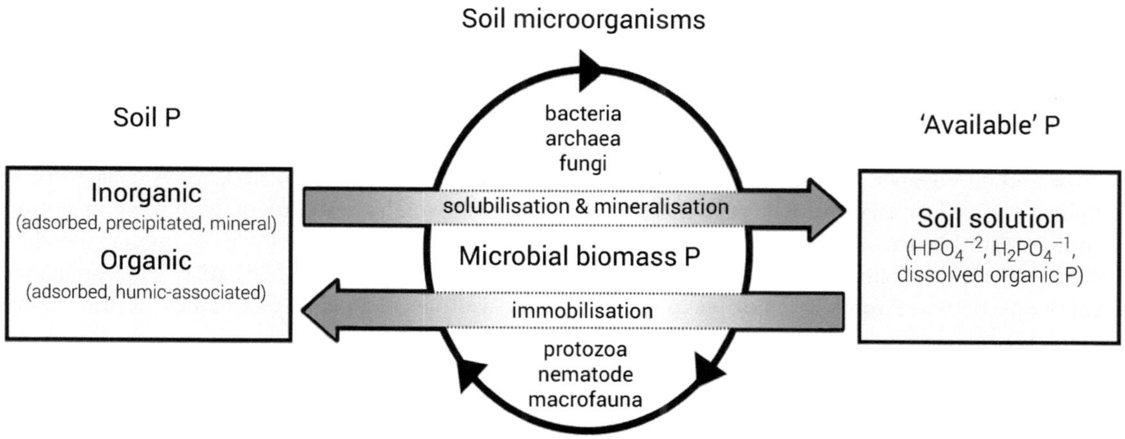

Figure 3.5 Schematic of the roles of microorganisms in regulating soil P availability. From Richardson and Simpson (2011) by permission of Oxford University Press.
Plant Physiol, Volume 156, Issue 3, July 2011, Pages 989–996, https://doi.org/10.1104/pp.111.175448.

Many soil microorganisms—referred to as P-solubilising microorganisms—have the capacity to solubilise these insoluble inorganic P forms by releasing organic acids (e.g., oxalic acid, citric acid, gluconic acid, lactic acid, and aspartic acid) that either directly dissolve precipitated P or chelate the Fe, Al, and Ca ions bound to phosphate, thereby releasing P (Sharma et al. 2013, Alori et al. 2017). The release of organic acids by microbes leads to acidification of microbial cells and their surroundings, which ultimately leads to the release of P ions from the P mineral by H^+ substitution for the cation bound to phosphate. For these reasons, the inoculation of agricultural soils with P-solubilising microorganisms to enhance P availability has attracted much recent attention (Sharma et al. 2013, Alori et al. 2017). However, while many bacteria, fungi, and actinomycetes are capable of solubilising inorganic P in laboratory settings and can improve plant P uptake in pot experiments, their importance for plant nutrition in the field remains uncertain (Richardson and Simpson 2011).

Mycorrhizal fungi and plant nutrient supply

Mycorrhizal fungi, which form symbiotic associations with plants, are ubiquitous in plant communities, but, as discussed in Chapter 2, there is a distinct pattern in the distribution of different mycorrhizal types according to biome, soil type, and limiting resources (Read 1991, Read and Perez-Moreno 2003): ERM fungi are prevalent in high latitude and high altitude ecosystems dominated by dwarf-shrub heathland vegetation on acidic soils of high organic matter content, in which the nutrients N and P are present mainly in combination with organic residues; ECM fungi dominate in boreal and temperate forest ecosystems that typically occur on acidic, organic soils, with most N and P in organic forms; and vegetation of tropical forest, grassland, and desert predominately have AM fungal associations. This pattern provides a simple framework to characterise the functional role of mycorrhizal fungi in relation to broad vegetation types and biogeochemical cycles.

The main benefit to plants of partnering with mycorrhizal fungi is enhanced nutrient (N and P) acquisition from soil, as they grow deep into the soil matrix, accessing nutrients that would otherwise not be available to roots. In line with this, plants typically allocate more C to their mycorrhizal partners when soil nutrients are limiting, and even preferentially allocate C to those fungal partners that best supply them with N and P (Bever 2015). Indeed, plants can be choosey about which fungi they trade with, seeking out those that give them the best deal in terms of nutrient supply and avoiding those that hoard nutrients rather than paying them back for the C they provide, implying a form of below-ground trading market (Kiers et al. 2011, Tedersoo et al. 2020). It has also been suggested that plants benefit from having multiple trading partners, with nutrients becoming 'cheaper' per unit of C with a greater number of fungal traders (Argüello et al. 2016).

In line with this, the idea that mycorrhizal fungi connect with and exchange nutrients and signalling compounds between roots of multiple plants of the same or different plant species through a common mycorrhizal network has become very popular (Figure 3.6). Evidence for this comes from field studies, including a classic study showing bidirectional C transfer between ECM tree species in forests (Simard et al. 1997), as well as work on orchids, which are at some stage of their life entirely dependent on receiving C via common mycorrhizal networks (Read et al. 2024). Also, by applying stable isotope C tracing techniques at the canopy

Figure 3.6 Hypothetical plant community including plant species that associate with different types of mycorrhizal fungi and which form three separate underground networks: (1) trees forming networks with ECM fungi (solid thin lines) are interconnected (arrow*); various plant species (2) and a tree (3) form AM networks and are also interconnected (dashed lines, arrow**); and (4) an orchid forms a third underground network. Different colours represent different mycorrhizal species of ERM fungi (solid thin lines) and AM fungi (dashed thin lines). Note that other combinations are possible (e.g., temperate forests with ERM trees often harbour an understorey of shrubs (e.g., *Vaccinium*) that form ECM associations). In these forests some fungi form both ERM and ECM associations, meaning that there might be interlinkages between the two networks. From Van der Heijden et al. (2015) by permission of John Wiley and Sons.
New Phytologist, Volume: 205, Issue: 4, Pages: 1406–1423. First published: 02 February 2015, DOI: (10.1111/nph.13288).

scale, C assimilated by individual spruce trees has been shown to be transferred to neighbouring trees, assisted by common mycorrhizal networks (Klein et al. 2016). But despite the evidence, the idea of resource sharing via fungal networks is not without controversy. Notably, it has been argued that popular claims that common mycorrhizal networks are widespread in forests are insufficiently supported, as are claims that mature trees preferentially transfer resources to offspring via these networks, thereby promoting their survival (Karst et al. 2023). So, while it is evident that different plants are connected by mycorrhizal fungi and that resources and chemical signals can be transferred belowground between plants, it remains unclear whether this is mediated by common mycorrhizal networks, and hence this warrants further study (Karst et al. 2023).

Types of mycorrhizal fungi differ in their nutrient acquisition strategies, which reflects, and contributes to, biogeochemical variation in ecosystems (Read and Perez-Moreno 2003). AM fungi, which dominate in more nutrient-rich ecosystems, characterised by high rates of N mineralisation and availability of mobile inorganic N, can uptake and transport a variety of chemical forms of N, both inorganic and organic (Hodge and Storer 2015), although they are believed to primarily use inorganic N forms (Smith and Smith 2011). Also, AM fungi have a high propensity for P acquisition by extending hyphae beyond the root zone into the soil matrix, and they produce phosphatase enzymes that release phosphate from organic matter, which is then taken up by the fungus and transported to the host plant (Read and Perez-Moreno 2003). In contrast, ECM and ERM fungi, which typically dominate in organic soils of low inorganic N availability, produce extracellular enzymes that degrade N-bearing organic compounds, such as chitin, lignin, and polyphenols, into constituent amino acids, which are assimilated by the fungus and transferred to the host plant (Read and Perez-Moreno 2003). Additionally, ECM fungi produce low molecular weight organic acids, such as oxalates, that increase P availability by weathering minerals and complexing cations (e.g., Ca, Fe, and Al) that would otherwise bind to phosphate, thereby taking it out of soil solution (Van Breeman et al. 2000). While it is generally thought that AM fungi lack saprotrophic capabilities, there is evidence that they can both enhance decomposition of and increase N capture from complex organic material in soil (Hodge et al. 2001, Leigh et al. 2009), but the mechanisms involved and their significance for plant nutrition remain elusive. Indeed, it has been suggested that AM fungal hyphae provide a 'highway' for bacteria to migrate to organic patches in soil to fulfil their role in nutrient mineralisation (Jansa and Hodge 2021). This has been shown to be the case for P-solubilising bacteria, which can swim in water films along AM fungal hyphae towards patches of organic P, which effectively extends the functional capability of the fungus to use otherwise inaccessible sources of P (Jiang et al. 2021).

Based on differences in the nutrient acquisition strategies of AM and ECM fungi, it has been proposed that forests dominated by AM- or ECM-associated trees have unique biogeochemical syndromes, or nutrient economies. This idea is conceptualised by the Mycorrhizal-Associated Nutrient Economy (MANE) framework for temperate forests, which predicts that: (i) an inorganic nutrient economy occurs in AM-dominated forest stands due to rapid decomposition of high quality litter and fast rates of C and N mineralisation; whereas (ii) in ECM-dominated forest stands, an organic nutrient economy is expected due to low quality litter pools and slow rates of C and N turnover, resulting in limited losses of inorganic N nutrients (Figure 3.7) (Phillips et al. 2013).

The MANE framework is supported by nutrient cycling data from temperate forest in the USA with a mixed assemblage of AM and ECM trees (Figure 3.8). Here, the ratio of organic N to inorganic N in surface soil was found to be positively related to the percentage of ECM trees at a plot scale, suggesting that organic N forms increase in importance as ECM trees become more abundant across the landscape. Conversely, rates of nitrification, a key process governing inorganic N cycling and N loss from soil by leaching, increase with the proportion of AM trees, whereas increasing abundance of ECM trees has little, if any, impact on rates of nitrification. This framework requires further testing across a range of ecosystems, but it provides a simple basis for understanding how forests dominated by

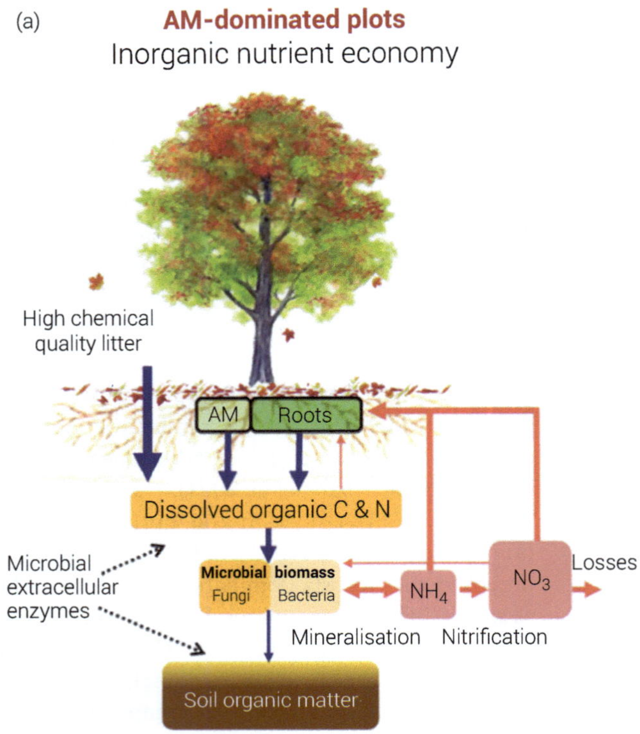

Figure 3.7 The Mycorrhizal-Associated Nutrient Economy (MANE) framework for temperate forests, which predicts an inorganic nutrient economy in AM-dominated forest stands as a result of the fast decomposition of high quality litter pools and elevated rates of C and N mineralisation, and an organic nutrient economy in ECM-dominated forest stands because of low quality litter pools and slow rates of C and N turnover, resulting in limited losses of inorganic nutrients. For each mycorrhizal group, the size of the boxes and arrows indicates the relative importance of pools and fluxes, respectively. Blue arrows relate to C fluxes and red arrows to N fluxes. From Phillips et al. (2013) by permission of John Wiley and Sons.
New Phytologist, Volume: 205, Issue: 4, Pages: 1406–1423. First published: 02 February 2015, DOI: (10.1111/nph.13288).

AM- or ECM-associated trees, and shifts in their relative abundance, contribute to spatial patterns of biogeochemical cycling at a landscape scale.

Rhizosphere priming

Another important mechanism by which plants access soil nutrients, especially N, is rhizosphere priming. This is defined as a short-term change in microbial mineralisation of SOM caused by the release of C-rich exudates from plants (Moreau et al. 2019). There is much uncertainty concerning the mechanisms by which root exudates influence microbial mineralisation–immobilisation N dynamics, and hence the supply of N to plants (Dijkstra et al. 2013, Moreau et al. 2019). But it is generally thought that in soils where N is limiting, root exudates stimulate the microbial production of extracellular enzymes (i.e., proteases and deaminases) to release N from N-containing polymers in SOM, such as proteins and chitin: this not only generates enough N to satisfy microbial N demand, but also enhances N mineralisation and N supply to plants (Zhu et al. 2014, Meier et al. 2017). Moreover, given that a large portion of organic N in soil resides in microbial residues formed following microbial death (i.e., microbial necromass), rhizosphere priming of microbial necromass recycling also represents a key strategy for plants to meet their N requirements (Pausch et al. 2024). In N-poor soils, however, rhizosphere microbes are more

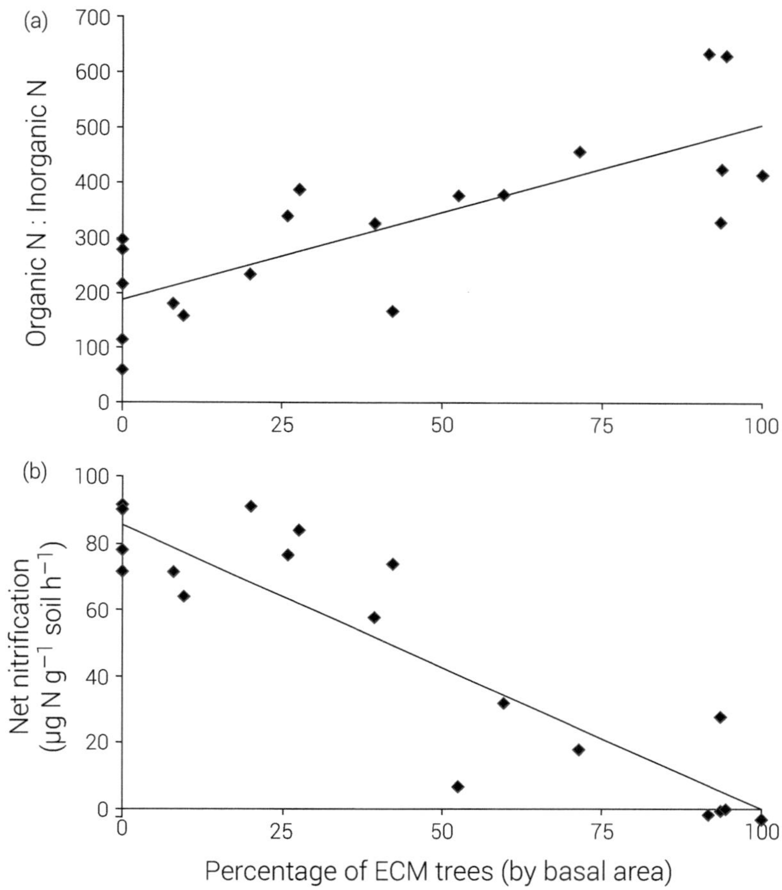

Figure 3.8 Relationship between the percentage of ECM trees and soil N cycling in a temperate forest in the USA. As the percentage of ECM trees increases, the ratio of organic N to inorganic N in the surface soil increases (a), while rates of nitrification decrease (b). Data indicate that forest plots dominated by ECM trees have soil pools dominated by organic N and low nitrification rates. From Phillips et al. (2013) by permission of John Wiley and Sons.
New Phytologist, Volume: 199, Issue: 1, Pages: 41–51. First published: 17 April 2013, DOI: (10.1111/nph.12221).

C than N limited and hence they use C-rich root exudates to meet their C and energy needs, rather than mining SOM for N, which leads to reduced mineralisation of soil organic N (Dijkstra et al. 2013). Whether priming has positive or negative effects on plant N supply also depends on other nutrients: in soils that are more P than N limited, root exudates are used by microbes to mobilise P, rather than to mine SOM for N, thereby reducing N mineralisation and plant N supply (Figure 3.9) (Dijkstra et al. 2013).

Many uncertainties remain about the significance of rhizosphere priming for plant N supply and how factors such as plant traits (e.g., plant N status), soil abiotic properties (e.g., moisture, texture, pH), and exudate quality and quantity influence priming (Moreau et al. 2019). Nevertheless, it is becoming clear that rhizosphere priming plays a critical role in driving microbial N-cycling processes as well as SOM decomposition, with consequences for C storage and loss (Kuzyakov 2010). Indeed, a synthesis of published studies revealed that, although highly variable, living roots can stimulate the rhizosphere priming effect by as much as 380% compared to rootless soil (Cheng et al. 2014), demonstrating its importance for SOM dynamics and storage.

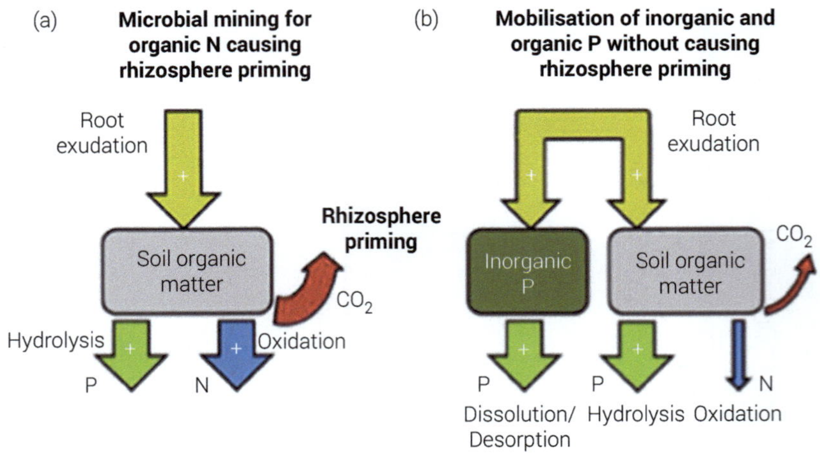

Figure 3.9 Schematic of how the availability of N and P in the soil can influence rhizosphere priming. When N availability is low, microbes use rhizodeposition to mine for N locked in organic matter, thereby increasing rhizosphere priming and the release of N (through oxidation of SOM) and P (mostly through hydrolysis of P-esters in SOM) (a). When P availability is low, rhizodeposition is used to mobilise P from inorganic and organic sources (through dissolution/desorption and hydrolysis, respectively), thereby increasing the release of P without affecting rhizosphere priming (b). From Dijkstra et al. (2013).

Further, as will be discussed later in Chapter 6, priming effects are of particular importance under global change, with studies showing that enhanced root exudation and the stimulation of extracellular enzymes involved in N mineralisation is a key pathway by which trees sustain long-term increases in growth under elevated CO_2 (Langley et al. 2009, Phillips et al. 2011). It has also been suggested that priming might be an evolutionary strategy developed between plants and rhizosphere microorganisms, whereby microbes benefit from plant-derived C; in turn, their plant host benefits through enhanced N supply via microbial mineralisation of soil organic N (Cheng et al. 2014).

Trophic interactions as control points of nutrient availability

The previous section considered how the mineralisation of nutrients is governed directly by the activities of microorganisms, principally bacteria and fungi. However, as identified in Chapter 2, the soil is also home to a rich and abundant community of invertebrates, including nematodes, collembolans, mites, and enchytraeids, which derive nutrients and C from roots and their exudates, but also from feeding on the microbes that live in the root zone (Pollierer et al. 2007, Gilbert et al. 2014, Bradford 2016). These faunal consumers of microbes, and their predators, strongly regulate soil nutrient cycling by altering the balance between microbial immobilisation and mineralisation, and hence the availability of nutrients for plant uptake. There are many ways that soil animals can affect microbial nutrient mineralisation, but for convenience they are typically divided into three broad pathways. First, soil animals affect nutrient cycling through their selective feeding on microbes, which alters microbial activity, abundance, and community structure. Second, soil animals affect nutrient mineralisation by altering the form of the organic matter in soil, in that they fragment and mix organic matter, increasing its susceptibility to microbial attack. Third, soil animals affect nutrient cycling via trophic cascades induced by higher-level predators that feed on animals of lower trophic levels that consume fungi, bacteria, and plant litter. Combined, these interactions between microbes and animals drive ecosystem-level processes of energy flow and nutrient cycling and contribute markedly to plant nutrient acquisition and growth. In the following section, the effects of these biotic interactions on the structure and activity of microbial communities and the consequences of this for plant nutrient availability are discussed.

Selective feeding

While many soil animals are generalist in their feeding behaviour (Ponsard and Arditi 2000, Scheu and Falca 2000, Maraun et al. 2003), there is also much evidence that they can be very selective in their feeding behaviour. It has long been known, for example, that nematodes have mouth parts that are adapted to feeding on either bacteria (bacterivores) or fungi (fungivores) (Yeates et al. 1993), and while not animals, certain protists even use chemical cues to discriminate between their bacterial and algal prey (Verity 1991). Despite feeding on a variety of food sources, studies reveal that soil microarthropods, especially Collembola, can be highly selective in their feeding behaviour, choosing to feed on some fungal species over others (Klironomos et al. 1992, Bardgett et al. 1993b, Jørgensen et al. 2005), often with fitness benefits in the form of increased growth and reproduction (Scheu and Simmerling 2004). Selective foraging is based on numerous factors, including the palatability of the fungus, with animals selectively feeding on metabolically active hyphae in preference to dead or inactive hyphae (Moore et al. 1985) and on N-rich regions of the fungal thallus (Leonard 1984). Preferential grazing can also arise from the avoidance of toxins produced by some fungal colonies (Parkinson et al. 1979), and some collembolan species can even locate and select their fungal food source by volatile compounds released from fungal mycelium (Bengtsson et al. 1988).

Through their selective grazing, soil invertebrates can exert strong selective pressures on fungal communities, with consequences for fungal community structure and diversity. Selective grazing by collembolans on the fungus *Marasmius androsaceus* colonising Sitka spruce (*Picea sitchensis*) needles, for example, reduced the abundance of this palatable fungus and increased the abundance of an unpalatable fungus *Mycena galopus*, which was also present on the needles (Newell 1984a, b). This change in fungal community structure also had consequences for ecosystem functioning, in that it reduced decomposition rates of coniferous litter since *M. galopus* decomposes litter at a slower rate than *M. androsaceus* (Newell 1984a, b). Selective grazing has also been shown to reverse the outcomes of competitive interactions of common decomposer fungi: by feeding selectively on the dominant wood-decaying basidiomycete fungus *Resinicium bicolor*, isopods (*Oniscus asellus*) prevented the competitive exclusion of other less competitive fungi of soil and wood, whereas grazing by nematodes reversed the outcomes of competitive interactions by stimulating growth of less competitive fungi (Crowther et al. 2011).

The above findings demonstrate two opposing mechanisms by which selective grazing by soil animals can influence fungal community composition: (i) the restriction of a more competitive fungal species through consumption; and (ii) the stimulation of a less competitive fungal species due to compensatory growth, which is likely due to the stimulation of new fungal growth (Hedlund et al. 1991) and/or mobilisation of storage compounds and increased nutrient uptake from organic substrates to counteract negative effects of grazing (Crowther et al. 2011). Indeed, it has long been known that grazing by collembolans can stimulate the growth and physiological activity of some fungal species (Hanlon and Anderson 1979, Bengtsson and Rundgren 1983, Bardgett et al. 1993b), but reduce the growth and physiological activity of others (Warnock et al. 1982). As such, grazing by invertebrates has the capacity to alter competitive interactions of soil fungi, with consequences for fungal community structure and the ecosystem functions that fungi drive.

Microbial-feeders, soil nutrient cycling, and plant growth

The effects of microbial-feeding organisms on nutrient mineralisation and plant growth are generally positive. The primary mechanism involved is the liberation of nutrients from microbes because of predation by microbial-feeding organisms, including microarthropods, nematodes, and protists, which excrete nutrients that are in excess of their own requirements into the soil and in forms that are available for plant uptake, i.e., the 'microbial loop' (Clarholm 1985). This process is widely recognised as a primary control point of nutrient availability in the rhizosphere (Moore et al. 2003, Bonkowski et al. 2009), and is not just driven by predatory animals and protists, but also by viral infection of microbes, which causes cell lysis and the release of considerable amounts of N into the soil environment (Kuzyakov and Mason-Jones 2018), and by bacterial predators that are known to be highly active in soil (Hungate et al. 2021, Petters et al. 2021, Groß et al. 2023). These biotic interactions are intimately linked to roots and their exudates that stimulate microbial populations in the rhizosphere, which increases the abundance of microbial consumers and their higher-level predators. The resulting stimulation of plant growth is due to two mechanisms: first, microbial consumers excrete excess N into the rhizosphere, making it available for plant uptake; and second, they stimulate microbial growth and turnover, which in turn stimulates consumers and their higher-level predators, thereby contributing to greater N supply to plants (Moore et al. 2003). As such, nutrient availability to plants is often controlled several trophic steps away from the root through 'top-down' control by trophic interactions in the rhizosphere.

There is a rich literature demonstrating that trophic interactions between animals and microbes play a key role in regulating nutrient availability in the rhizosphere, with consequences for plant nutrition and growth. A classic study by Ingham et al. (1985) showed that the addition of microbial-feeding nematodes to soil enhanced N uptake by, and growth of, a grass species (Figure 3.10a). Similarly, when acting together, Collembola and nematodes were found to increase concentrations of NH_4^+ in the soil solution, leading to increased plant nutrient uptake by the grass *Nardus stricta* (Bardgett and Chan 1999). Classic studies of tree responses to microbial-feeding animals show similar responses. For example, leaf, stem, and shoot biomass of birch seedlings (*Betula pendula*) was found to increase when grown in the presence of a diverse fauna of boreal forest soil (Figure 3.10b) (Setälä and Huhta 1991), and faunal grazing on ECM fungi associated with *B. pendula* enhanced nutrient uptake and growth of *B. pendula*, despite reductions in the biomass of ECM (Setälä 1995). While not classified as animals, protists also play a crucial role in nutrient cycling via their predation on bacteria and other soil microbes, which can liberate as much as a third of the consumed N into the rhizosphere (Geisen et al. 2018). There is also evidence that protists facilitate plant N uptake indirectly by enhancing N uptake by root-associated AM fungi, which are very efficient in the uptake of N released from bacteria by bacterial-feeding protists (Herdler et al. 2008, Koller et al. 2013a, b).

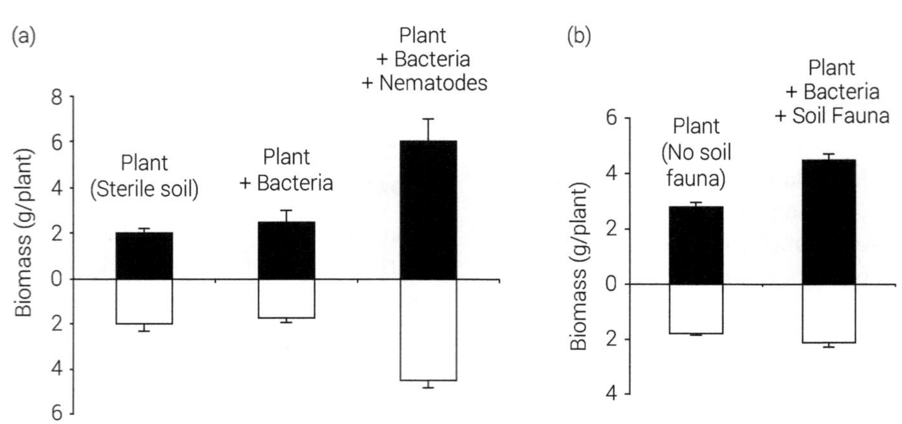

Figure 3.10 Influence of soil trophic interactions on plant growth. Adding bacterial-feeding nematodes to soil increased growth shoot and root growth of the perennial grass *Bouteloua gracilis* (a). Addition of diverse fauna to soil increased root and shoot biomass of birch (*Betula pendula*) seedlings in boreal forest soil (b). From Bardgett (2005) based on data from (a) Ingham et al. (1985) and (b) Setälä and Huhta (1991) by permission of Oxford University Press.

Microbial predation can also influence the performance of above-ground organisms, especially herbivores that benefit from enhanced plant nutrition. For example, plant-sucking aphids were found to perform better when the host plant was grown in the presence of microbial-feeding Collembola or earthworms, as compared to when these soil organisms were absent (Scheu et al. 1999). Similarly, bacterial-feeding protists were found to indirectly increase aphid abundance on barley shoots through their positive effects on soil N turnover and plant nutritional status (Bonkowski et al. 2001). These examples indicate that indirect linkages and feedbacks operate between consumer organisms of both above-ground and below-ground food webs.

Beneficial effects of microbial-feeding soil organisms on plant growth are mostly assigned to nutritional effects via the 'microbial loop'. However, predation of microbes in the rhizosphere can also have non-nutritional effects on plant growth via inducing the production of plant growth-promoting hormones by bacteria. Several studies show that plants grown in the presence of microbial-feeding protists develop extensive and highly branched root systems, and that these effects resemble those caused by plant growth-promoting hormones, such as auxin (Bonkowski and Brandt 2002, Bonkowski 2004). Whilst is has long been known that protists can themselves release auxins, selective feeding by protists in the rhizosphere can also stimulate the growth of auxin-producing bacteria, which has been proposed as a key mechanism by which bacterial-feeding protists impact root growth (Bonkowski and Brandt 2002, Bonkowski 2004) (Figure 3.11). It should be noted, however, that changes in root architecture may be due to other effects of protist feeding on bacterial communities. Stimulation of nitrifying bacteria, for example, may lead to hot spots of NO_3^- concentration in soil; as well as being a source of N for plant growth, NO_3^- acts as a signal for lateral root elongation and may aid direct root growth towards nutrient patches (Zhang and Forde 1998). Nevertheless, what these studies reveal is that microbial feeders in the rhizosphere impact plant growth through both nutritional and non-nutritional responses to microbial consumption (Bonkowski et al. 2009).

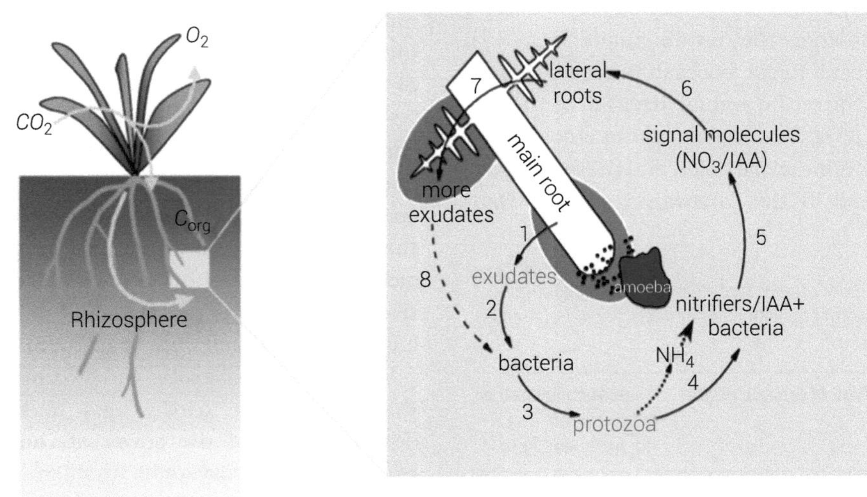

Figure 3.11 Schematic of microfaunal-induced hormonal effects on root growth, whereby root exudation (1) stimulates growth of a diverse bacterial community (2) and subsequently of bacterial-feeders such as protozoa (3). Ammonia is excreted by protozoa and selective grazing favours nitrifiers and indole-3-acetic acid (IAA+)-producing bacteria (4). The release of signal molecules (5), such as NO_3 and IAA, induces lateral root growth (6), leading to release of more exudates (7), and subsequent bacterial growth (8). From Bonkowski (2004) modified from Bonkowski and Brandt (2002) by permission of John Wiley and Sons.
New Phytologist, Volume: 162, Issue: 3, Pages: 617–631. First published: 13 April 2004, DOI: (10.1111/j.1469-8137.2004.01066.x.

Consumption of litter and production of faecal pellets

Soil fauna affect ecosystem processes of nutrient cycling via the activities of litter transformers, such as microarthropods and some macrofauna, that consume plant detritus and egest this material into soil as faecal pellets. Such faecal pellets have a higher surface-to-volume ratio compared with the original leaf litter (Webb 1977), and they also provide a highly favourable environment for microbial growth, which together enhances their decomposition and the release of nutrients (Zimmer and Topp 2002). Decomposition of dead plant litter is also enhanced through the action of enodosymbiotic microbes that live in the guts of many soil animals, such as earthworms, isopods, and termites. These enodosymbiotic microbes produce extracellular enzymes that degrade cellulose and phenolic compounds, which further enhances the degradation of ingested material (Zimmer and Topp 1998). Some animals, for instance many termites, have gut bacteria that fix atmospheric N and can potentially act as an ecosystem N input (Abe et al. 2000).

Earthworms consume vast amounts of litter fall in many ecosystems, but especially those with mull-type soils where they are abundant (Table 3.1). Indeed, in mixed forest ecosystems, earthworms can consume the entire annual litter fall of the forest (Satchell 1967). This organic matter is consumed along with soil mineral particles, and these two fractions are mixed in the earthworm gut and then egested as surface or subsurface casts. Estimates from various habitats around the globe show that earthworms produce between 2 and 250 tonnes of cast ha^{-1}, whereas typical values for many temperate ecosystems probably range from 20 to 40 tonnes ha^{-1} (Edwards and Bohlen 1996, Edwards 2004). These casts contain significantly greater numbers of microbes and have higher enzyme activities than surrounding soil, either because of enrichment of organic matter with microbes in the worm's intestine or because the cast provides a better substrate for microbial growth, being rich in organic matter and available nutrients. As a result, rates of organic matter decomposition are typically enhanced due to the presence of earthworms (Cortez et al. 1989).

Table 3.1 Amount of organic matter ingested or incorporated into soil by earthworm populations of different habitats (adapted from Bohlen 2002).

Ecosystem	Type of organic matter	Amount consumed or incorporated (kg ha^{-1} year^{-1})
Maize field	Maize residues	840
Orchard	Apple leaves	2000
Mixed forest	Canopy tree leaves	3000
Oak forest	Oak leaves	1071
Alfalfa field	Alfalfa residues	1220
Tallgrass prairie	Total organic matter	740–8980
Savanna	Total organic matter	1300

Similarly, rates of N mineralisation are greater in earthworm casts relative to surrounding soil, largely due to them being enriched with inorganic N and N-rich excretory products and mucus from the earthworm (Lavelle and Martin 1992), and P availability is also greater in casts than surrounding soil, largely due to the stimulation of phosphatase activity (Sharpley and Syers 1976). Whilst these positive effects of casting on microbial activity are often transitory, their net effect, along with other earthworm activities, is the stimulation of soil nutrient availability and enhanced plant growth. Indeed, a meta-analysis of studies exploring effects of earthworm presence on plant growth in agroecosystems found that, on average, the presence of earthworms increases crop yields by 25%, largely due to enhanced rates of soil N cycling (Van Groenigen et al. 2014a). Moreover, it has been estimated that earthworms contribute to roughly 6.5% of global grain production, and 10% of grain production in Sub-Saharan Africa and 8% in Latin America and the Caribbean (Fonte et al. 2023).

The invasion of exotic earthworms into regions that lack native earthworms provides a clear demonstration of the power of these organisms to transform ecosystems (Bohlen et al. 2004). For example, the invasion of exotic European earthworms into large parts of North America that lack native earthworm species—due to Pleistocene glaciations eliminating previous earthworm species—has had substantial impacts on forest ecosystems through stimulating organic matter breakdown, increasing soil C loss and leaching of nutrients, and making soils more vulnerable to erosion (Bohlen et al. 2004, Eisenhauer et al. 2019). Moreover, while the enhancement of organic matter mineralisation by invasive earthworms may have short-term benefits for plant nutrition and growth (Scheu and Parkinson 1994), in the longer term it can lead to complete removal of the surface soil organic layer, adversely affecting native plant species (Gundale 2002, Frelich et al. 2006) and ecosystem functioning, and promoting invasion by alien plant species (Bohlen et al. 2004).

Dung beetles, which are widespread in tropical forests and savannas, have significant impacts on soil nutrient cycling through the movement and burial of large quantities of mammalian faeces in soil, which they do for feeding and nesting purposes (Hanski and Cambefort 1991, Nichols et al. 2008). Numerous studies show that this burial of faeces stimulates microbial activity and rates of nutrient cycling, with notable benefits for the growth and nutrient content of plants (Nichols et al. 2008). Further, they play an important role in secondary seed dispersal, through relocating seeds in mammalian dung to beneath the soil surface, which can benefit seed germination and seedling establishment, because seeds are placed in a more suitable microenvironment where they are less vulnerable to predation (Shepherd and Chapman 1998, Andresen and Levey 2004, Griffiths et al. 2016). Through such roles, dung beetles not only act as key regulators of ecosystem processes, but also influence vegetation dynamics (Nichols et al. 2008).

Trophic cascades and soil functioning

The above examples focus mostly on microbial consumers and how they influence rhizosphere nutrient cycling and plant growth. However, soil food webs also include higher-level predators that consume fauna of lower trophic levels, such as those that feed on fungi, bacteria, and plant litter. These higher-level predators can induce 'trophic cascades' through changing the density of prey fauna, and thereby indirectly influence the abundance of soil microbes. Trophic cascades have long been recognised by soil ecologists to indirectly influence below-ground processes driven by soil microbes. For instance, the experimental reduction of predatory mites was shown to enhance densities of bacterial-feeding nematodes (their main prey source), which in turn reduced rates of microbial decomposition of plant litter (Santos et al. 1981). Similarly, reducing the abundance of large arthropod predators, including spiders, centipedes, beetles, and predatory larvae, was found to negatively affect the abundances of lower trophic levels in the soil food web and reduce litter decomposition in a dry beech forest: a 50% reduction in the density of large predators was associated with a 50% decline in rates of leaf litter decomposition (Melguizo-Ruiz et al. 2020).

The above studies demonstrate that predator-induced trophic cascades can be important drivers of processes of decomposition and nutrient cycling, although the nature of responses can vary considerably depending on environmental conditions. For instance, indirect effects of higher-level predators (i.e., spiders) on rates of plant litter decomposition on the forest floor have been shown to depend on moisture regime: at a dry site, spiders had no indirect effect on the rate of litter decomposition, whereas at a moist site, they accelerated litter decomposition under reduced rainfall, but had no effect under high rainfall (Lensing and Wise 2006). Further, pronounced indirect and negative effects of large-predator reductions on leaf litter decomposition have been demonstrated in dry forests, but not in wet forests (Melguizo-Ruiz et al. 2020). These and other studies showing that indirect effects of higher-level predators on ecosystem functioning depend, for example, on temperature (Koltz et al. 2018) and soil fertility (Lenoir et al. 2007) indicate that the nature and importance of trophic cascades in soil food webs depends strongly on environmental context and, as such, are responsive to environmental change.

There is evidence that non-consumptive trophic interactions may play an important role in shaping soil food webs and their functioning. It is well known that fear of predation can induce behavioural and phenotypic responses in a variety of organisms, with consequences for ecosystem nutrient cycling (Schmitz et al. 2010, Hawlena et al. 2012), and it has been proposed that such predator-induced responses could play a role in soil (Hawlena and Zaguri 2016). Experimental evidence for the occurrence of non-consumptive trophic interactions in soil is, however, scant. But studies show that collembolans, and presumably other soil fauna, change their behaviour in responses to cues of predator risk, especially necromones from conspecifics (Sitvarin et al. 2015), and that collembolans reduce their activity under risk of predation, resulting in lower rates of soil respiration (Sitvarin and Rypstra 2014). It is also known that termites respond to predation risk from ants by changing their foraging behaviour (Korb and Linsenmair 2002), which will likely impact soil functioning, and that risk of predation can induce bacterial defences with potential to alter soil processes (Jousset 2012). As such, there is clearly potential for non-consumptive interactions to cascade through soil food webs with consequences for soil functions.

A note on fungal and bacterial energy channels

Another approach to studying how soil food webs influence nutrient cycling is through exploring their energy channels. The general idea here is that although the detrital food web is complex, it consists of two broad assemblages of species, dubbed energy channels: the bacterial energy channel, composed of protists, rotifers, bacterial-feeding nematodes, and some microarthropods, and the fungal energy channel, consisting of saprophytic fungi, fungal-feeding nematodes, and microarthropods (Figure 3.12) (Moore et al. 2003). It has been proposed that these two energy channels differ in ways that influence soil nutrient cycling, with the bacterial energy channel representing a 'fast' cycle, characterised by rapid rates of nutrient mineralisation, and the fungal energy channel representing a 'slow' or 'conservative' cycle with low rates of nutrient mineralisation and greater nutrient retention (Coleman et al. 1983, Moore and Hunt 1988). This general idea is supported by knowledge that decomposer fungi can immobilise substantial quantities of nutrients in their hyphal networks, whereas bacteria have a more exploitative nutrient use strategy through rapid use of labile substrates. Moreover, because bacteria have low C:N ratios and concentrations of defence chemicals relative to fungi, bacteria are generally consumed by their predators to

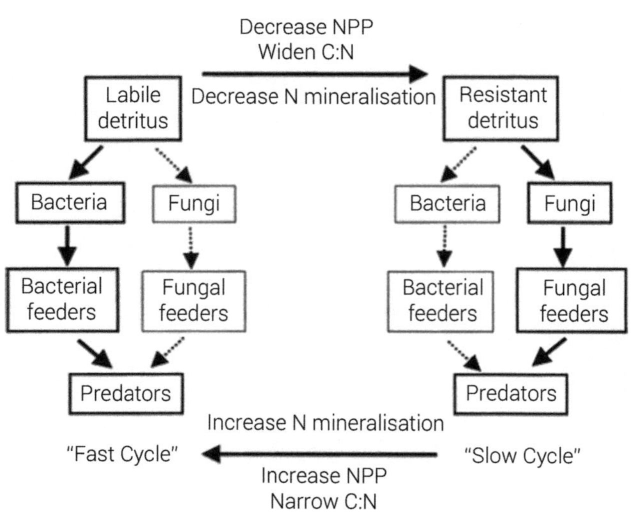

Figure 3.12 Simplified bacterial and fungal energy channels of the below-ground food web. The bacterial energy channel represents the 'fast cycle' due to the higher turnover rates of bacteria and their consumers relative to the fungi and their consumers, whereas the fungal energy channel represents the 'slow cycle'. Changes in the C:N ratio of the detritus, net primary productivity (NPP), disturbance, or rates of N mineralisation have been linked to shifts in the relative dominance of one channel relative to the other. From Moore et al. (2003) by permission of John Wiley and Sons.

a greater extent, which leads to greater nutrient turnover and release via the 'microbial loop' (Wardle 2002).

Following these ideas, it might be expected that ecosystems dominated by the bacterial energy channel will have high rates of nutrient cycling and nutrient availability and narrow C:N ratios, whereas those dominated by the fungal energy channel will have high organic matter and low resource availability and higher C:N ratios. In support, a global meta-analysis revealed that the soil fungal to bacterial ratio was positively related to the soil C:N ratio (Figure 3.13), being greatest in coniferous forest soils with high organic matter content and high C:N ratios and lowest in deserts and grasslands with low C:N ratios (Fierer et al. 2009). Consistent with this, a recent global analysis of the biogeography of the relative abundance of soil fungi versus bacteria revealed a striking latitudinal trend of increasing fungal dominance in cold and high latitude environments with large soil C stocks, and of bacterial dominance in warmer regions with high primary productivity (Yu et al. 2022). Moreover, studies done at a field scale reveal that increases in the relative dominance of the fungal over the bacterial channel, for instance resulting from reduced tillage or extensification of agricultural land management, are associated with a slowing of C and N cycling (Hendrix et al. 1986, Bardgett and McAlister 1999, De Vries et al. 2012a) and greater microbial retention of C and N, leading to reduced N loss in leachates (Gordon et al. 2008, De Vries et al. 2012a, b) and increased C storage potential (Six et al. 2006, Malik et al. 2016). Other soil physical and chemical factors might contribute to these patterns, but available evidence points to soils with fungal-dominated microbial communities being more efficient in retaining nutrients and having greater soil C stocks than their bacterial-dominated counterparts.

While the bacterial–fungal energy channel model provides a broad conceptual framework for exploring how shifts in soil food webs impact ecosystem nutrient and C cycling, it does have issues. One issue is that bacterial and fungal energy channels operate simultaneously, and hence nutrient

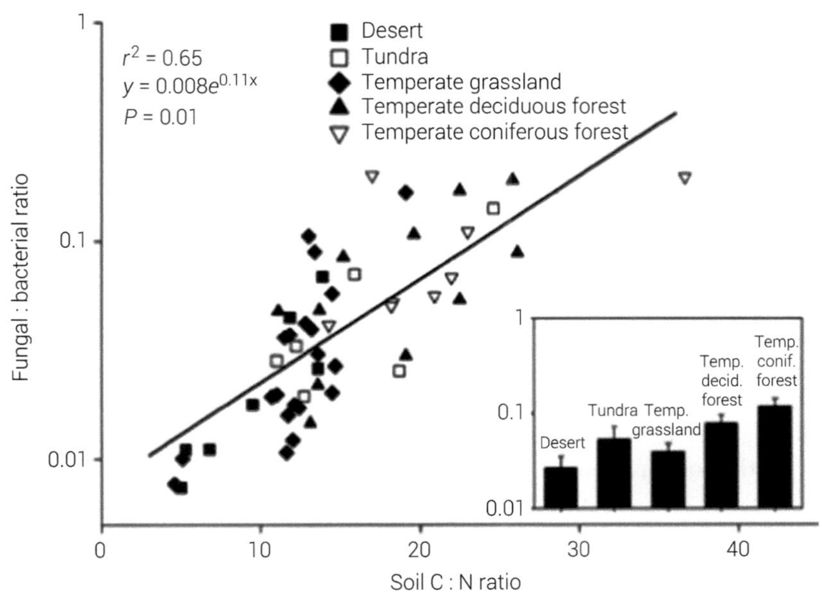

Figure 3.13 Relationships between fungal to bacterial ratios estimated using quantitative polymerase chain reaction and soil C:N ratios across five biomes. A fungal:bacterial ratio of 1 means that fungal and bacterial rRNA gene copies are in equal abundance. The bar chart in the insert shows averages and standard errors of the fungal:bacterial ratios determined for each of the five biomes. From Fierer et al. (2009) by permission of John Wiley and Sons.

availability to plants will be controlled by their relative dominance, rather than by the influence of one channel or the other. Following this, it has been proposed that plant growth is regulated simultaneously by top-down control through predation affecting nutrient mineralisation via the bacterial energy channel, and bottom-up control influencing nutrient immobilisation through the fungal energy channel (Moore et al. 2003). Another issue is that a clear split of the bacterial and fungal energy channels does not exist in nature; rather, soil food webs are highly complex and interconnected. As an example, protists are commonly treated as bacterivores that control the bacterial energy, but they display enormous functional versatility, feeding on fungi, bacteria, and nematodes, and acting as parasites of animals and plants (Geisen 2016). Also, the bacterial–fungal energy channel model assumes that bacteria consume labile substrates in soil at a much faster rate than do fungi, whereas there is evidence to suggest that this is not the case; rather, fungi use substantial amounts of labile C (De Vries and Caruso 2016). Given these issues, it has been argued that the bacterial–fungal energy channel model is oversimplified and that models need to be adapted to incorporate emerging understanding of the complexity of the soil food web (De Vries and Caruso 2016, Geisen 2016).

Soil biota contributions to soil biophysical properties

As well as directly regulating organic matter decomposition and nutrient transformations, soil biota indirectly influence ecosystem processes and plant growth by modifying their surrounding soil physical environment. As discussed in Chapter 2, the structural complexity of the soil environment strongly shapes the structure, function, and distribution of soil biota. But this relationship is reciprocal, as soil organisms also shape the soil physical matrix, with feedback consequences for the hydrological properties of soil, nutrient cycling processes, gas fluxes, and the growth of plants. Most literature in this area has focussed on the role of bacteria and fungi in the formation of soil aggregates, which are the basic building blocks of soil structure (Phillipot et al. 2023). A recent global meta-analysis, for example, showed that fungi contribute mostly to the formation of macroaggregates (i.e., those larger than 250 μm), whereas bacteria contribute to both macro- and microaggregate (i.e., those smaller than 250 μm) formation (Lehmann et al. 2017). Fungi contribute to macroaggregate formation due to their filamentous growth form, with fungal hyphae enmeshing mineral particles and microaggregates together. However, fungi also produce extracellular polymeric substances, such as polysaccharides, which act as aggregate binding agents, and hydrophobins (Rillig and Mummey 2006), which influence the wettability of aggregates, potentially helping to stabilise them (Lehmann et al. 2020a). Bacteria similarly promote aggregate formation through producing aggregate binding agents, but, also, filamentous bacteria that share 'fungal-like' characteristics, such as actinomycetes, likely enmesh mineral particles together in a similar way to fungal hyphae, thereby contributing to aggregate formation (Lehmann et al. 2017).

Soil fauna affect ecosystem properties by acting as ecosystem engineers, substantially modifying the physical structure of the soil profile, and hence the habitat and activities of other organisms and the passage of materials through soils. Engineers tend to be macrofauna, which exploit plant litter at the soil surface, such as earthworms and termites, and create macropores and channels because of their feeding and burrowing activities. Macrofauna also improve soil porosity and drainage through their burrowing activities. For example, earthworms increase the proportion of macropores and channels in soil (Knight et al. 1992, Lavelle et al. 1997), which enhances the movement of both water and soluble nutrients through soil and to interconnected waterways (Bardgett et al. 2001a). Indeed, studies that have eliminated earthworms from pasture using pesticides have reported dramatic declines in water infiltration of up to 93% (Sharpley et al. 1979) and, conversely, the inoculation of agricultural soils with earthworms caused a 4- to 12-fold increase in leachate volumes and a 10-fold increase in the amount of N contained within them (Subler et al. 1997). The inoculation of earthworms (*Aporrectodea caliginosa*) into limed coniferous forest soils likewise enhanced the amount of N in leachates by 50-fold, thereby increasing potential nutrient loss from these soils (Robinson et al. 1992, 1996). Such effects of earthworms on water infiltration and leaching are likely to have strong influences on nutrient availability to plants, and on the transfer of water and nutrients to groundwaters and adjacent ecosystems, with potential consequences for water quality (Bardgett et al. 2001a).

In tropical soils, the main ecosystem engineers are termites, which can reach densities upward of 7,000 individuals m^{-2} and make up some 95% of the total below-ground insect biomass (Aber et al. 2000). Termites feed mainly on decaying organic matter and roots, and through their mound building and gallery excavating activities they have substantial impacts on soil nutrient cycling and the physical structure and hydrology of soils (Aber et al. 2000, Hyodo et al. 2000). In doing so, termite mounds not only serve as foci for nutrient and water redistribution in many landscapes (Spain and McIvor 1988, Aber et al. 2000), but also strongly affect vegetation structure, especially in 'termite savanna' ecosystems, a term coined for vegetation formations in Africa and South America consisting of grasslands with discrete islands of woodland based on large termite mounds (Harris 1964). In these situations, a vegetation succession is triggered when mounds become stable, for example following colony death: grasses first colonise the mound, which are then replaced by shrubs and finally trees (Harris 1964).

Ants have substantial impacts on the soil physical environment through the formation of nests. Most ants build nests within soil, where they concentrate organic material and create hot spots or 'oasis' of high nutrient availability (Hölldobler and Wilson 1990, Moutinho et al. 2003, Verchot et al. 2003). These nests also create extensive macroporosity in soil, which enhances infiltration rates of water and soluble nutrients and provides an important route for the transmission of water down the soil profile, especially in arid and semi-arid regions (Lobry De Bruyn and Conacher 1994). Similarly, rates of water infiltration in soils with nest entrances of funnel ants (*Aphaenogaster barbigula*), which can reach densities of up to 88,000 ha^{-1}, were found to average 23.3 mm min^{-1}, compared to just 5.9 mm min^{-1} in nest-entrance-free soil (Eldridge 1993). It is of note that some ants, mostly of the genus *Formica* that are common in temperate and boreal forests (Hölldobler and Wilson 1990), build their nests on the soil surface rather than in soil. These surface mound-building ants can have substantial impacts on ecosystem processes, which largely operate through the enrichment of organic matter and nutrients in mounds relative to adjacent soil (Jurgensen et al. 2008) and through multitrophic effects on soil food webs (Wardle et al. 2011b), rather than through modifying soil physical properties.

Functional consequences of soil biodiversity

The last few decades have witnessed an explosion of studies examining the ecosystem consequences of declines in biological diversity, largely provoked by concerns over current unprecedented rates of global species extinction (Barnosky et al. 2011, Pimm et al. 2014). Although experiments examining biodiversity–function relationships have generated much debate, there is now a consensus that biodiversity loss reduces the efficiency by which ecological communities capture resources, produce biomass, and recycle biologically essential nutrients (Cardinale et al. 2012). Nevertheless, the bulk of studies on diversity–function relationships have been done on plant and aquatic communities, with soil biological diversity and its impact on ecosystem functioning receiving much less attention. This situation is, however, changing, and scientific interest in the functional importance of soil biodiversity has boomed (Bardgett and Van der Putten 2014), as has interest among policy makers and land managers in the importance of soil biodiversity for sustainable management, human health, and ecosystem services (Wall et al. 2015). This section considers the available evidence for relationships between soil biological diversity and ecosystem functioning and addresses the question: does soil biodiversity matter? The section focusses mainly on the role of soil biodiversity in regulating processes of decomposition and nutrient cycling, whereas its role as a driver of vegetation dynamics will be considered in Chapter 4.

It is well known that soil species, both within and across different trophic groups, differ in their effects on soil functioning. But what is less well known is how changes in the diversity and composition of soil communities influence the functioning of soil. Several experimental studies considered this issue in the 1990s and 2000s, and a synthesis of them concluded that diversity effects on processes of soil nutrient and C cycling are highly variable, but that effects of species loss are greatest at the low end of the diversity spectrum (Nielsen et al. 2011). While this later finding lends support to the idea that there is a high degree of functional redundancy in soil communities (i.e., many species in soil are replaceable with others without influencing general soil functions, such as decomposition and nutrient cycling), the synthesis also revealed that changes in the composition of soil communities and the variety of organism functional traits and biotic interaction, are of more importance for soil functioning than are changes in species richness per se (Nielsen et al. 2011). For example, using microcosms, functional dissimilarity among detritivores (i.e., the degree to which component species differ in their effects on ecosystem processes) was shown to be a better predictor of litter decomposition than was species richness (Heemsbergen et al. 2004). Also, selective removal of certain groups of organisms—using sieves of different sizes to eliminate different groups of soil organisms—from complex soil communities was found to impair ecosystem multifunctionality, based on the simultaneous quantification of multiple ecosystem functions related to soil nutrient cycling and plant productivity (Figure 3.14) (Wagg et al. 2014). Another microcosm study showed that higher species richness and increased functional dissimilarity of faunal communities increases CO_2 but decreases nitrous oxide (N_2O) emissions; this was attributed to an enhanced rate of litter decomposition in the presence of larger fauna within mixtures, especially earthworms (Lubbers et al. 2020). However, studies also show that rare soil microbes can have an important functional role, for instance in plant defence (Hol et al. 2010), and that changes in the functional complexity of soil communities elicit divergent responses of individual ecosystem processes. This suggests that the net effect on ecosystem functioning is an aggregate of both positive and negative relationships for different processes (Bradford et al. 2014).

The above-mentioned studies were all done under controlled laboratory conditions or in glasshouses, mostly using structurally simplistic soils and narrow diversity gradients, which bear little resemblance to the highly diverse and complex communities found in heterogeneous field soils. However, a new generation of studies have attempted to overcome these limitations by testing for soil biodiversity–function relationships in situ across a broad range of soils and ecosystem types. In a cross-biome experiment, spanning sites from

Figure 3.14 Soil biodiversity and community composition determine ecosystem multifunctionality. (a) A gradient of increasing simplification of soil communities was created by sieving soil through different meshes and using that as an inoculum in micrososms. (b) Positive relationship between ecosystem multifunctionality and soil biodiversity index, created using soil communities of increasing simplification. From Wagg et al. (2014).

the subarctic to the tropics, for example, the loss of key components of the decomposer community consistently slowed rates of litter decomposition and nutrient cycling, thereby demonstrating negative effects of soil biodiversity loss on these functions across a range of ecosystem types (Handa et al. 2014). A study of soils taken from different countries across Europe also found that certain soil food web properties strongly and consistently predict processes of C and N cycling, suggesting that soil food web parameters can be important predictors of biogeochemical cycles (De Vries et al. 2013).

Several studies have sought to test for soil biodiversity–function relationships using molecular approaches in combination with the measurement of multiple ecosystem functions across different ecosystem types and biomes. In one such study, soil microbial diversity (bacteria and fungi) was found to positively relate to ecosystem multifunctionality, based on the simultaneous quantification of multiple ecosystem functions related to soil nutrient cycling and plant productivity, across a broad range of ecosystem types in Scotland and a global network of drylands (Delgado-Baquerizo et al. 2016). Further, the positive relationship between microbial diversity and ecosystem multifunctionality was detectable even when accounting for the influence of other factors known to regulate ecosystem functions, especially climate, soil abiotic factors, and spatial predictors. In a subsequent study, positive relationships between various elements of soil biodiversity (including bacteria, fungi, protists, and invertebrates) and ecosystem multifunctionality were detected across a globally distributed set of sites, spanning five continents and a wide range of climates, soils, and vegetation types (Figure 3.15) (Delgado-Baquerizo et al. 2020b). Notably, these relationships were backed up by controlled laboratory studies and were especially strong for soils with a greater diversity of large invertebrates and a larger number of dominant phylotypes within highly connected ecological networks. Together, these studies point to the importance of diverse soil communities and ecological networks for maintaining multiple ecosystem functions across ecosystems and biomes, but also indicate that some soil organisms play a greater role than others.

Some recent studies suggest that greater soil biodiversity enhances the stability of ecosystem functions, consistent with the insurance hypothesis, i.e., that high diversity insures against declines in ecosystem function because many species provide greater guarantees that some will maintain their function if others do not (Yachi and Loreau 1999). This topic is considered in greater detail in Chapter 6 in the context of below-ground responses to climate extremes, but two studies are worth highlighting here. The first by Wagg et al. (2021) used sieves of varying size to create an experimental gradient of declining soil bacterial and fungal diversity in grassland mesocosms, and measured how a range of ecosystem functions, including plant production, litter decomposition, and soil C assimilation, varied over time in response to a variable watering regime, including periods of drought. Under these fluctuating environmental conditions, declining bacterial and fungal diversity was associated with reduced stability of ecosystem functions (assessed in terms of their invariability over time), which was taken to suggest that impaired soil microbial diversity destabilises ecosystem functions in response to environmental change (Figure 3.16). The second study by Liu et al. (2022) tested the same idea in the field, using survey data of soil fungal diversity along with satellite-derived temporal measurements of plant productivity as a measure of ecosystem stability. They found that diversity–stability relationships for soil fungi varied depending on fungal functional group: the richness of fungal decomposers was consistently and positively associated with ecosystem stability worldwide, whereas the opposite pattern was found for the richness of fungal plant pathogens, especially in grasslands. The mechanistic basis for these positive soil microbial diversity–stability relationships in both studies remain unresolved, but it is likely related to asynchrony in species responses to environmental change, which is greater in high diversity soils. However, what these studies do suggest is that soil microbial diversity enhances not only ecosystem functioning, but also the stability of ecosystem functions in response to environmental change.

While our understanding of diversity–function relationships in soil is growing, many uncertainties remain. One such challenge is the need to

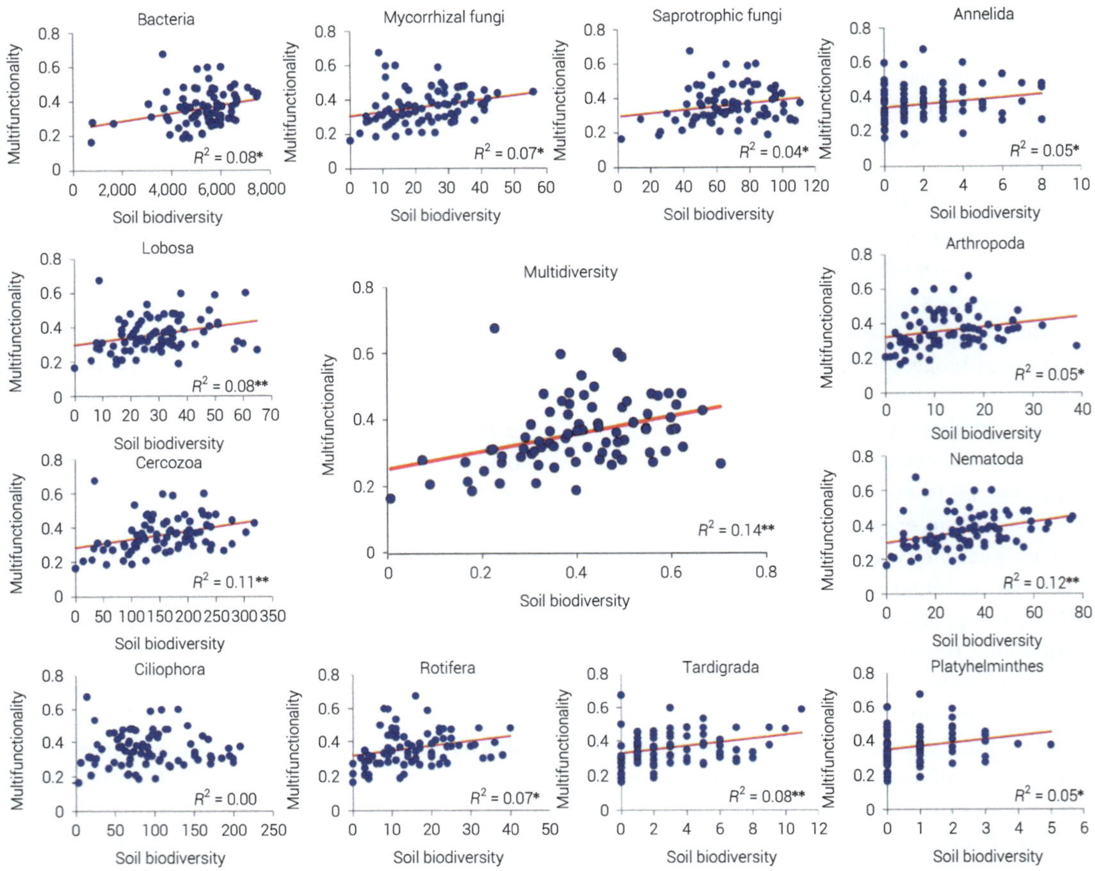

Figure 3.15 Relationships between multifunctionality, based on measures of 11 ecosystem functions, and the biodiversity of selected groups of soil organisms, measured as species richness and multidiversity for the entire community (standardised between 0 and 1), across 81 sites globally covering multiple biomes. From Delgado-Baquerizo et al. (2020b) by permission of Springer Nature.

better understand the mechanisms by which different facets of soil biological diversity influence ecosystem functioning, considering the many ecological interactions that operate between soil organisms, both within and between different functional groups, as well as the enormous structural and chemical complexity of soil. For instance, it remains unclear how the importance of soil biological diversity varies for different ecosystem functions and the extent that diversity effects are regulated by the traits of dominant organisms rather than species diversity per se. It is also becoming increasingly clear that relationships between soil biological diversity and ecosystem function are strongly context dependent, varying from one ecosystem to another and with abiotic factors such as moisture and temperature. For example, in a study of 125 sites across eastern Australia, the richness of soil bacteria, fungi, invertebrates, and protists was found to be a strong predictor of ecosystem functions in non-forested ecosystems, but not in forested ecosystems, where plant richness was a stronger predictor (Ding et al. 2021). Also, abiotic factors, such as moisture and temperature, were found to be considerably more important determinants of soil functions in a boreal forest than were variations in the structure and biomass of the faunal community in soil (Liiri et al. 2002). Further, in an experimental study using a model soil, bacterial diversity was a strong predictor of C use efficiency

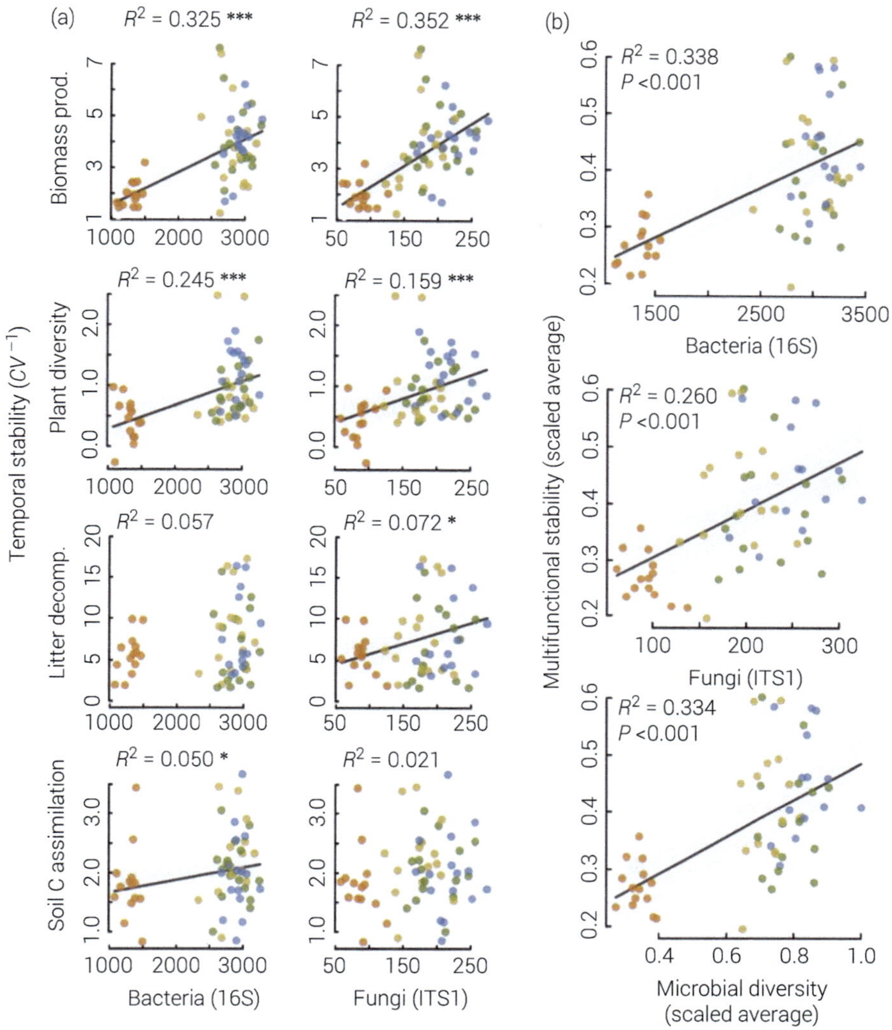

Figure 3.16 The relationship between (a) temporal stability (the inverse coefficient of variation, CV^{-1}) of each of four functions (i.e., plant biomass, plant diversity, litter decomposition, and soil C assimilation) and the richness of bacteria and fungi; and (b) multifunctional stability, calculated as the scaled average of the stability of each of the four ecosystem functions shown in relation to the richness of bacteria, fungi, and microbial diversity (the scaled average of fungal and bacterial richness). Regression lines and 95% confidence bands are shown for significant relationships ($P < 0.05$). From Wagg et al. (2021).

(i.e., the fraction of C taken up by microbial cells and retained in biomass as opposed to being respired) in moist soils, but not in dry soils, indicating that relationships between microbial diversity and ecosystem function are strongly moderated by soil abiotic properties (Domeignoz-Horta et al. 2020). Relationships between soil microbial diversity and multifunctionality were also found to be moderated by moisture availability along a 4,000-km aridity gradient in northern China, although they became stronger in more arid regions (Hu et al. 2021b), and in a global-scale decomposition experiment, soil animals increased decomposition rates in temperate and wet tropical climates but had neutral effects in ecosystems where temperature or moisture constrain biological activity (Figure 3.17) (Wall et al.

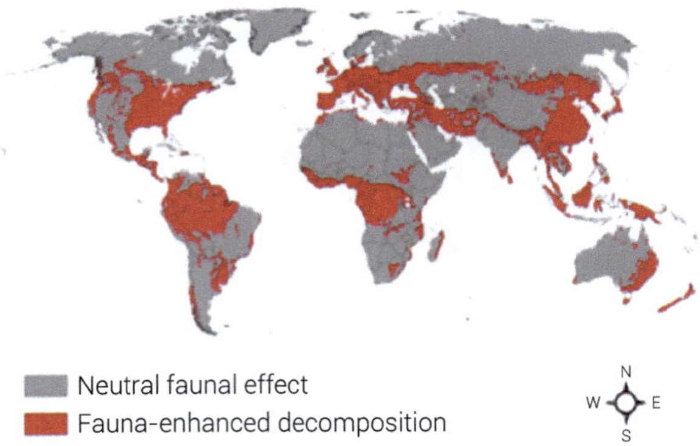

Figure 3.17 Map showing climatic regions where soil animals were found to enhance decomposition rates, adapted from Wall et al. (2008). Results from temperate and wet tropical climatic regions (red) show fauna to increase decomposition rates, but to have neutral effects in other regions (grey). This suggests that for regions predicted to be warmer and wetter, enhanced decomposition rates may result from greater effects of soil animals and other biota. Results are based on the Global Litter Invertebrate Decomposition Experiment (GLIDE). From Wall et al. (2008) by permission of John Wiley and Sons.
Global Change Biology, Volume: 14, Issue: 11, Pages: 2661–2677. First published: 22 October 2008, DOI: (10.1111/j.1365-2486.2008.01672.x.

2008). Together, these findings suggest that relationships between soil biodiversity and ecosystem functioning depend on environmental context and therefore change along environmental gradients.

Conclusions

Microbes have many important roles to play in the operation of ecosystem processes of decomposition and nutrient cycling, and these activities can determine the availability of nutrients to plants. However, to fully understand the role of soil biota in driving these processes requires consideration of the interactions that occur between microbes and other organisms within the soil food web. There is ample evidence, for example, that the feeding activities of microbial-feeding fauna can greatly influence the abundance and activities of microbes, ultimately affecting nutrient flux. Also, higher-level consumers can have cascading effects on soil food webs that ultimately affect microbes and the processes that they drive. Soil animals greatly modify the physical structure of soil, which shapes the habitat and activities of other fauna and microbes and influences the physical movement of water and solutes through soil. Whilst variations in soil food web diversity and complexity clearly influence ecosystem functioning, it is evident that relationships vary considerably depending on biotic factors, such as the traits or dominant species and shifts in species composition and relative abundance, as well as abiotic factors, such as temperature, moisture, and nutrient availability. Much is still to be learnt about the functional role of biological diversity and ecological networks in soil, especially of microbes, and the relative importance of these biotic forces compared to abiotic ones, as drivers of ecosystem function.

CHAPTER 4

Feedbacks between plants and soil biota

Key messages

- Plant species and shifts in plant diversity and composition have considerable impacts on soil biota and ecosystem functioning.
- Plant species effects on soil biota induce negative and positive feedbacks that shape plant population and community dynamics, and ecosystem functioning.
- Plant associations with soil food webs and their functioning are broadly related to the functional traits of dominant plant species, especially those of roots.
- Major gaps in knowledge exist over the role of root traits in modifying soil biota, their feedbacks to plants, and ecosystem functioning.

Soil organisms are intimately linked to plants. Not only do plants provide C and other nutrients to the decomposer community, but also plant roots form intimate associations with soil organisms, which can be both beneficial (i.e., symbionts) and detrimental (i.e., herbivores and pathogens) to plant growth. Soil organisms, in turn, influence plant growth indirectly by recycling dead plant material and making nutrients available for plant uptake and directly through the action of the root-associated organisms that selectively influence the growth of plants, both positively and negatively, thereby influencing plant population and community dynamics (Figure 4.1). It is now widely accepted that these biotic interactions or feedbacks between plants and soil organisms are important drivers of plant population and community dynamics, and ecosystem functioning (Van der Heijden et al. 2008, Bardgett and Wardle 2010, Van der Putten et al. 2013, Bennett and Klironomos 2019).

This chapter examines the way that plants and soil biota influence each other in both positive and negative ways, and considers the ecological role of these biotic interactions in regulating plant community dynamics and ecosystem functioning. First, the main mechanisms by which individual plants and changes in the diversity and composition of plant communities shape soil biological communities and their functioning is considered. Then, how soil organisms in turn influence plant growth and act as agents of plant population and community dynamics is considered. The goal of this chapter is to illustrate the nature and ecological significance of feedbacks between plants and soil biota at the population, community, and ecosystem scale.

Plant species effects on soil biota

The principal route by which plants impact belowground organisms and communities is through modifying the amount and quality of organic matter returned to soil. Plant species differ greatly in their functional traits, defined as morphological, biochemical, structural, physiological, or behavioural characteristics that impact the fitness of individual species via their effects on growth, reproduction, and survival (Violle et al. 2007). This variation in functional traits is reflected in differences in the quantity and quality of resources that they produce, in terms of both litter inputs (shoot and root) and root exudates, which shape the structure and function of soil food webs. Plant inputs to soil also vary greatly in the relative portions of different C constituents that they contain, which

The Ecology of Soil. Richard D. Bardgett, Oxford University Press. © Richard D. Bardgett (2025). DOI: 10.1093/9780191858710.003.0004

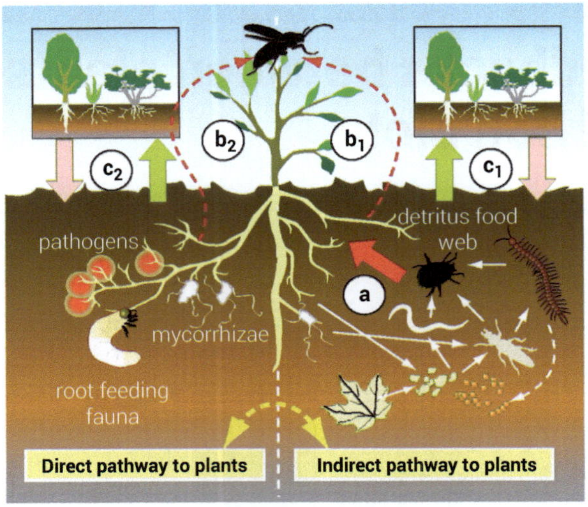

Figure 4.1 Above-ground communities are affected by both direct and indirect consequence of soil food web organisms. Right-hand half of figure: Feeding activities in the detritus food web (slender white arrows) stimulate nutrient turnover (thick red arrow), plant nutrient acquisition (a), and plant performance, and thereby indirectly influence above-ground herbivores (red broken arrow) (b_1). Left-hand half of figure: Soil biota exert direct effects on plants by feeding on roots and forming antagonistic or mutualistic relationships with their host plants. Such direct interactions with plants influence the performance not only of the host plants themselves, but also that of the herbivores (b_2) and potentially their predators. Further, the soil food web can control the successional development of plant communities both directly (c_2) and indirectly (c_1), and these plant community changes can in turn influence soil biota. From Wardle et al. (2004a). Ecological Linkages Between Aboveground and Belowground Biota, Volume: 304, Issue: 5677, Pages: 1629–1633, DOI: (10.1126/science.1094875).

degrade at different rates, reflecting their resource value to decomposer organisms (Figure 4.2). These can be broadly characterised into three C components of differing degradability: (1) the most easily degraded C is the labile, low molecular weight fraction that includes sugars and amino acids, which provide a soluble and readily usable resource for microbes; (2) the intermediate fraction includes moderately labile substances such as cellulose and hemicelluloses, which are the most abundant constituents of plant residues; and (3) the most recalcitrant fraction are structural materials, such as lignin. The extent that plant species differ in their influence on soil organisms and their activities therefore depends on interspecies differences the quantity and quality of plant-derived resources entering the soil from litter and root exudates. Plant species also influence soil biota through modifying the soil physical environment, and hence the soil habitat, which is largely due to differences among species in root traits that modify soil structural properties.

It has long been known that individual plant species modify below-ground communities and processes in their zone of influence. The mull and mor theory of Müller (1884), which was discussed in Chapter 1, recognised the role that plant species have in determining the community composition of soil fauna with consequences for nutrient cycling. Also, during the 1950s it was demonstrated experimentally that the dwarf-shrub *Calluna vulgaris* produces tannin–protein complexes that inhibit N mineralisation, and hence the supply of nutrients to coexisting plant species (Handley 1954, 1961). Following this, several pot experiments were performed during the 1990s that revealed a high degree of divergence in soil microbial communities and their consumers (e.g., nematodes) among plant species, which has largely been attributed to variations in root exudation patterns (e.g., Griffiths et al. 1992, Grayston et al. 1998, Bardgett et al. 1999b). Effects of individual plant species on microbes and their consumers have also been detected in the field. In tallgrass prairies of North

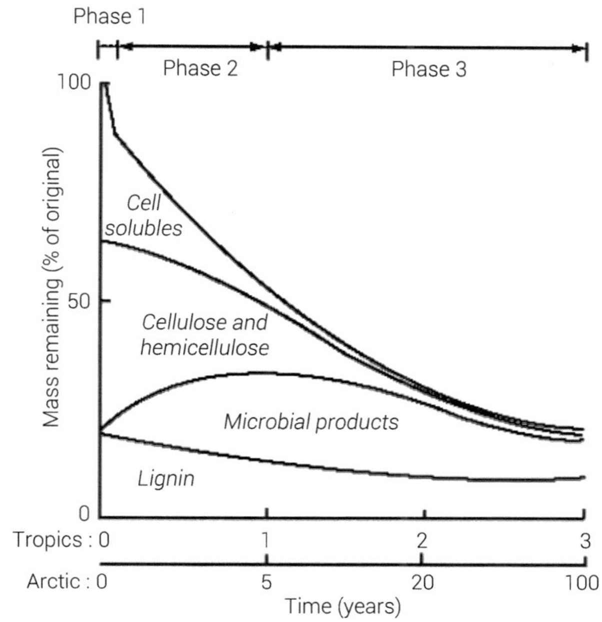

Figure 4.2 Time course of leaf litter decomposition in relation to C components of differing degradability and the timescales commonly found in warm tropical and cold arctic conditions. Phase 1 is dominated by leaching of soluble compounds, whereas phase 2 involves degradation of labile, low molecular weight fraction, such as sugars and amino acids, and then of moderately labile substances such as cellulose and hemicelluloses, which are the most abundant constituents of plant residues. Finally, the most recalcitrant fraction, such as lignin, is degraded. From Chapin et al. (2002) by permission of Springer Nature.

America, for example, sampling of soils under various C3 and C4 grass species revealed that some bacterial and nematode groups were responsive to the presence of certain plant species (Porasinski et al. 2003). Also, experiments in New Zealand pastures showed that the community structure of soil microbes, microbe-feeding nematodes, and herbivorous nematodes and arthropods were all responsive to the removal of subsets of flora from the plant community (Wardle et al. 1999a). Such effects are likely due to differences in root turnover and root exudation patterns of grassland plants, but also to variations in resource quality of litter inputs from roots and shoots.

Knowledge of how plants shape rhizosphere microbial communities has advanced significantly in recent years with the use of molecular techniques, which has revealed a vast diversity of rhizosphere microbes involved (Mendes et al. 2011). It is now well established that rhizosphere microbial communities, including protists, and endophytic and AM fungi display considerable variation in their diversity and composition among plant species (e.g., Fitzpatrick et al. 2018, Leff et al. 2018, Kivlin et al. 2022) and genotypes (e.g., Lundberg et al. 2012, Edwards et al. 2015) and are highly dynamic as plants develop (Ceja-Navarro et al. 2021). Moreover, this interspecies variation in the recruitment of rhizosphere microbes has significant consequences for plant growth: plant species that accumulate beneficial rhizosphere microbes, such as N-fixing bacteria or mycorrhizal fungi, can improve their growth, whereas the accumulation of pathogens suppresses plant growth (Fitzpatrick et al. 2018, Hu et al. 2018, Semchenko et al. 2018). Plants also release a wide variety of compounds into the rhizosphere that serve to recruit protective microbes or directly repel, inhibit, or kill pathogenic microbes, thereby conferring defence against soil-borne pathogens (Berendsen et al. 2012, Baetz and Martinoia 2014). This has led to the rhizosphere being referred to as a preventive microbial buffer zone or 'front line' of defence against pathogen attack (Figure 4.3). The release of volatile organic compounds (i.e., methyl

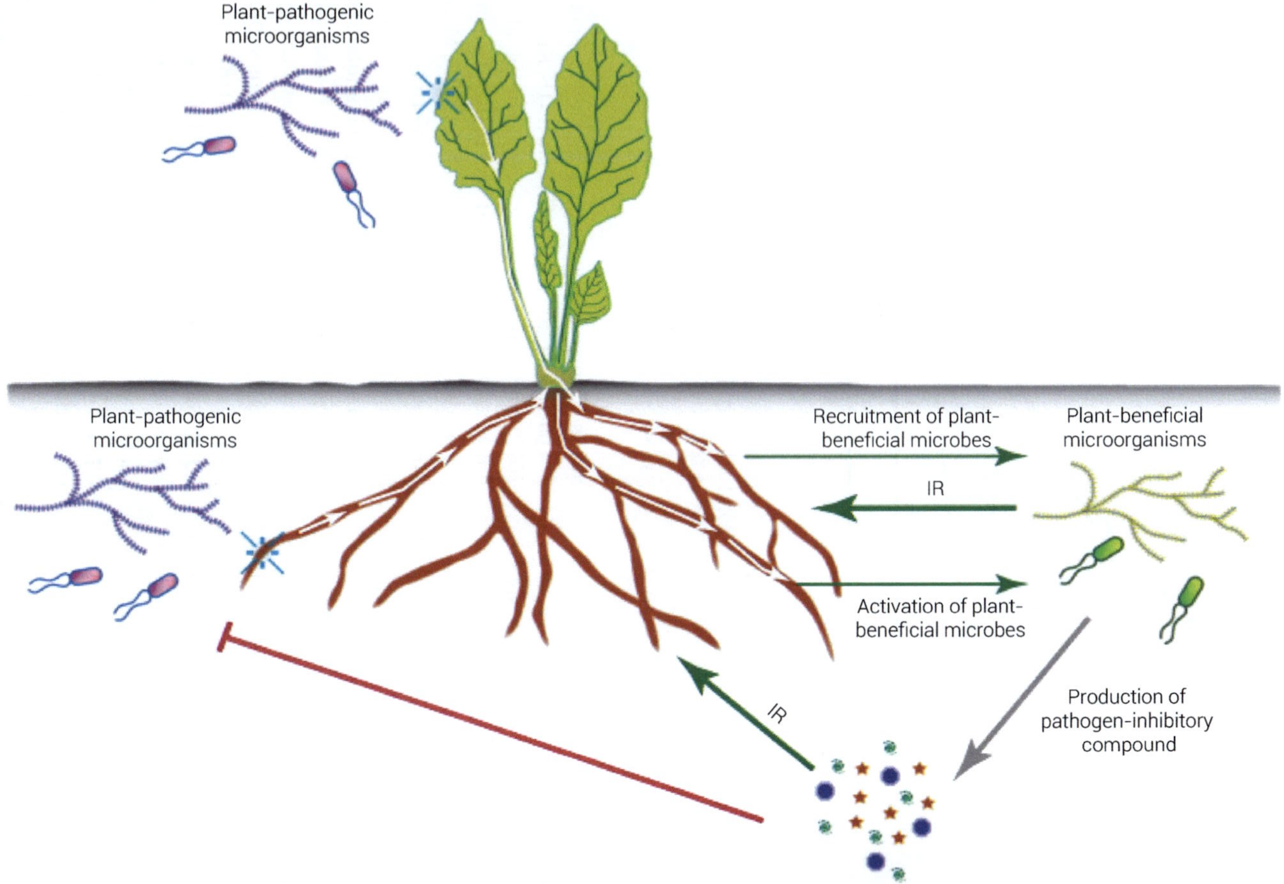

Figure 4.3 Schematic of recruitment and activation of beneficial microbes by the plant upon attack from plant pathogenic microbes. Infected plants perceive pathogen invasion in their roots or shoots, and in response increase the secretion of microbe-stimulatory compounds in non-infected roots. These stimulants can recruit and activate plant-beneficial microbes, which can induce resistance (IR) directly or produce pathogen-inhibitory compounds. Some pathogen-inhibitory compounds are known to induce resistance themselves. From Berendsen et al. (2012) by permission of Elsevier.

jasmonate) from roots has also been shown to trigger the formation of plant growth-promoting soil microbial biofilms that extend into soil beyond the immediate rhizosphere, which supports the suggestion that plants form an 'extended rhizosphere' of beneficial microbial function (Kulkarni et al. 2024). It has also been suggested that plants capable of shaping their rhizosphere microbiome in a way that is beneficial to their reproductive success will be favoured during evolutionary selection (Berendsen et al. 2012).

Other plants, namely parasitic plants that depend partly or entirely on their host for nutrients, C, and water, can indirectly shape the structure and function of soil microbial communities through their influence on their host plants (Figure 4.4). There are around 4,000 species of parasitic plants that occur across most major biomes and they typically have negative impacts on their host plants (Press and Phoenix 2005). However, they can have positive effects on their hosts and/or co-occurring plant species via modifying soil microbial communities and nutrient cycling. In model grassland communities, for example, the widespread parasitic plant *Rhinanthus minor*, which preferentially infects fast-growing grasses, was found to have marked indirect effects on the composition of soil microbial communities, increasing the abundance of bacteria relative to fungi, which in turn increased rates of N cycling (Bardgett et al. 2006). The exact mechanism involved is not known, but parasite-driven changes in root turnover and exudation patterns of host plants, which alter C supply to rhizosphere microbes, are likely the cause (Bardgett et al. 2006). Also, in subarctic plant communities, hemiparasitic plants accumulate nutrients in their leaves, thus producing high quality plant litter which releases nutrients that would otherwise remain in the host plant or in slowly decomposing plant litter (Quested et al. 2002). This in turn enhances the rate of

(a) (b)

Figure 4.4 Parasitic plants such as *Rhinanthus minor* in meadow grasslands (a) and *Bartsia alpine* in subarctic communities (b) depend partly or entirely on their host for nutrients, C, and water and indirectly impact soil microbial communities and ecosystem function through their influence on their host plants. Images by Richard Bardgett and Helen Quested, respectively.

decomposition of litter from non-parasitic plants (Quested et al. 2003a) and increases soil nutrient availability and plant growth (Quested et al. 2003b).

Home-field advantage

An extension of the idea that plants shape soil communities is the concept of home-field advantage. This refers to the notion that plant species may select for decomposer communities that preferentially colonise and decompose their own litter, leading to the expectation that litter decomposition, and thus rates of nutrient cycling, is faster in close vicinity of the plant from which it originates (Austin et al. 2014, Fanin et al. 2021). Several studies support the idea of home-field advantage and show that effects can be substantial. For example, litter from three coexisting *Nothofagus* species in mixed old-growth forest in Patagonia was found to decompose much faster when placed under its own species than when under one of the other species (Vivanco and Austin 2008) (Figure 4.5). Similarly, litters of three dominant tree species of high elevation forest in Colorado, USA—namely trembling aspen (*Populus tremuloides*), lodgepole pine (*Pinus contorta*), and Engelmann spruce (*Picea engelmannii*)—were found to decompose faster when they were incubated with their own soil biota (Ayres et al. 2009). Moreover, this home-field advantage was most pronounced for lodgepole pine, which produced the lowest quality litter (i.e., high litter lignin: N) (Ayres et al. 2009), thereby supporting the idea that home-field advantage increases with decreasing litter quality, as has been shown by others (e.g., Hunt et al. 1988, Strickland et al. 2009, Milcu and Manning 2011). This litter quality effect on home-field advantage has been explained by the need for specialist decomposer microorganisms, especially saprotrophic fungi, to break down more recalcitrant litter, which contrasts with high quality litter that can be degraded by many microorganisms. As such, home-field advantage has been shown to be less pronounced for

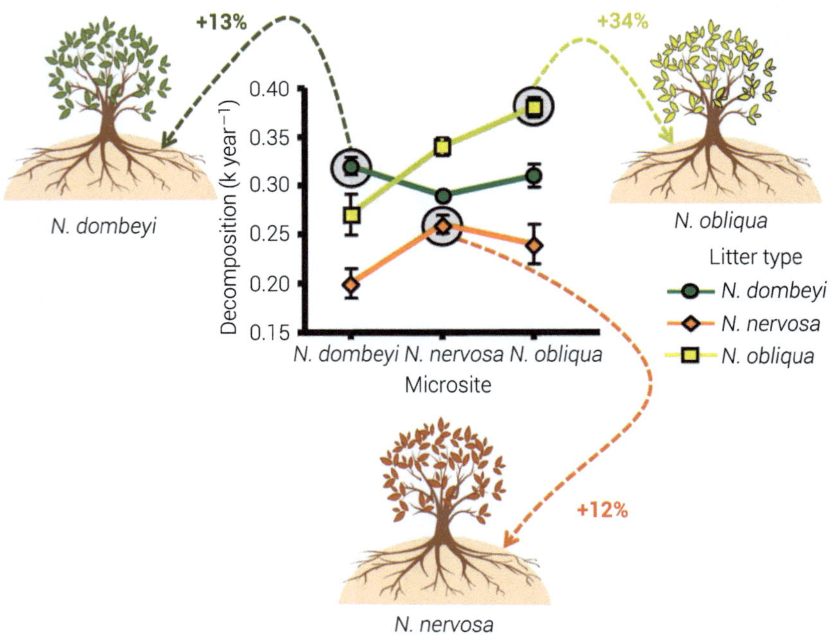

Figure 4.5 Home-field advantage of three southern beech species (*Nothofagus* spp.) in mixed old-growth forest in Patagonia, showing that leaf litter from a particular tree species is decomposed faster beneath that species (i.e., in its home environment) than beneath a different species (i.e., away from its home environment). Grey circles indicate home-field advantage and the litter affinity effect for each species. From Austin et al. (2014) by permission of John Wiley and Sons.
New Phytologist, Volume: 204, Issue: 2, Pages: 307–314. First published: 08 August 2014, DOI: (10.1111/nph.12959).

plants with high quality litter and more pronounced for those that produce poor quality litter requiring specialist decomposers (Ayres et al. 2009).

While several studies support the idea of increased decomposition of litter in its home environment, others do not (e.g., Ayres et al. 2006, Gießelmann et al. 2011, St John et al. 2011). Moreover, a synthesis of published data from 35 studies showed that while on average litter decomposition was 7.5% faster at home, there was considerable variation in the strength and direction of home-field advantage effects (Veen et al. 2015), suggesting that home-field advantage is not universal. This inconsistency has been attributed to several factors, one being variation in climate and soil abiotic properties, which in some situations overwhelms home-field advantage (Austin et al. 2014). In arid ecosystems, for example, water availability is by far the dominant control on rates of litter decomposition (Austin 2011), and in old-growth forest in Patagonia, N addition has been shown to eliminate home-field advantage observed with *Nothofagus* tree species under ambient conditions (Vivanco and Austin 2011). Plant species also differentially shape decomposer communities, and hence rates of decomposition, via mechanisms other than those related to their leaf litter, such as through root traits (see below) or species-specific associations with mycorrhizal fungi, which can have negative and positive effects on decomposer communities and decomposition (Hättenschwiler et al. 2011, Nottingham et al. 2013). Very little is known about how such plant species-specific differences in root traits and symbiotic associations may translate to influence home-field advantage, but it is a topic deserving further study.

Another factor contributing to variability in home-field advantage is the legacy of green leaves, or the microbial community on foliage before litter fall: the phyllosphere community. The issue here is that the phyllosphere microbial community, which includes a diverse range of bacteria, fungi, and protists, may persist through the process of leaf senescence and litter fall, thereby influencing the decomposition of litter in or on the surface of soil (Austin et al. 2014). Evidence for this is scant, but an analysis of data from published studies revealed that home-field advantage was smaller when phyllosphere communities were removed from leaves by sterilisation, suggesting that these 'green leaf hitchhikers' (*sensu* Austin et al. 2014) play an important role in increasing rates of decomposition at home (Fanin et al. 2021). Very little is known about the mechanisms by which phyllosphere microorganisms influence home-field advantage, but it has been suggested that they contribute to litter degradation through conditioning leaf tissue during an initial stage of decomposition following leaf death, which modifies the subsequent succession of 'home' microbial communities in soil (Fanin et al. 2021). In other words, this conditioning of leaf litter by phyllosphere microorganisms may play a role in stimulating the home-field advantage brought about by a range of 'home' soil decomposers. Future work is needed to identify the specific mechanisms involved, but the concept of home-field advantage clearly needs to be extended to include phyllosphere microorganisms and their role in conditioning leaf material and in shaping the 'home' soil decomposer community (Austin et al. 2014, Fanin et al. 2021).

Trait-based frameworks

Plant ecologists have long recognised that plant species differ in their functional traits (Grime 1979, Chapin 1980) and have identified a strikingly universal spectrum of leaf traits—the leaf economics spectrum—that balances efficient resource conservation against rapid resource acquisition and growth potential (Wright et al. 2004, Reich 2014, Diaz et al. 2016). In brief, a rapid return on investment at the exploitative end of the spectrum (i.e., 'fast' economic strategy) is associated with leaves with shorter longevity, higher nutrient concentrations, high gas exchange, and thinner and/or less dense leaves, whereas a slower return on investment at the conservative end of the spectrum (i.e., 'slow' economic strategy) is associated with the opposite set of traits (Reich 2014). There is also strong evidence that the leaf economics spectrum is an important driver of leaf litter quality, with important implications for soil biological properties, especially the rate of decomposition of litters produced by plant species (Box 4.1). Plant species at the conservative end of the spectrum, which are adapted to nutrient-poor conditions, generally produce litter of lower nutrient contents and higher concentrations of structural carbohydrates (e.g., cellulose and lignin) and defence compounds (e.g., polyphenols) than exploitative species adapted to high resource conditions (Cornelissen and Thompson 1997, Wardle 2002). Reflecting this, litter from resource-conservative coniferous species decomposes more slowly than that from more exploitative woody angiosperm species, which in turn breaks down more slowly than that from herbaceous species (Cornelissen 1996).

Box 4.1 Factors controlling decomposition rate of litter

A hierarchy of abiotic and biotic factors controls decomposition. At the highest level of the hierarchy, climate determines the rate of decomposition, in that it is the primary determinant of soil moisture and temperature, which both affect rates of the main physical, chemical, and microbiological reactions that control decomposition. In general, decomposer organisms and their enzymes are more active in warm, moist conditions, such as in the tropics, whereas they are less active under excessively cold and/or wet conditions, such as found in northerly ecosystems where peat accumulates. At the second level in the hierarchy, the landscape, the nature of parent material largely influences decomposition rates. For example, all else being equal, decomposition is faster in neutral than acidic soil, and an abundance of clay minerals tends to reduce rates of decomposition, due to increased water-holding capacity and, in some cases, waterlogging of soil. The third level in the hierarchy is the quality and quantity of the organic matter produced, which depends on the nature and composition of plant communities. The release of energy and nutrients stored in dead organic matter depends strongly on the proportion of support tissues rich in lignin and lignocellulose, the proportion of secondary chemical compounds, such as tannins or polyphenols, and the ratio of nutrients to C. Whilst the relative importance of these factors is not always clear, certain key measures of litter quality are commonly used as predictors of decomposition. Of these, the C:N ratio is often used as an index of litter quality, in that litter of a lower C:N generally decomposes faster than that with a high ratio. In more recalcitrant litters, the lignin:N ratio is often a good predictor of decomposition rate.

The recognition of such relationships between plant traits and decomposition processes has led to the suggestion that the leaf economics spectrum could provide a framework for understanding plant species associations with soil food webs and their functioning (Figure 4.6) (Wardle et al. 2004a, De Deyn et al. 2008, Bardgett 2017). The general idea is that plant species at the conservative end of the leaf economics spectrum, with low relative growth rate and low specific leaf area, produce litter of low resource quality that promotes the growth of fungi and their consumers (e.g., fungal-feeding collembolans and mites), which slows down rates of nutrient and organic matter decomposition and enhances nutrient retention and C sequestration. In contrast, fast-growing exploitative plant species, with high relative growth rate and specific leaf area, produce high quality resource inputs in the form of litter and root exudates, which promotes the bacterial component of the microbial community and its consumers (e.g., bacterial-feeding protists and nematodes), leading to enhanced organic matter decomposition and nutrient turnover. Through these linkages, this framework also suggests that there is a form of mutualism between the plant and soil community, in that certain plant species can modify soil food webs and decomposition processes in a way that creates a more favourable habitat for that species, thereby increasing its competitive status.

Several studies provide support for the above-mentioned idea that plant traits associate with soil microbial communities in a way that is broadly consistent with the leaf economics spectrum. A comparative field-based study of grassland plant species grown in monoculture for 7 years, for example, showed that exploitative grassland species with high leaf and litter quality were associated with elevated biomass of bacteria relative to fungi, and high rates of N mineralisation and availability of inorganic N and P in soil (Orwin et al. 2010). Similarly, the abundance of archaeal ammonia oxidisers, which play an important role in the process of nitrification (Chapter 3), was found to be greater in soil modified by exploitative or 'fast' grassland plant species of high N demand than conservative or 'slow' grassland plants species (Thion et al. 2016). Similar trait associations have been found

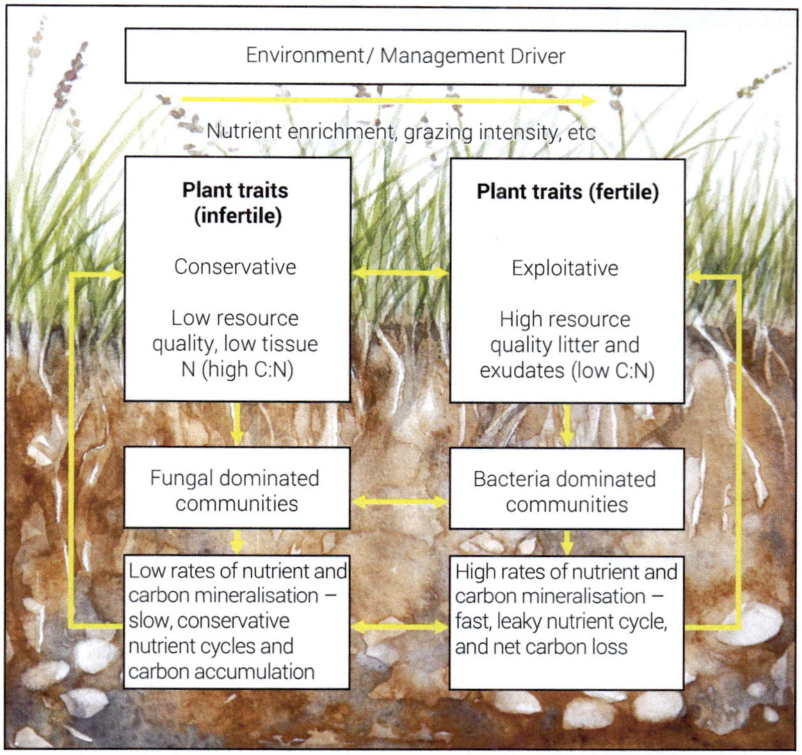

Figure 4.6 Hypothetical influence of plant traits on soil fungal and bacterial energy channels in soil food webs and their relation to soil C and N cycles. From Bardgett (2017) adapted from Wardle et al. (2004a) by permission of the Royal Irish Academy. Image by Jill Colquhoun Bardgett.

at larger scales. In a study of a ponderosa pine (*Pinus ponderosa*) forest ecosystem, understorey vegetation dominated by fast-growing species of high leaf N and specific leaf area were found to be associated with soils that have high nitrification potential, presumably due to the promotion of nitrifying bacteria and archaea (Laughlin 2011). Also, across a wide range of grasslands in Europe under different management regimes, dominance of exploitative plant species with high leaf N and specific leaf area was consistently associated with bacterial-dominated microbial communities, rapid rates of microbial activity, and low soil C stocks and N retention. In contrast, grasslands dominated by conservative species were found to be associated with fungal-dominated microbial communities, reduced bacterial activity, and greater soil C storage and N retention (De Vries et al. 2012c, Grigulis et al. 2013). These findings are broadly consistent with the mass ratio hypothesis (Grime 1998), which states that ecosystem functioning is chiefly determined by trait values of the dominant contributors to plant biomass, in this case the dominance of plant species with leaf traits associated with different resource acquisition strategies.

There is evidence that traits of subordinate species (i.e., those that contribute to only a small proportion of the total biomass) can punch above their weight in terms of their impact on belowground communities and functions. For example, the removal of subordinate species from a semi-natural grassland of the Jura Mountains, Switzerland, was found to strongly impact soil microbial communities, plant litter decomposition, and soil nutrient cycling (Mariotte et al. 2013). Similarly, during primary succession, low biomass invasive species were shown to have disproportionately large effects on soil microbial and nematode communities relative to their contribution to total plant biomass (Peltzer et al. 2009), and in boreal forest, dwarf shrubs, which make up only a small portion of the total plant biomass, had much greater effects on soil microbial biomass, litter decomposition, and nutrient availability than did the more abundant trees and mosses (Wardle and Zackrisson 2005). Together, these studies indicate that in addition to the traits of dominant species, those of subordinate species also matter for below-ground communities and processes, although the mechanisms involved remain poorly understood (Mariotte 2014).

Finally, it should be noted that not all studies have found relationships between leaf traits and soil microbial communities: studies in hyper diverse tropical rainforests (Barberán et al. 2015) and temperate grassland (Leff et al. 2018) show that plant taxonomy is a much better predictor of soil microbial communities than leaf traits. However, as noted by the authors of these studies, unmeasured plant traits, especially those of roots, will likely have played an important, but undetected role (Barberán et al. 2015, Leff et al. 2018). Nevertheless, for coarse measures of soil communities and their functions, it appears that the leaf economics spectrum provides a general framework for understanding plant species associations with soil biota at both the individual plant and plant community scale.

Incorporating root traits

Plant roots form intimate associations with soil organisms. It therefore makes sense that the traits of roots, which display tremendous variation among plant species and are extremely plastic in response to changing soil resource availability, serve as important drivers of rhizosphere communities and ecosystem functions (Figure 4.7) (Bardgett et al. 2014, Freschet et al. 2021). While there is still much uncertainty about the mechanisms by which different root traits impact soil organisms and the functions they drive, they clearly play a central role. For instance, variation in root traits, especially root exudation, is widely proposed to be a dominant factor explaining plant species-specific associations with rhizosphere communities and C and N cycling (Jones et al. 2004, Moreau et al. 2019), and several studies have revealed strong relationships between certain root traits and particular groups of rhizosphere microbes. Comparative studies of grassland plants, for example, found that the abundance of archaeal ammonia oxidisers is positively related to root N content (Thion et al. 2016), and that the diversity of fungal pathogens is negatively related to root diameter, i.e., plant species with finer root systems accumulate a greater diversity of fungal pathogens (Semchenko et al. 2018). The composition of rhizosphere fungal communities was also found to be related to root traits, especially root diameter, specific root length, and root N across a wide range of temperate grassland species, with relationships being especially strong for AM fungi (Sweeney et al. 2021). The abundance of mycorrhizal fungi has been shown to be positively related to root diameter across a wide range of plant species and biomes (Kong et al. 2014, Ma et al. 2018), leading to the suggestion that root traits have evolved along a spectrum bounded by two contrasting strategies of root life: an ancestral 'conservative' strategy in which plants with thick roots depend on symbiosis with mycorrhizal fungi for soil resources; and a more-derived 'opportunistic' strategy in which thin roots enable plants to leverage photosynthetic C for soil exploration (Ma et al. 2018).

A growing focus on root traits has led to the question of whether the plant economic spectrum holds for roots as it does for leaf traits, i.e., that roots follow a one-dimensional gradient in trait syndromes from fast foraging and short lifespan (i.e., acquisitive strategy) to slow foraging and long lifespan (i.e., conservative strategy) (Freschet et al. 2010, Reich 2014). Based on this hypothesis, plant species from the resource-acquisitive end of the trait spectrum will exhibit high specific root length, thinner roots, high N content, and low root lifespan, whereas roots at the resource-conservative end of the spectrum will exhibit the opposite traits. However, evidence for the existence of such a simple one-dimensional 'fast–slow' root economics spectrum is mixed (Kramer-Walter et al. 2016, Weemstra et al. 2016, Ma et al. 2018, Kong et al. 2019, Carmona et al. 2021), which likely reflects the multidimensional nature of roots and wide range of below-ground resource uptake strategies expressed by plants. Indeed, plants have the capacity to 'outsource' resource acquisition via associations with mycorrhizal fungi, rather than rely entirely on

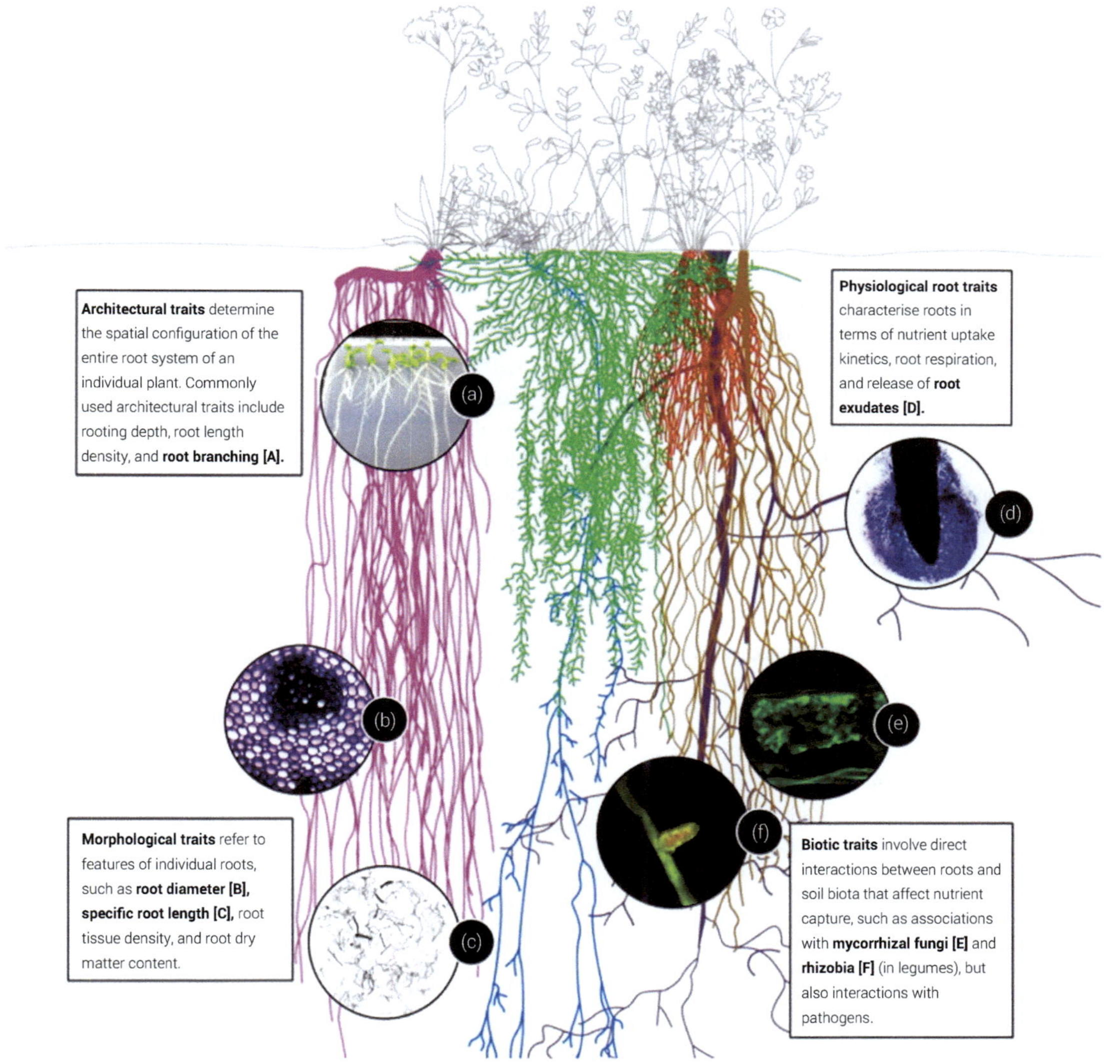

Figure 4.7 Root trait categories with potential to impact ecosystem processes, including architectural, morphological, physiological, and biotic traits. The general idea behind a trait-based framework is that variation in root traits across individual plant species, communities, and ecosystems explains variation in a range of ecosystem processes, including C and nutrient cycling, and the structural stability of soil. Images represent the different root trait categories with potential to impact ecosystem processes: (A) root systems of *Leucanthemum vulgare* growing on agar plates by Liesje Mommer, Wageningen University (WU), representing architectural traits; (B) cross section of root of *L. vulgare* stained with toluidine blue by Eric Visser, Radboud University, Nijmegen; (C) root length scan of washed roots from a diverse grassland community by Natalie Oram (WU), both representing morphological traits; (D) root nodule in *Medicago truncatula* by Erik Limpens (WU); (E) arbuscules of AM fungi in *M. truncatula* root by Rik Huiskes (WU), both representing biotic traits; and (F) rhizodeposition around *Zea mays* root by Davey Jones, Bangor University, representing physiological traits. From Bardgett et al. (2014) by permission of Elsevier.

plant organs, which has led to the suggestion that root–mycorrhizal fungal associations short-cut the one-dimensional economic spectrum. This idea is supported by a recent global analysis that showed most root trait variation is explained by two orthogonal gradients depict different plant strategies for acquiring soil resources represent: the ‘collaboration’ gradient, which ranges from ‘do-it-yourself’ resource uptake associated with high specific root length and root branching density, to 'outsourcing' of resource uptake by mycorrhizal fungi, associated with thicker roots and high mycorrhizal colonisation; and the 'conservation' gradient, which is similar to the 'fast–slow' root economics spectrum, with plants that invest in high root tissue density and longer root lifespan on one side of the gradient, and plants with high root N and root turnover on the other (Figure 4.8) (Bergmann et al. 2020).

Root traits related to the plant economic spectrum also act as important drivers of C and N dynamics via influencing SOM mineralisation, i.e., priming (see Chapter 3). It has been proposed that root traits that stimulate the growth of bacteria over fungi, especially high rates of root exudation, promote faster soil C and N dynamics (De Deyn et al. 2008). Evidence for this idea is scant, but recent comparative studies of plant species lend some support. For example, a study of coexisting tree species of deciduous–evergreen mixed forest showed that root exudation was positively associated with competitive root traits (root respiration and root N) and negatively with a conservative trait (root tissue density), which was taken to suggest that increasing nutrient availability by root exudation might be a complementary strategy of fast-growing trees (Sun et al. 2021a). Similarly, comparative studies of grassland plants have demonstrated that root exudation rate is positively correlated with root diameter, which is associated with 'outsourcing' of resource uptake (Williams et al. 2022), and that acquisitive and conservative species are associated with high and low rates of rhizodeposition, respectively, which in turn led to fast versus slow rates of SOC turnover (Henneron et al. 2020a). Acquisitive grassland plant species, with higher rates of photosynthesis and rhizodeposition than conservative species, were found to induce stronger acceleration of soil N cycling via rhizosphere priming of SOM (Henneron et al. 2020b). It is well known that the release of root exudates can enhance microbial mineralisation of organic matter via the process of priming, which increases nutrient availability to plants (Chapter

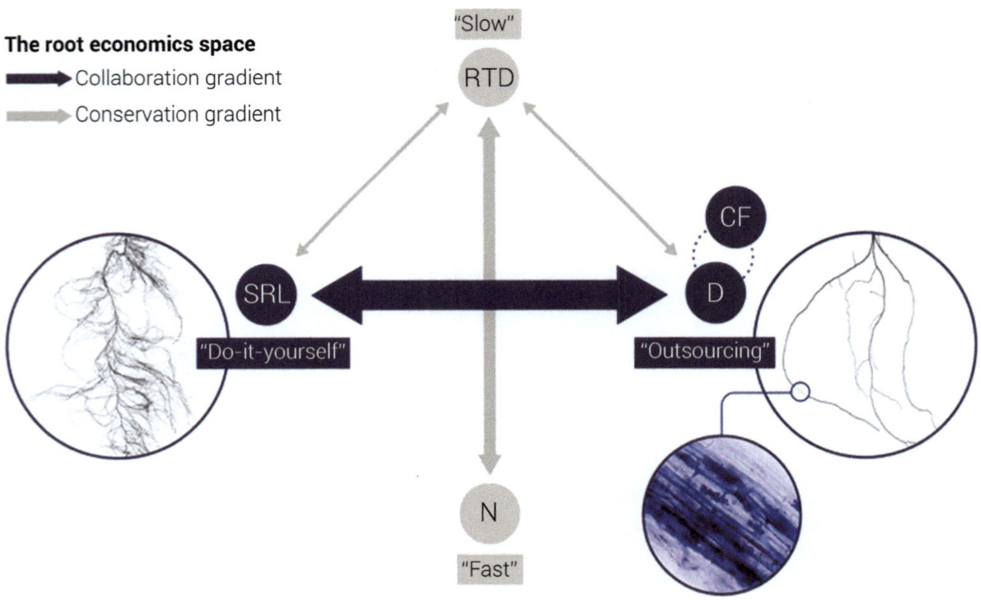

Figure 4.8 Hypothetical framework of root economics space, showing: (i) a collaboration gradient, which ranges from 'do-it-yourself' soil exploration by high specific root length (SRL) to 'outsourcing' by investing C into mycorrhizal partners and hence extraradical hyphae, which requires a large cortex fraction (CF) and root diameter (D); and (ii) a conservation gradient ranging from roots with high root tissue density (RTD) and a slow resource return on investment, but that are long lived and well protected, to fast roots with high N content and metabolic rate for fast resource return on investment, but a short lifespan. Arrows indicate negative correlations between the single traits. From Bergmann et al. (2020), by permission of AAAS Science Advances.

3). However, these studies go a step further by revealing a coupling between plant growth strategies and soil biogeochemical processes driven by rhizodeposition.

A role for plant polyphenols

Plant polyphenols, the most widely distributed class of plant secondary metabolites, have traditionally been considered in relation to their role in anti-herbivore compounds, defending plants against herbivore attack, both above- and below-ground (Hartley and Jones 1997). However, it has become increasingly clear that these compounds also play an important role in modifying soil biota and nutrient availability, often with consequences for the growth and competitive status of the plant that produced them (Hättenschwiler and Vitousek 2000). Plant polyphenols are broadly divided into low molecular weight compounds (e.g., simple phenols, phenolic acids, and flavonoids) and high molecular weight oligomers and polymers, such as tannins, which mostly enter soil via foliar leaching and plant litter (Box 4.2). However, root exudates also contain a wide range of phenolic compounds, such as flavonoids, phenols, and chalcones, which act as signalling molecules as well as a substrate or toxin to microbes, with consequent impacts on the structure and functioning of the rhizosphere microbiome (Badri et al. 2013, Zwetsloot et al. 2018). Phenolic profiles of root exudates display much interspecies variability (Zwetsloot et al. 2018), thereby potentially acting as an important root trait explaining species-specific effects of plants on rhizosphere microbial communities.

Plant-produced polyphenols impact processes of soil nutrient cycling via a variety of mechanisms. These include inhibition of groups of microbes involved in decomposition and nutrient transformations (Baldwin et al. 1983, Schimel et al. 1998), suppression of enzyme activities (Fierer et al. 2001, Triebwasser et al. 2012), and formation of tannin–protein complexes, which inhibit N mineralisation and the production of inorganic N (Northup et al. 1995, Bragazza et al. 2013).

Box 4.2 The fate of polyphenols in soil

Phenolic compounds have many roles in plant biology, including UV protection, herbivore and pathogen defence, and as contributors to plant colouration. They are defined by the presence of at least one aromatic ring bearing one (phenol) or more (polyphenols) hydroxyl substitutions. Polyphenols are often divided into two groups: (i) low molecular weight compounds, such as simple phenols, phenolic acids, and flavonoids; and (ii) oligomers and polymers of relatively high molecular weight, such as tannins. Low molecular weight phenolics occur universally in higher plants, whereas higher molecular weight compounds are most abundant in woody plants and often absent in herbaceous plants. Total polyphenol concentration of plants varies considerably among species, as does the relative polyphenol content of shoots and roots (Munakata et al. 2019). They enter the soil via two broad pathways: (i) as leachates from leaf and root material; and (ii) within leaf and root litter. There is also growing evidence that root exudates represent a significant source of phenolic compounds in soil and that the phenolic composition of root exudates varies considerably among plant species (Zwetsloot et al. 2018). Little is known on the relative importance of these routes, but it has traditionally been assumed that significantly more polyphenols enter soil via decomposing litter than through leachates, although the relative contribution of root exudates remains poorly understood. On entering the soil, soluble polyphenols are faced with a variety of fates: they may be degraded and mineralised by heterotrophic microbes; they can be transformed into insoluble and recalcitrant humic substances; they might be adsorbed to clay minerals or form chelates with Fe and Al ions; or they might remain dissolved and be leached out of the soil and into waterways as dissolved organic material. Insoluble polyphenols enter the soil as litter and are usually the main form of polyphenol that enters soil. These are slowly decomposed by microbes, contributing to the soluble polyphenol fraction. Some soil animals can digest polyphenols in their guts (e.g., earthworms), thereby enhancing their breakdown or mixing them with minerals to form organo-mineral complexes.

Through these roles, plant species-specific variations in polyphenol production can have marked effects on nutrient cycles and plant competitive interactions. A study in Alaskan taiga, for example, showed that polyphenols produced by balsam poplar (*Populus balsamifera*) inhibit N fixation by alder (*Alnus tenuifolia*) and increase microbial immobilisation of N. This in turn caused a decline in alder growth with increasing dominance of poplar, thereby contributing to species replacement (Schimel et al. 1998) (Figure 4.9). Plant polyphenols have also been shown to play an important role in N-limited forests on the coast of northern California (Northup et al. 1995). Here, tannins released from decomposing Bishop pine litter (*Pinus muricata*) were found to inhibit N mineralisation and increase DON release from the pine litter, the latter being the preferred N source of this tree species (Figure 4.10). This finding was taken to suggest that trees in N-limited ecosystems, such as Bishop pine, benefit from increasing DON availability relative to inorganic N because it short-cuts the microbial mineralisation step by allowing increased uptake of DON via ECM associations (Northup et al. 1995). Indeed, it is well known that ECM fungi produce extracellular enzymes that degrade N-bearing organic compounds, such as tannin–protein complexes, into constituent amino acids, which are then assimilated by the fungus and transferred to the host plant (Read and Perez-Moreno 2003). As such, high polyphenol production by tree species on infertile soils might represent an adaptive strategy to regulate the fate of N and influence the plant's competitive advantage for uptake of organic N (Northup et al. 1995, Wurzburger and Hendrick 2009).

Plant phenolics also play an important role in soil N cycling and plant competition in alpine ecosystems. An example of this comes from studies done on co-occurring plant species of alpine moist meadows in Colorado: the slow-growing alpine herb *Acomastylis rossi*, which produces litter rich in phenolics, and the fast-growing, N-demanding grass *Deschampsia caespitosa*, which should theoretically outcompete *Acomastylis*. In a microcosm study, phenolic compounds in litter of *Acomastylis* were found to enhance microbial activity in soil, as they acted as a C source for soil microbes, which in turn led to net N immobilisation and reduced soil N availability (Bowman et al. 2004). *Acomastylis* litter also reduced the uptake of N by and growth of *Deschampsia* relative to when it was grown with its

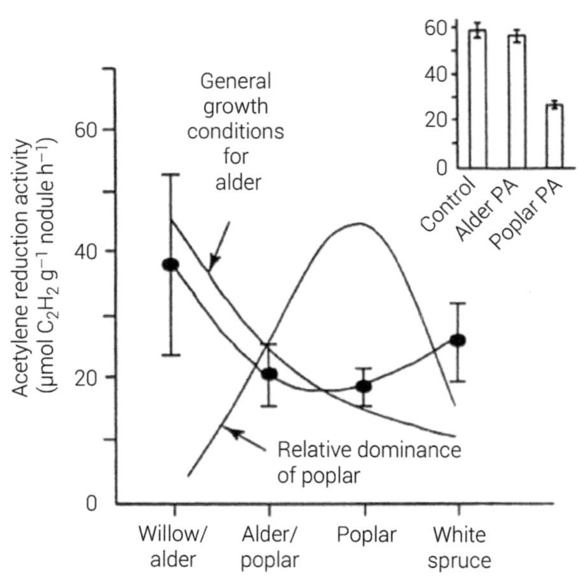

Figure 4.9 Changes in N_2 fixing activity, measured as acetylene reduction activity of nodules, and general growth conditions of alder (*Alnus tenuifolia*) along a successional sequence of Alaskan taiga forest. N_2 fixing activity and growth of alder declined with increasing dominance of poplar, which was attributed to a negative effect of polyphenols (PA) produced by poplar on alder, as observed in the laboratory (see insert graph). From Hättenschwiler and Vitousek (2000) based on Schimel et al. (1998) by permission of Elsevier.

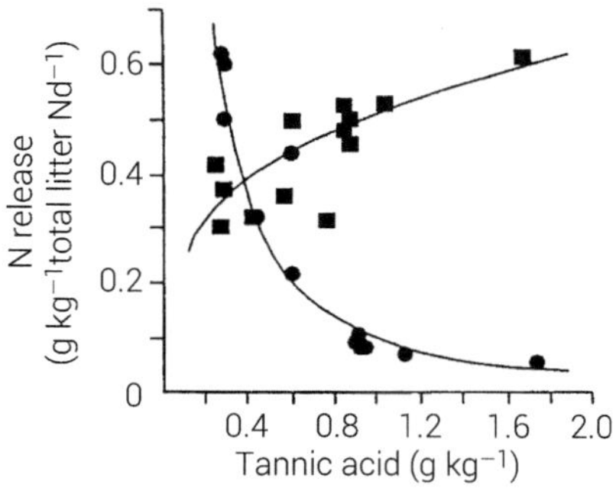

Figure 4.10 Release of N as either inorganic (•) or dissolved (■) organic N from Bishop pine litter (*Pinus muricata*) as a function of the concentration of total phenolics in litter (tannic acid equivalents) during a 3-week laboratory incubation study. Data show that high levels of polyphenols released from pine litter inhibit N mineralisation but increase the release of DON from pine litter. From Hättenschwiler and Vitousek (2000) based on Northup et al. (1995) by permission of Elsevier.

own litter. Coupled with this, studies done in the field revealed that rates of N mineralisation were 10-fold lower in soils dominated by *Acomastylis* relative to those dominated by *Deschampsia* (Steltzer and Bowman 1998). Taken together, these findings suggest that *Acomastylis* litter influences soil N availability in a way that facilitates its persistence in alpine meadows. In other words, because *Acomastylis* can tolerate low N availability due to its slow growth and conservative N use, it gains a competitive advantage over its faster-growing, N-demanding neighbour, *Deschampsia*, which is negatively affected by a reduction in resource supply. More broadly, these results, and those of the other highlighted studies, indicate that in addition to their role in deterring herbivores, phenolic compounds play an important role in regulating plant interactions through their effect on soil biogeochemical processes mediated by soil microbes.

Invasive plants and soil biota

The invasion of exotic plant species into new habitats is a major driver of biodiversity loss and habitat degradation worldwide (Vilà et al. 2011, Simberloff et al. 2013, Essl et al. 2020). The scale of this issue is enormous: invasive plants are almost everywhere on Earth and a total of 13,168 plant species have established self-sustaining populations somewhere outside of their native range, often posing a serious threat to resident plant species and having major impacts on below-ground communities and ecosystem functioning (Van Kleunen et al. 2015). As such, it would be amiss if the topic of how invasive plant species impact soil biota was not considered here.

Invasive plants impact indigenous soil biota and the functions they drive via a variety of routes, but the main pathway is via the introduction of novel plant traits into an ecosystem. As discussed above, plant traits represent a major driver of below-ground communities and their functioning, and the scale of impact of invasive plants on soil biota is generally related to the novelty of their traits relative to those of native plant species: the greater the difference in traits between invaders and native species in invaded areas, the greater the impact on soil biota and ecosystem functioning (Wardle et al. 2011a). Moreover, novel traits of invasive plant species can broadly impact soil biota and ecosystem functions in invaded areas via two broad, simultaneous pathways: the litter pathway, driven by leaf trait differences between invaders and native plant species, and the rhizosphere pathway, driven by differences in the root traits of invaders and native plant species (Figure 4.11) (Zhang et al. 2019b).

Regarding the litter pathway, it is generally the case that leaf traits of invasive plant species are positioned further along the leaf economics spectrum towards exploitative strategies, with higher foliar N and P concentrations and leaf area than native plant species (Leishman et al. 2007). In line with this, invasive plants typically produce greater amounts of leaf litter than their native counterparts (Liao et al. 2008), and this litter is of higher quality (i.e., lower C:N and/or lignin:N ratio) and decomposability (Lee et al. 2017). This increase in the supply of high quality leaf litter from invasive plants in turn generally benefits the decomposer community, as shown by a recent global meta-analysis of studies exploring the effects of invasive plants on soil biota via the litter pathway: across a broad range of studies, plant litter from invasives was shown to, on average, increase the abundance of detritivores, as well as bacteria and their consumers (i.e., microbial feeders), relative to litter from natives (Zhang et al. 2019b). Associated effects of the invasive plant litter pathway on soil N cycling were not detected in this meta-analysis, which was explained by the short-term nature of most studies on this topic

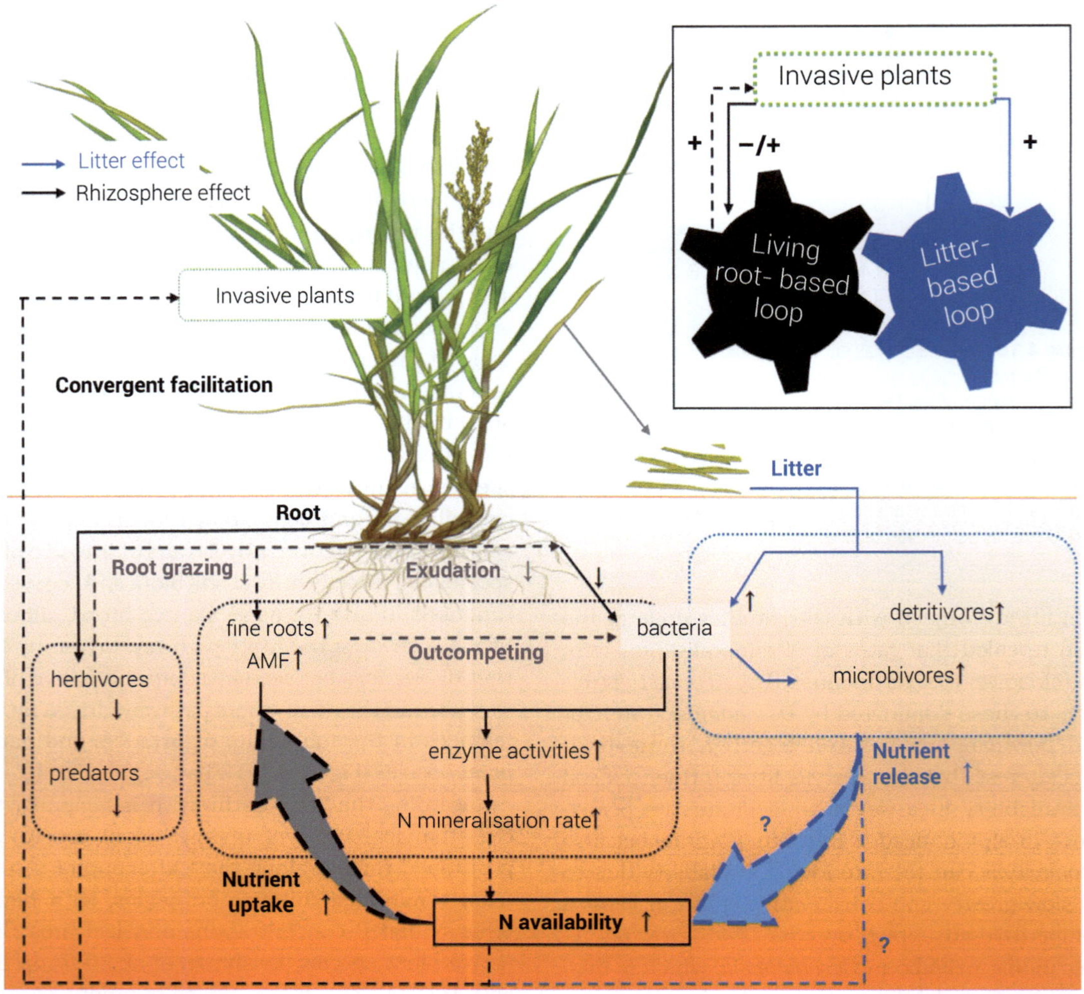

Figure 4.11 Schematic illustrating how differential litter and rhizosphere effects of invasive plants on soil biota converge to facilitate plant invasion. Blue and black arrows mean litter and rhizosphere effects, respectively, and solid and dashed arrows represent experimentally determined and potential effects/feedbacks, respectively. '↑', positive effect; '↓', negative effect; '?', unknown; AMF, arbuscular mycorrhizal fungi *In face of plant competition for N, bacteria may have to produce more exoenzymes to degrade N-containing organic compounds, which will increase the N enzymatic activities. From Zhang et al. (2019b) by permission of John Wiley and Sons. *Ecology Letters*, Volume: 22, Issue: 1, Pages: 200–210. First published: 20 November 2018, DOI: (10.1111/ele.13181).

and the likelihood that nutrient cycling responses lag changes in decomposer communities (Zhang et al. 2019b). However, as discussed earlier in this chapter, it follows that stimulation of decomposer community through the input to soil of high quality litter will enhance rates of N cycling and plant N supply in the longer term, potentially enforcing the dominance and spread of invasives (Zhang et al. 2019b).

The other pathway by which invasives impact soil biota in their new range, and which operates simultaneously, is the rhizosphere pathway (Zhang et al. 2019b). The rhizosphere pathway is largely driven by differences in biochemical root traits of exotic and native plant species, and especially the exudation of phytotoxic compounds that negatively impact native plants and soil biota. This biochemical suppression of native plants and their associated soil biota is a key basis of the novel weapons hypothesis, which posits that the success of many invasive plants is due to their production of phytotoxic biochemicals which are not produced by natives in the invaded range (Callaway and Ridenour 2004). Much evidence for the novel weapons hypothesis comes from classic studies on the exotic forb *Centaurea maculosa* (spotted knapweed), a European invader of North American grasslands that produces root exudates containing the polyphenol catechin, which has an allelopathic effect on native plants in its invaded range (e.g., Callaway et al. 2005, Inderjit et al. 2008, Thorpe et al. 2009).

The exudation of phytotoxins by invasive plants has also been shown to disrupt mycorrhizal associations of native plants, thereby reducing their capacity to uptake nutrients from soil and suppressing their growth. For example, the release of antifungal compounds in root exudates of *Alliaria petiolata* (garlic mustard), a European invader of North American forests, has been shown to disrupt mycorrhizal associations of native tree seedlings, thereby contributing to its success in the invaded range (Stinson et al. 2006, Callaway et al. 2008). While this provides a compelling example of how the exudation of phytotoxins by invaders can disrupt mutualistic associations in the places they invade, many uncertainties remain. For instance, it is unclear whether allelopathic effects of invaders on native soil biota persist in the longer term (Lankau et al. 2009, Lankau 2011) or whether they even reach sufficient soil concentrations to exert negative effects (Blair et al. 2005, Perry et al. 2007, Duke et al. 2009). Moreover, in some situations, invaders can enhance the abundance and diversity of mycorrhizal fungi in their invaded range rather than suppress them (Lekberg et al. 2013, Zhang et al. 2019b) and/or benefit from exploiting mycorrhizal associations of native species (Reinhart and Callaway 2004, Sun and He 2010, Menzel et al. 2017). Despite these uncertainties, the exudation of novel compounds by roots represents an important mechanism by which invasive plants disrupt below-ground communities and their associations with native plants, and thereby modify invaded ecosystems.

Another route by which invasive species influence below-ground properties that sustain native species is through modifying the availability of growth-limiting resources. A classic example of this comes from pioneering studies on the effects of the invasion of the N-fixing actinorhizal shrub *Myrica faya* into young, N-limited *Metrosideros polymorpha* forests in Hawaii, which previously lacked symbiotic N fixers (Vitousek and Walker 1989). This N-fixing shrub was introduced into the Hawaiian Islands from the Azores and Madeira in the 1800s as an ornamental or medical plant. Due to its prolific seed production and effective dispersal by exotic birds, and rapid growth rate in open-canopied sites, it has spread extensively into young, N-limited natural forests of Hawaii. This has led to four-fold increases in soil N inputs due to its N-fixing activities, hence greatly increasing soil N availability. This enrichment of native forest soils with N, the main nutrient that they lack, has potentially dramatic consequences for these ecosystems. Not only does it increase their productivity, but also it can potentially feedback by creating a more favourable environment for invasion by a broader range of exotic species (Vitousek and Walker 1989). Indeed, sites with fertile soils tend to be invaded more successfully by a broader range of plant species than sites with infertile soils on Hawaii (Gerrish and Mueller-Dombois 1980) and elsewhere (Pickard 1984). Overall, this example demonstrates how biological invasions can greatly modify below-ground properties in such a way that substantially alters the structure and productivity of natural ecosystems.

As a final point, invasive organisms can also impact soil biota indirectly by modifying disturbance regimes. This is best illustrated by the introduction of fire-enhancing invasive grasses into ecosystems that increase fire frequency, a classic example being the invasion of the C4 perennial grass *Schizachyrium condensatum* into submontane forest in Hawaii (D'Antonio and Vitousek 1992). Here, *S. condensatum* has spread into the canopies of native shrubs of submontane forest, creating a continuous layer of fuel and increasing the frequency and intensity of fire. This has not only killed native trees and shrubs, but also has enabled *S. condensatum* and other grasses to invade burned areas, further increasing fire risk (D'Antonio and Vitousek 1992). In other words, grass invasion has set in motion a feedback cycle that has transformed non-flammable, mostly native-dominated forest into highly flammable, low diversity, alien-dominated grassland.

There are numerous other examples of invasive grasses altering fire regimes across ecosystems, and a recent analysis of invasive grasses in the USA identified substantial regional-scale alteration of fire regimes: eight invasive grass species have increased fire occurrence by up to 230% and fire frequency by up to 150% across the USA (Fusco et al. 2019). As will be discussed in Chapter 6, such alteration of fire regimes has potential to have dramatic, yet largely untested, below-ground impacts and feedback consequences for plant community dynamics and ecosystem functioning.

Plant diversity, soil biota, and ecosystem functioning

A wealth of studies investigating biodiversity–ecosystem function relationships point towards multiple benefits of higher levels of plant diversity, including increased plant productivity, recycling of essential nutrients, and resistance of plant production to extreme climatic events (Tilman and Downing 1994, Cardinale et al. 2012, Isbell et al. 2015). The mechanisms underlying such positive biodiversity–function relationships remain a topic of debate, but they are generally attributed to: (i) complementarity effects, whereby niche complementarity, or differences in the way species use limiting resources, leads to more efficient resource capture and hence higher plant productivity; and (ii) selection effects, due to an increased probability of including a plant species with specific characteristics in high diversity mixtures, such as a highly productive species that becomes dominant, or N fixers that increase N supply (Loreau and Hector 2001). How plant diversity impacts the abundance, diversity, and functioning of below-ground communities remains less certain and varies depending on a host of factors, including facets of plant diversity (i.e., species and functional diversity), taxonomic groups of soil organisms, environmental context, and spatial scale. Nevertheless, a large amount of evidence points to changes in plant diversity and composition acting as important determinants of below-ground communities, especially at a local scale.

Hypothetically, positive effects of plant species diversity on soil biota and processes should operate via a range of mechanisms including: (i) increased supply and heterogeneity of plant-derived resources entering soil; (ii) increased probability of including plants with specific functional traits in more diverse mixtures, such as specific root exudation profiles or root morphological traits; and (iii) increased niche space for plant species-specific microbial associations, especially symbionts (Wardle and Van der Putten 2002) (Figure 4.12). The prevalence of positive effects of plant diversity on soil organisms and their activity at a local scale was supported by a global meta-analysis of 106 studies revealing that plant species mixtures increased soil microbial biomass, the fungal to bacterial biomass ratio, and respiration across different terrestrial ecosystems, and that these positive effects of plant diversity on soil microorganisms increased with the number of species in mixtures and stand age (Chen et al. 2019a). Likewise, synthesis of 623 paired observations of plant mixtures and corresponding monocultures from 40 studies revealed that, on average, soil animal diversity was 10% higher in plant mixtures than in corresponding monocultures, and that the positive mixture effect strengthens with higher plant species richness and time (Zhang et al. 2022). Data from 12 different studies that manipulated grassland plant species diversity also showed that higher plant diversity enhanced soil microbial biomass, with diversity effects becoming stronger over time (Thakur et al. 2015). Individual studies also show that plant mixtures generally increase soil phosphatase activity and available soil P across various ecosystems relative to monocultures (Chen at al. 2022), and that the abundance and richness of soil microfauna and mesofauna is positively related to plant diversity due to increased root biomass and rhizodeposition, and hence resource supply to microbes and their consumers (Eisenhauer et al. 2013). In line with this, plant diversity effects on soil biota have been shown to operate across trophic levels, cascading from decomposers and herbivores to carnivores and omnivores (Scherber et al. 2010).

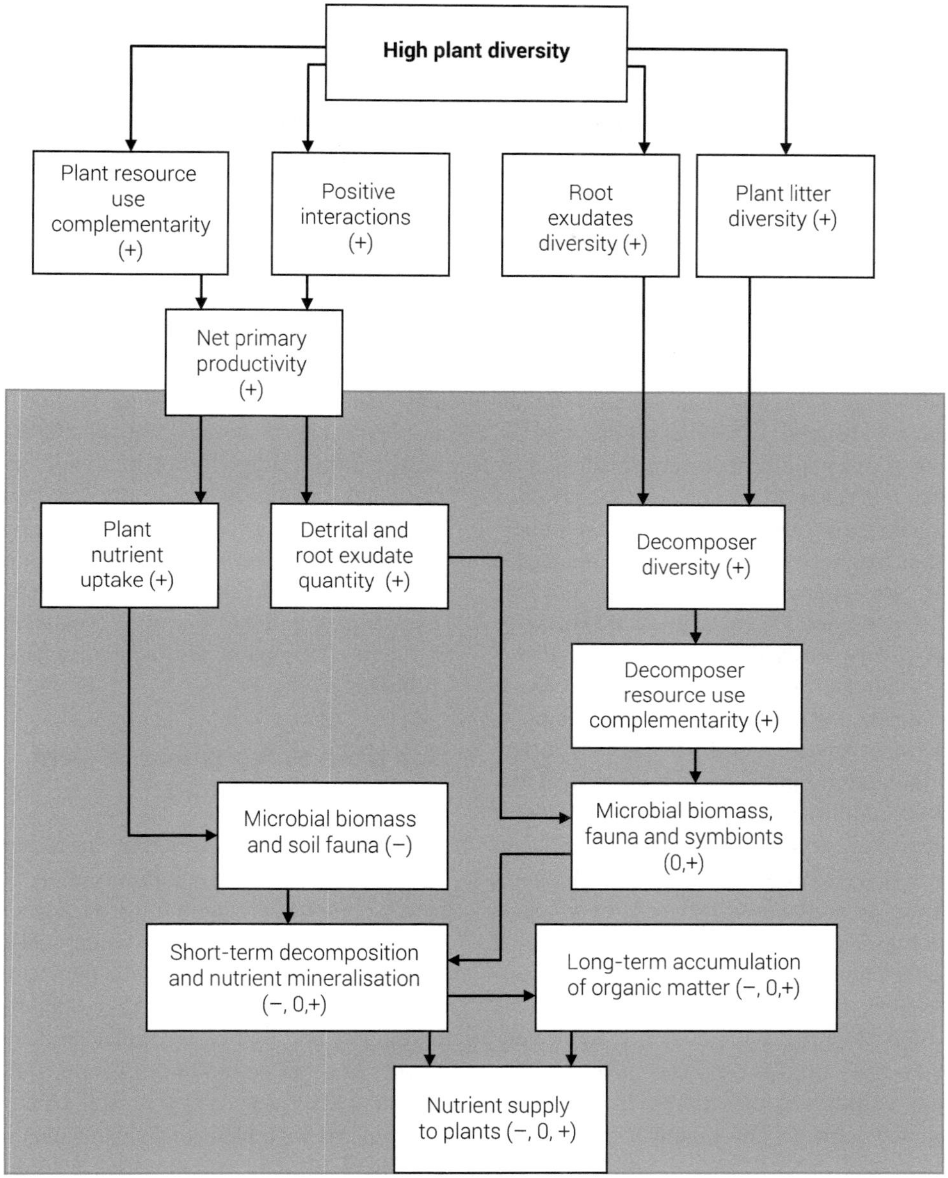

Figure 4.12 Hypothetical mechanisms by which changes in plant species richness may affect decomposer-mediated processes. +, 0, and − indicate positive, neutral, and negative effects, respectively. From Wardle and Van der Putten (2002) by permission of Oxford University Press.

Not all studies have detected positive effects of plant diversity on soil biota, with many showing that plant species identity is a more important driver than is diversity per se or that diversity effects on soil biota are unpredictable. A field study in New Zealand grassland, for example, found no predictable effect of manipulating grassland plant diversity, using in situ plant removals, on the diversity of soil microbes, microbe-feeding nematodes, predatory nematodes, or herbivorous nematodes (Wardle et al. 1999a). Also, a glasshouse study in which three plant species of three functional groups (grasses, N-fixing legumes, and forbs) were grown in monoculture and in mixtures found that plant species identity, rather than plant species or functional group richness, had strong effects on the biomass or population densities of primary consumers (microbes and herbivorous nematodes) and secondary consumers (microbe-feeding nematodes and enchytraeids) (Wardle et al. 2003b). The diversity of Collembola was also found to be unresponsive to plant species and functional group richness in a grassland diversity experiment in Switzerland, whereas collembolan diversity increased with the presence of legumes due to the production of high litter quality and higher microbial biomass in the rhizosphere of this plant functional group (Salamon et al. 2004). Individual plant species that affected resource quality were also found to be more important than plant diversity as a driver of soil nematode community structure and diversity in an outdoor grassland experiment (De Deyn et al. 2004), again pointing to plant species identity rather than diversity per se as the main driver of soil biota. It should be noted that studies also show that plant species richness, identity, and productivity differentially influence groups of soil organisms. In model grassland communities, for example, increases in the abundance of saprotrophic fungi, bacteria, and actinomycetes were driven by larger plant biomass in high diversity treatments rather than by plant species richness per se, whereas the abundance of AM fungi increased in response to increasing plant species richness (De Deyn et al. 2011). This suggests that AM fungal abundance is more sensitive to changes in plant species diversity per se than are abundances of saprophytic microbes, which respond more to the quantity of resource supply.

It is arguably unsurprising that plant diversity effects on soil biota are so variable given the wide variability among studies in the way they are designed, the levels of plant diversity and plant species used, and the fact that relationships can vary across different taxonomic groups, habitats, and spatial scales (Bardgett and Wardle 2010). Indeed, an extensive analysis of soils taken from grasslands worldwide found few significant relationships between plant and soil microbial diversity at the plot scale (i.e., alpha diversity), suggesting plants and soil microorganisms respond to different environmental drivers (Prober et al. 2015). Further, a comparative study assessing a range of chronosequences worldwide found that the diversity of plants and soil organisms (bacteria, fungi, protists, and invertebrates) was unrelated to each other during ecosystem development, suggesting that temporal changes in plant diversity are not associated with corresponding changes in soil biodiversity (Delgado-Baquerizo et al. 2019). A recent global meta-analysis of 84 sites, covering 3,900 samples, also found that correlations between plant and soil microbial diversity vary considerably across biomes, as does the proportion of studies showing positive, negative, and neutral correlations (Liu et al. 2020). Together, these studies suggest that relationships between plant and soil microbial diversity detected at a local scale (i.e., in studies at single sites or in pots) do not persist across larger spatial and temporal scales as they become overridden by environmental factors (e.g., climate and soil abiotic properties) that more strongly shape plant and soil microbial diversity (Prober et al. 2014).

Soil biota as regulators of plant community dynamics

Understanding the factors that drive change in the composition of plant communities has long been a major theme of community ecology. A variety of abiotic factors, such as climate, disturbance, competition for light and nutrients, and biotic interactions such as those that operate between herbivores and plants, are considered to be major drivers of plant community dynamics. However, it is now widely accepted that soil biota and their interactions with plants also act as major structuring forces in plant communities (Van der Heijden et al. 2008, Bardgett and Wardle 2010, vam der Puttem et al. 2013, Bennett and Klironomos 2019). There are three major types of biotic interactions between plants and soil biota that impact plant population and community dynamics: (i) interactions between dead plant materials and the decomposer subsystem, leading to changes in soil nutrient availability; (ii) effects of biotrophic interactions between living plant roots and their symbionts, herbivores, and pathogens; and (iii) microbial-mediated niche partitioning of nutrient use by plants. How decomposer organisms act on plant growth through influencing the availability of nutrients in soil has been covered in Chapter 3. This section focuses on the latter two processes as drivers of vegetation dynamics, namely interactions between plant and root-associated organisms, and microbial mediation of resource partitioning. In doing so, recent developments in understanding of the role of plant–soil feedback in plant community dynamics are considered, which is based on the notion that a given plant species or genotype may increase (positive feedback) or decrease (negative feedback) its growth rate relative to other plants via its impact on the soil community, with consequences for spatial and temporal patterns of vegetation. Several mechanisms contribute to plant–soil feedbacks, including changes in nutrient availability and the accumulation of symbionts and natural enemies in the root zone of host plants. While these mechanisms are strongly interrelated, here, their roles are considered individually.

Mycorrhizal fungi

There is an abundance of evidence for mycorrhizal fungi benefiting the performance of their host plants, largely through improved plant uptake of nutrients and water and increased tolerance to biotic and abiotic stress (Smith and Read 2010). However, whilst mycorrhizal fungi associate with most plants within any community, they confer different degrees of benefit on certain plant species, thereby indirectly influencing the structure and dynamics of plant communities. Furthermore, as discussed in Chapter 3, they form extensive underground common mycorrhizal networks among plant species that facilitate C and nutrient transfers between individual plants (Simard et al. 1997, Klein et al. 2016) and act as a conduit for chemical signals, for instance warning neighbouring plants of insect attack (Babikova et al. 2013). Indeed, there is now a large body of literature showing that mycorrhizal fungi can act as structuring forces of plant communities at both local and landscape scales by amplifying or relaxing interspecific plant competition, with negative and positive effects on plant species coexistence and diversity, respectively (Tedersoo et al. 2020).

Several studies have demonstrated that AM fungi can enhance plant species coexistence by benefiting the growth of weak competitors that are more dependent on mycorrhizal fungi than stronger competitors (Grime et al. 1987, Gange et al. 1993, Stanescu and Maherali 2017). Such responses may be due to interplant transfers of C and nutrients via mycorrhizal networks leading to more even distribution of resources within the plant community, reducing the ability of certain species to monopolise resources (Grime et al. 1987). However, they have also been attributed to the dual effects of less competitive plant species having a greater colonisation and dependency on AM fungi than competitive species, and competitive species suffering more than less competitive species from negative soil feedback due to a greater accumulation of fungal pathogens (Tedersoo et al. 2020). Indeed, it was shown that grassland plants with resource-acquisitive traits, such as high shoot N concentrations and thin roots, attract more diverse communities of fungal pathogens and a lower diversity of AM fungi than do those with conservative traits, resulting in strong plant growth suppression (i.e., negative feedback) of the former (Semchenko et al. 2018). Plant responses to soil organisms have also been shown to vary according to their nutrient-acquisition strategy in hyperdiverse Australian shrublands, with ECM plants displaying positive feedback, and N-fixing and non-mycorrhizal plants experiencing negative feedback, which contributed to species coexistence of these shrublands (Teste et al. 2017).

Mycorrhizal fungi do not always promote plant species coexistence and diversity. Hartnett and Wilson (1999) found that suppression of AM fungal infection in tallgrass prairie, achieved using the fungicide Benomyl, led to a large increase in plant species diversity, which was due to a reduced dominance of obligately mycotrophic C4 tall grasses and consequent competitive release of subordinate facultative mycotrophs. This suggests that active AM fungal associations reduce floristic diversity in tallgrass prairie due to the promotion of competitive dominance of obligately mycotrophic species. Furthermore, a comprehensive study of North American temperate tree species revealed that AM trees accumulate more pathogens and display negative feedback, whereas ECM trees display positive feedback, presumably due to increased nutrient capture and protection from pathogens, which promotes their dominance in forest ecosystems (Bennett et al. 2017). Similarly, studies in subtropical forests reveal that ECM trees tend to exhibit positive feedback (Chen et al. 2019b), growing better in conspecific than heterospecific soil, and that this might contribute to the clustering of ECM seedlings around adult ECM trees (Johnson et al. 2018, Sasaki et al. 2019) and, over time, to monodominance in forest ecosystems (Peh et al. 2011).

Some plant species can also gain an advantage through their litter chemistry, which maximises the acquisition of nutrients by a host's mycorrhizal roots, while hindering nutrient acquisition by mycorrhizal roots of co-occurring plants. The expansion of ECM-associated willows into alpine meadows, for example, has been shown to be associated with reduced AM fungal colonisation and growth of alpine herbaceous plants because willow leaf litter, and presumably the organic substrates contained within it, increases ECM activity and resource acquisition to the detriment of AM fungi and their host plants (Becklin et al. 2012). Similarly, ERM plants produce phenolic-rich leaf litter that can retard the growth of both AM and ECM fungi (Kohout et al. 2011, Teste et al. 2017), and in hardwood forests, the expansion of rhododendron has been attributed to its ERM roots being better able to access N from its own tannin-rich litter than do roots of ECM and AM plants (Wurzburger and Hendrick 2009). The expansion of some non-mycorrhizal plants, such as the invasive tamarisk (*Tamarix* sp.), or salt cedar, which has replaced historically dominant native cottonwoods (*Populus fremontii*) along rivers and streams in southwestern USA, has also been linked to changes in soil chemistry that disrupt mycorrhizal associations (Meinhardt and Gehring 2012). Here, it has been proposed, but not fully tested, that tamarisk negatively affects cottonwood mycorrhizal fungi through increasing soil salinity and producing phytochemicals in its litter and root exudates that act as 'biochemical weapons' against mycorrhizal fungi of the native cottonwoods (Meinhardt and Gehring 2012). These are just a few examples, but they demonstrate the existence of intricate feedbacks between plants, mycorrhizal fungi, and plant litter chemistry that can maximise the acquisition of nutrients by one plant species over another, thereby altering plant competitive interactions and species distributions.

Mycorrhizal diversity also impacts plant communities, with studies revealing a general trend of positive effects of both ECM and AM fungal diversity on plant productivity (Hazard and Johnson 2018, Powell and Rillig 2018, Neuenkamp et al. 2019). However, responses are quite inconsistent, with mycorrhizal diversity effects varying depending on factors such as host identity and soil conditions. For instance, while positive effects of AM fungal richness (1–14 species) on plant diversity and productivity were detected in model grasslands (Figure 4.13), which was attributed to more efficient exploitation of soil P reserves (Van der Heijden et al. 1998), positive effects of ECM species richness on biomass of *Betula pendula* were only detected in low fertility substrate, whereas ECM species richness negatively affected *Pinus sylvestris* growth in highly fertile substrate (Jonsson et al. 2001). Furthermore, there is a general lack of understanding regarding the mechanisms that underpin diversity effects of mycorrhizal fungi on plant communities, and whether they are explained by the inclusion of fungal species that differ in resource acquisition strategies in more plant species-rich communities (i.e., complementarity effects), or fungal species that are especially competitive in terms of their capacity to exploit soil resources for their host plant (i.e., selection effects). This issue was tested by Wagg et al. (2011) who observed a positive relationship between AM fungal richness and plant productivity in a simple grass–clover model system, as found in other studies (e.g., Van der Heijden et al. 1998, Vogelsang et al. 2006, Maherali and Klironomos 2007), and that this was driven by complementarity effects among fungi in one soil type and by a selection effect resulting from the fungal species that stimulated plant productivity the most in another soil type. These findings, and those of other studies (e.g., Jonsson et al. 2001), indicate that diversity effects of mycorrhizal fungi are highly context dependent and vary depending on soil conditions and the identity of the fungal species and plant partners involved. Future work is clearly needed to unravel the mechanisms that underpin diversity effects of mycorrhizal fungi on plant communities

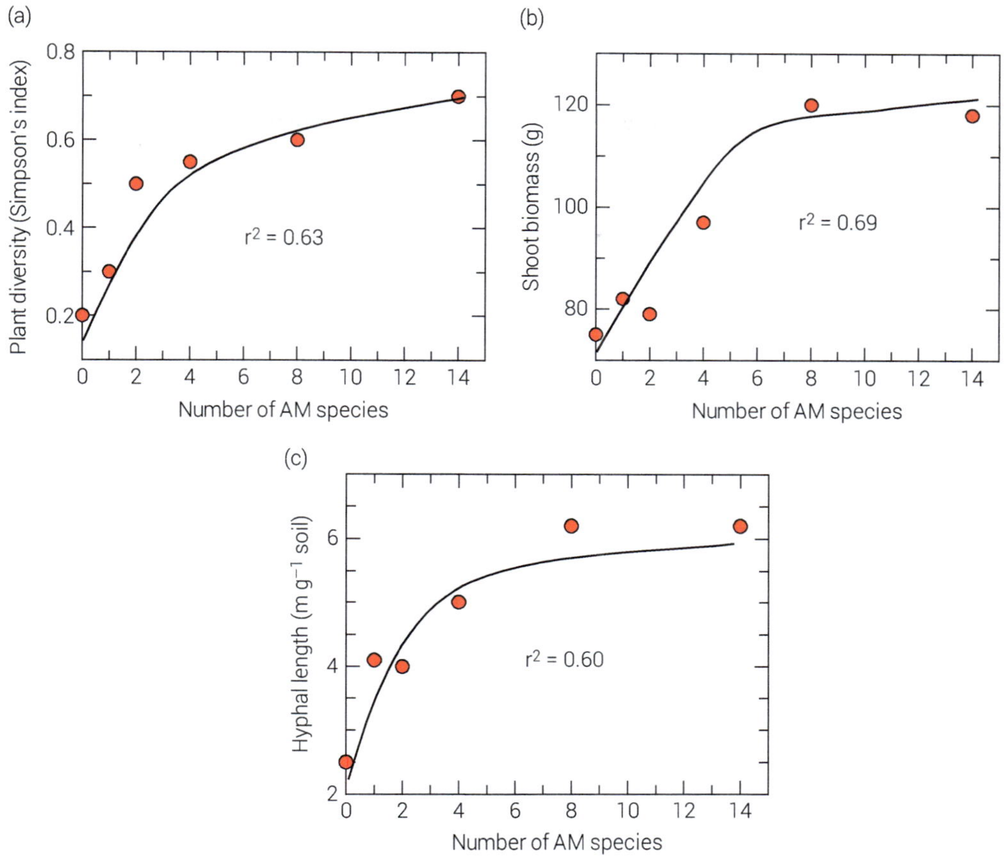

Figure 4.13 Effects of manipulating the diversity (1–14 species) of native arbuscular mycorrhizal fungi on (a) plant diversity (Simpson's index), (b) shoot biomass, and (c) total fungal hyphal length in experimental grassland units. From Bardgett (2005) using data from Van der Heijden et al. (1998) by permission of Oxford University Press.

under different environmental conditions, but it is evident that they play an important role.

N-fixing organisms

Vascular plants with N-fixing symbionts can have profound effects on vegetation dynamics, especially through acting as facilitators of plant succession. N fixers are especially abundant during early stages of primary succession when N is typically limiting, for instance on newly exposed land following glacial retreat or volcanic eruptions, or on river terraces and sand dunes (Walker 1993, Chapin et al. 1994, Bellingham et al. 2001, Crews et al. 2001, Menge and Hedin 2009). In these situations, the low availability of N favours N fixers that can acquire atmospheric N_2 to sustain their growth, which in turn raises soil N to levels needed to support other later successional species (Walker and Del Moral 2003). The mechanisms by which N fixers increase soil N availability include the release of N-rich root exudates into soil and the build-up and decomposition of shoot and root litter of high N content (Walker and Del Moral 2003). Furthermore, associations between N-fixing plants and mycorrhizal fungi can aid the transport of N from N fixers to neighbouring plants via hyphal networks (Haystead et al. 1988), and low amounts of root herbivory of N-fixing legumes has been shown to increase the transfer of C and N to neighbouring grasses (Bardgett et al. 1999a, Ayres et al. 2007). However, the importance of these below-ground mechanisms of

N transfer between neighbouring plants for plant succession remains untested.

N-fixing plants have been shown to influence the course of vegetation succession in a wide range of ecosystems (Figure 4.14). For example, in river terraces in New Zealand, the N-fixing shrub *Carmichaelia odorata* facilitated seedling growth of some dominant trees via improved N supply (Bellingham et al. 2001). In alpine ecosystems in Norway, the presence of the legume *Oxytropis lapponica* was found to be associated with greater plant diversity due to an increase in N supply in these N-limited ecosystems (Olsen et al. 2013). Also, in sand dunes in the Netherlands, plant biomass was found to increase following a stage dominated by the N-fixing shrub *Hippophaë* (Olff et al. 1993). Some of the earliest evidence for facilitation by N fixers, however, comes from the studies of primary succession on substrates exposed by glacial retreat. In a now classic study, Crocker and Major (1955) documented primary succession at Glacier Bay, southeast Alaska, and reported that N-fixing alder shrubs (*Alnus sinuata*) increased soil N and SOM and decreased soil pH, creating a more favourable environment for *Picea sitchensis* (sitka spruce), the next successional dominant. While this view has since been challenged on the basis that other factors contribute to the replacement of *Alnus* by *Picea*, such as competitive interactions between these species (Chapin et al. 1994), it still represents an important demonstration of the ability of N fixers to facilitate succession and hence act as important determinants of vegetation dynamics. N-fixing tree species have also been shown to play a key role in tropical secondary succession in that they provide a large fraction of the N needed to support rapid forest recovery following disturbance (Batterman et al. 2013).

Plants with N-fixing symbionts also impact vegetation dynamics via the amelioration of local conditions, especially in extreme environments. For example, in semi-arid ecosystems, the leguminous, N-fixing shrub *Retama sphaerocarpa* was found to facilitate understorey herbs and increase overall plant diversity and biomass due to the amelioration of local soil conditions (Pugnaire et al. 1996). Not only did this shrub increase local soil N and organic matter content, but also it helped intercept dust and water, which further improved soil conditions, creating 'islands of fertility' in an otherwise barren environment (Pugnaire et al. 1996). These 'islands of fertility' were found to facilitate the growth and

Figure 4.14 Vascular plants with N-fixing symbionts are extremely widespread as early colonists in primary succession. This figure shows dominant N-fixing plants of different primary seres: (a) *Alnus sinuata*, which forms thickets on deglaciated terrain at Glacier Bay, southeast Alaska; (b) *Coriaria arborea* colonising recently deglaciated terrain at the Fox Glacier, South Island, New Zealand; and (c) *Coriaria arborea* colonising floodplains near Kaikoura, South Island, New Zealand. Images by Richard Bardgett.

nutrition of understorey herbs, thereby increasing overall plant diversity. But they also created a more favourable environment for decomposer organisms, which further enhanced nutrient cycling and supply to neighbouring plants (Pugnaire et al. 1996). It should be noted that while leguminous shrubs such as *R. sphaerocarpa* can facilitate the establishment and growth of understorey plants, associated improvements in soil conditions can increase competition among understorey plant species, leading to displacement of certain species. In mesic environments of higher fertility, for example, *R. sphaerocarpa* was found to reduce plant species richness of understorey plants due to higher soil N and increased competition (Rodríguez-Echeverría and Pérez-Fernández 2003), which is counter to its positive effects on understorey plants in harsher semi-arid environments (Pugnaire et al. 1996).

N-fixing symbionts occur later in succession. For example, cyanobacteria in moss carpets fix significant amounts of N in boreal forests, contributing as much as 50% of total N input in late successional and old-growth forests (Lagerström et al. 2007, DeLuca et al. 2008, Gundale et al. 2010). N-fixing trees are also abundant in many late successional tropical forests (Crews 1999, Hedin et al. 2009), although rates of N fixation are typically low in these ecosystems and are much less than in early successional or secondary forests (Vitousek et al. 2013). Furthermore, while symbiotic N fixation is typically high when N is in short supply, it can be constrained in some N-limited ecosystems, thereby limiting its impact on N cycling and plant community dynamics. For example, low light availability in the forest understorey can constrain symbiotic N fixation despite low levels of N availability in soil (Taylor and Menge 2018), and preferential grazing on N-rich plants with N-fixing symbionts in grasslands can reduce N inputs and sustain N limitation (Vitousek et al. 2013).

Root pathogens

Farmers have long known that the repeated cultivation of a particular crop leads to the build-up of crop-specific soil-borne pathogens, which causes declines in yield; this is the basis of crop rotation, which prevents the build-up. However, soil-borne pathogens also play an important role in regulating vegetation dynamics in natural ecosystems via negative feedback on their host plant. Some of the earliest studies on this topic demonstrated that the growth of certain grassland plants was reduced when grown with their own 'cultured' soil community relative to when grown with a soil community of another species due to the accumulation of host-specific pathogens (Bever 1994, Bever et al. 1997). Also, studies done in sand dunes in the Netherlands revealed that the build-up of root pathogens and root-feeding nematodes in the root zone of early coloniser Marram grass (*Ammophila arenaria*) led to its degeneration and replacement by other species, namely *Festuca rubra*, which was not susceptible to these pathogens (Van der Putten et al. 1993) (Figure 4.15). Through this mechanism, negative feedback was shown to facilitate species replacement and the course of plant succession.

A plethora of studies have since identified soil-borne pathogens as important drivers in several areas of plant ecology, including species coexistence and the maintenance of plant diversity. The general idea here is that negative feedback caused by the accumulation of host-specific pathogens prevents any single plant species from being dominant, thereby promoting species coexistence. Evidence for this mostly comes from studies in tropical forests, where the accumulation of host-specific fungal pathogens in soil beneath adult trees limits the establishment and growth of conspecific seedlings close to the adult trees, thereby putting heterospecific seedlings at an advantage and facilitating tree species coexistence and diversity (Packer and Clay 2000, Mangan et al. 2010, Liu et al. 2015b). This effect, termed the Janzen–Connell effect (Janzen 1970), has also been shown to occur in temperate forests where plant-associated fungi reduce conspecific seedling recruitment and survival at high adult conspecific density, thereby potentially increasing plant diversity (Jia et al. 2020). Soil-borne pathogens have also been shown to contribute to spatial mosaics of plants (Blomqvist et al. 2000) and to fluctuations in species abundance over time (in't Zandt et al. 2021), due to spatial and temporal variation among species in the extent that they display negative feedback (Chung 2023).

(a) (b)

Figure 4.15 In sand dune systems in the Netherlands, the build-up of both root pathogens and root-feeding nematodes in the root zone of Marram grass (*Ammophila arenaria*) led to the degeneration of this plant and its replacement by other species, namely *Festuca rubra*, which was not susceptible to these pathogens. The images show Marram grass in its vigorous stage of growth (a) and in its degraded form, being replaced by *Festuca rubra* (b). Images by Wim van der Putten.

The invasion of exotic plant species into new territories, which was considered earlier in this chapter, is strongly influenced by negative feedback. Rare plants, for example, tend to accumulate pathogens that limit their growth (Xu et al. 2015, Reinhart et al. 2021), whereas abundant and invasive species accumulate pathogens more slowly, thereby allowing them to proliferate (Klironomos 2002). However, in some situations, negative plant–soil feedback can disproportionally harm more-abundant species, thereby keeping them in check and helping maintain species coexistence (LaManna et al. 2016, Maron et al. 2016). The invasion of exotic plants into new territories is facilitated by an 'escape' from soil-borne pathogens that suppress their growth in the native range: the so-called enemy release hypothesis. For example, soil microbes from the home range of the invasive exotic spotted knapweed (*Centaurea maculosa*), a native of western Europe, were found to have a stronger inhibitory effect on its growth than soil microbes from where the weed had invaded in North America, which contributes to its success as an invasive species (Callaway et al. 2004). Further, the invasion of black cherry trees (*Prunus serotina*) into northwestern Europe has been shown to be facilitated by an 'escape' from virulent soil fungal pathogens that hamper its growth in its native range in the USA (Reinhart et al. 2003, 2010). Similarly, range expansion of plants originating from southern Europe into northwestern Europe has been attributed to them experiencing fewer pathogenic effects in their root zone than native plant species (Van Grunsven et al. 2007, Engelkes et al. 2008). Such release of invasive plants from their natural soil-borne pathogens may also favour the selection of genotypes that invest less in defence and more in growth, thereby further increasing their competitiveness over native plant species (Dawson and Schrama 2016).

The effects of soil-borne pathogens and strength of negative feedback vary greatly among plant species, which is consistent with there being considerable interspecific variability in the composition of rhizosphere fungal communities (Leff et al. 2018, Mommer et al. 2018, Sweeney et al. 2021). Also, some plant species have the capacity to actively suppress fungal pathogens in their root zone, for instance through releasing antifungal compounds in root exudates (Baetz and Martinoia

2014, Li et al. 2018a) or by stimulating specific antagonists (Berendsen et al. 2012). The roots of neighbouring plants can also influence the strength of negative feedback, for instance by creating a barrier to fungi pathogens navigating to their host (Van Ruijven et al. 2020). The spread of the soil-borne disease *Phytopthora* in pepper, for example, was reduced when it was intercropped with maize, especially when root intermingling was at its highest (Yang et al. 2015). Negative feedback also varies with plant life history strategy—plants with an annual life cycle tend to experience stronger negative feedback than do perennials (Kulmatiski et al. 2008), as do early compared to late successional plant species (Kardol et al. 2007)—and can also be neutralised by abiotic factors, such as reduced water and high nutrient availability (Fry et al. 2018a, in't Zandt et al. 2019).

Evidence also suggests that plants have evolved traits that govern rhizosphere microbiome assembly and hence negative feedback. Grassland plant species with resource-acquisitive traits, for example, appear to suffer more from negative feedback than do those with conservative traits (Cortois et al. 2016). This has been linked to plants with exploitative traits (e.g., rapid growth rates, high shoot N, and thin roots) attracting more diverse communities of fungal pathogens in their root zone (Semchenko et al. 2018) (Figure 4.16), which may also be linked to differences in root exudate composition (Yacine et al. 2024). These antagonistic relationships between soil fungi and grassland plants appear to be strongly influenced by soil fertility, with more fertile soils promoting negative feedback, due to shifts in plant traits from conservative towards more exploitative, and an increase in the diversity of soil-borne fungal pathogens (Semchenko et al. 2018). Indeed, the interplay between soil nutrient availability and soil microorganisms is increasingly recognised as an important determinant of plant–soil feedback. For example, nutrient-poor soils might favour positive feedback mediated by mycorrhizal fungi due to improved plant nutrient acquisition, whereas nutrient-rich soils could promote negative feedback by increasing the accumulation of pathogens and/or reducing the benefits of symbionts (Semchenko et al. 2018). However, increased nutrient availability could dampen or neutralise negative feedback by increasing plant health and/or the production of plant defences (Bennett and Klironomos 2018). Although the mechanisms underpinning plant–soil feedback responses to soil nutrient availability remain unclear, it is evident that both abiotic and biotic soil properties should be considered when studying plant–soil feedback.

As a final note, it has been proposed that soil-borne pathogens contribute to positive diversity–productivity relationships in plant communities. The rationale behind this is that at low plant

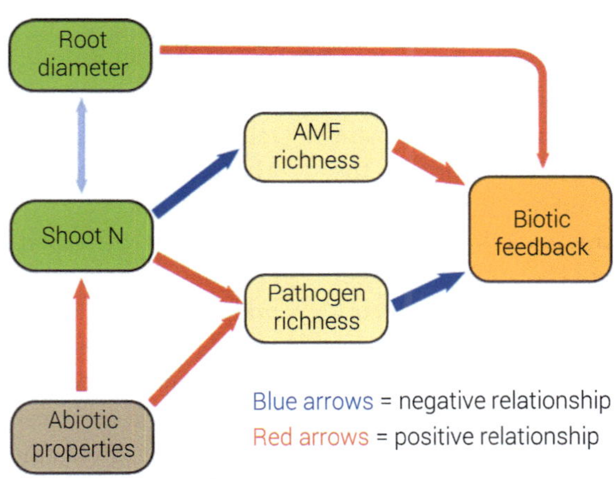

Figure 4.16 Fast-growing grassland plant species with higher shoot N and finer roots accumulate fewer beneficial AM fungi and attract a more diverse pathogen community, resulting in net negative interactions with soil microbiome. From Semchenko et al. (2018).

diversity, the density of individual plant species is high and conspecifics grow in proximity, so host-specific fungal pathogens strongly reduce plant growth. At high plant diversity, however, the density of a particular plant species will be lower and conspecifics will be further apart, so negative effects of soil fungal pathogens on plant growth are dampened (Van Ruijvan et al. 2020). This idea, the so-called dilution effect, was tested by Maron et al. (2011) who created plant communities of varying diversity and crossed them with a fungicide to suppress soil pathogens. In non-fungicide-treated plant communities, they detected a strong positive relationship between plant diversity and aboveground plant biomass, whereas in fungicide-treated communities, where fungal pathogens were suppressed, this relationship disappeared. Moreover, this dampening, or flattening, of the diversity–productivity relationship in fungicide-treated plots occurred because plant production was boosted much more in the low diversity than in the high diversity plant communities. While there is potential for the fungicide to have side-effects on other groups of soil organisms, such as mycorrhizal and saprotrophic fungi, these findings, and those of other studies (Schnitzer et al. 2011, Kulmatiski et al. 2012), provide support for the idea that plant diversity can dilute impacts of host-specific fungal pathogens on plant growth, thereby acting as a driver of diversity–productivity relationships.

Root-feeding fauna

A wide range of root-feeding fauna that live belowground, including nematodes and the larvae of beetles, flies, and moths. These fauna can shape plant populations and communities directly through the consumption of roots and indirectly via interactions with other soil organisms, including microbial pathogens and symbionts and above-ground insect herbivores. At its most dramatic, root herbivory causes reductions in crop yield, or in some cases the complete loss of crops, with major economic consequences (Blackshaw and Kerry 2008). The role of root-feeders in natural ecosystems is less studied, but they similarly have negative direct impacts on plant growth by causing damage to or the death of roots. Moreover, by selectively feeding on the roots of different plant species, based on host specificity and/or nutritional quality, they can impact plant communities and the course of succession. For example, insect root herbivores were found to accelerate secondary succession in grassland by selectively feeding on early successional forbs, reducing their colonisation and persistence (Brown and Gange 1992) and facilitating the growth of late successional species (Schädler et al. 2004). Similarly, in model grassland communities, soil invertebrates were found to enhance secondary succession and local plant diversity through selective feeding on dominant plant species, which enabled late successional subordinate plant species to proliferate (De Deyn et al. 2003). The addition of macrofauna to soils containing micr- and mesofauna was also found to cause shifts in vegetation composition in model grassland communities, albeit increasing the dominance of grasses at the expense of herbs and legumes due to preferential feeding by large insects on N-rich herbs and legumes (Bradford et al. 2002).

How root-feeding insects influence plants also depends on interactions with other organisms in soil, a topic that has been extensively reviewed by Johnson and Rasmann (2015). For example, studies suggest that root herbivores perform worse on plants with mycorrhizal fungi due to improved plant nutrition, which makes them more tolerant to root herbivory, but also mycorrhizal fungi can affect the production of plant defensive compounds and defence signalling (Johnson and Rasmann 2015). Furthermore, some specialist root-feeders, such as weevils (*Sitona* spp.), can actively seek out and feed on root nodules of N-fixing plants, thereby selectively affecting the performance of these plants in mixed communities (Johnson and Rasmann 2015). The abundance of root-feeders, and hence their potential to impact vegetation dynamics, is also affected by predation by other arthropods, such as mites, ants, beetles, and spiders, and by natural enemies. The most abundant and ubiquitous natural enemies are entomopathogenic nematodes, which ambush or actively search for root-feeding insects using chemical cues (Campbell et al. 2003), and entomopathogenic fungi, which occur in soils globally (Vega et al. 2009). Both entomopathogenic nematodes and fungi are effective in controlling populations of root-feeders and therefore indirectly moderate their potential to impact plant growth and community dynamics.

Effects of root herbivores on plant communities may manifest themselves through their influence on above-ground consumers of plant material, and vice versa. It is generally thought that insect herbivores that feed on the same plant have negative effects on each other, which has been attributed to the production of plant secondary compounds in response to the initial attacking species that negatively impact the next species (Kaplan and Denno 2007). However, studies that have examined how root-feeders impact above-ground herbivores reveal contrasting responses, reflecting the range of mechanisms and organisms involved (Soler et al. 2012). For example, positive effects of root-feeding insects on phloem-feeding aphids have been attributed to a stress response, whereby an impaired capacity of roots to acquire water and nutrients leads to the accumulation of amino acids and carbohydrates in above-ground plant tissues, thereby enhancing aphid performance (Gange and Brown 1989, Masters et al. 1993). In contrast, negative effects of root-feeding insects on leaf-chewers have been attributed to increased production of plant secondary compounds, such as terpenes and nicotine, which negatively impact leaf-chewers, but not phloem-feeding aphids (Bezemer and Van Dam 2005, Soler et al. 2012). Moreover, and in contrast to insect root-feeders, the feeding by nematodes on roots has been found to affect phloem-feeding aphids consistently and negatively, which has been explained by the triggering of defence signalling pathways and/or competition for plant assimilates in the phloem (Bezemer et al. 2005, Kaplan et al. 2011). Finally, it should be noted that root herbivory has the potential to impact host–parasitoid interactions above-ground (Soler et al. 2012), which further emphasises the multitrophic nature of interactions between root-feeders and above-ground organisms and their potential to shape plant populations and communities.

Below-ground resource partitioning

Primary productivity in many ecosystems is limited by N, and under such situations it is generally assumed that plants and microbes compete for soil N. However, it is now evident that plants and microbes can partition the N pool both temporally and spatially to alleviate competition for N. Studies in alpine ecosystems, for example, reveal a strong temporal partitioning of N use by microbes and plants that is strongly linked to seasonal dynamics of microbial and plant communities. In these highly seasonal ecosystems, marked shifts in microbial communities and their functioning occur between winter and summer, with an almost complete turnover of the microbial community (Figure 4.17) (Schadt et al. 2003, Broadbent et al. 2021). These seasonal dynamics in microbial communities are in turn associated with a strong temporal partitioning of N between plants and soil microbes: plants take up N maximally after snowmelt, and microbes immobilise N most strongly in the autumn following plant senescence, retaining this N throughout the winter (Jaeger et al. 1999, Bardgett et al. 2002). Moreover, the release of microbial N for plant uptake in spring appears to be facilitated by a marked decline in microbial biomass after snowmelt, leading to a pulse of soluble N into the soil, in the form of protein (Lipson and Schmidt 2004). This pulse coincides with a peak in soil protease activity, which facilitates the supply of amino acids for plant uptake at a time of high plant N demand (Raab et al. 1999). Together, these studies indicate that in highly seasonal, N-limited ecosystems, the cycling of labile N pools relies on intimate, temporal partitioning of N between plants and microbes based on their seasonal resource demands (Figure 4.18) (Bardgett et al. 2005a).

Soil nutrients exist in a variety of chemical forms in soil, both inorganic and organic, and it has been proposed that the partitioning of the soil N pool based on plant uptake of different N form could provide a mechanism for plant species coexistence, and hence the maintenance of botanical diversity (Figure 4.19). Evidence for such partitioning of the soil N pool based on chemical form is controversial. On the one hand, there is overwhelming evidence that plants from across a range of ecosystems can take up a diversity of chemical forms of N, both organic and inorganic (e.g., Schimel and Chapin 1996, Näsholm et al. 1998, Bardgett et al. 2003, Harrison et al. 2008. Andersen et al. 2017), and that plant species vary in their capacity to take

Figure 4.17 In high alpine grasslands, such as at Hohe Mut (2,650 m) near Obergurgl, Austria (lat. 46.84862, long. 11.02957), there are marked taxonomic and functional differences between winter (a) and summer (b) soil microbial communities, and spring snowmelt triggers an abrupt shift in soil microbial community composition and functioning that is closely linked to biogeochemical pools and fluxes, and the partitioning of N between plant and microbial pools. Images by Richard Bardgett.

up different forms of N, pointing to the existence of fundamental niches based on N chemical form (Miller and Bowman 2002, 2003, Weigelt et al. 2005). Moreover, studies using in-situ ^{15}N labelling techniques demonstrate that these chemical niches are realised in the field, especially in strongly N-limited ecosystems. In arctic tundra, for example, coexisting plant species were shown to display differential uptake of chemical forms of N, which was proposed to contribute to species coexistence (McKane et al. 2002). Similarly, in temperate grasslands, plants and soil microbes have been shown to display chemical niche partitioning over soil depths and time (Liu et al. 2016) and between plant functional groups (Kahmen et al. 2006), thereby alleviating competition for N.

On the other hand, there are studies that do not lend support to the idea that resource partitioning based on the chemical form of N contributes to plant species coexistence. Studies of semi-natural temperate grasslands, for example, show that while coexisting plant species take up a variety of N forms, including inorganic N, amino acids, and peptides, they have a strong preference for inorganic over organic N, and for simple over more complex organic N forms (Harrison et al. 2007, Wilkinson et al. 2015). Studies of tropical plants similarly provide evidence for a high degree of flexibility in uptake of N forms among plant species, but that only a few species display weak preferences for a specific N form (Andersen et al. 2017). While these studies challenge the view that differences in the availability of inorganic and organic N facilitate resource partitioning by coexisting plants, they do support the emerging view that organic N forms, including amino peptides, represent a significant, but often neglected, component of the terrestrial N cycle (Farrell et al. 2011).

A similar niche partitioning model may hold for soil P, given that it also occurs in a diversity of chemical forms, both organic and inorganic. P limitation is also pervasive across terrestrial ecosystems (Hou et al. 2020), and a wide range of plant strategies exist to access soil P (Zemunik et al. 2015), including plastic adjustments in mycorrhizal and root exudation traits among coexisting plant species (Zhu et al. 2023); as such, P resource partitioning provides a plausible explanation for the maintenance of high levels of plant species diversity found in P-limited ecosystems. Only a handful of studies

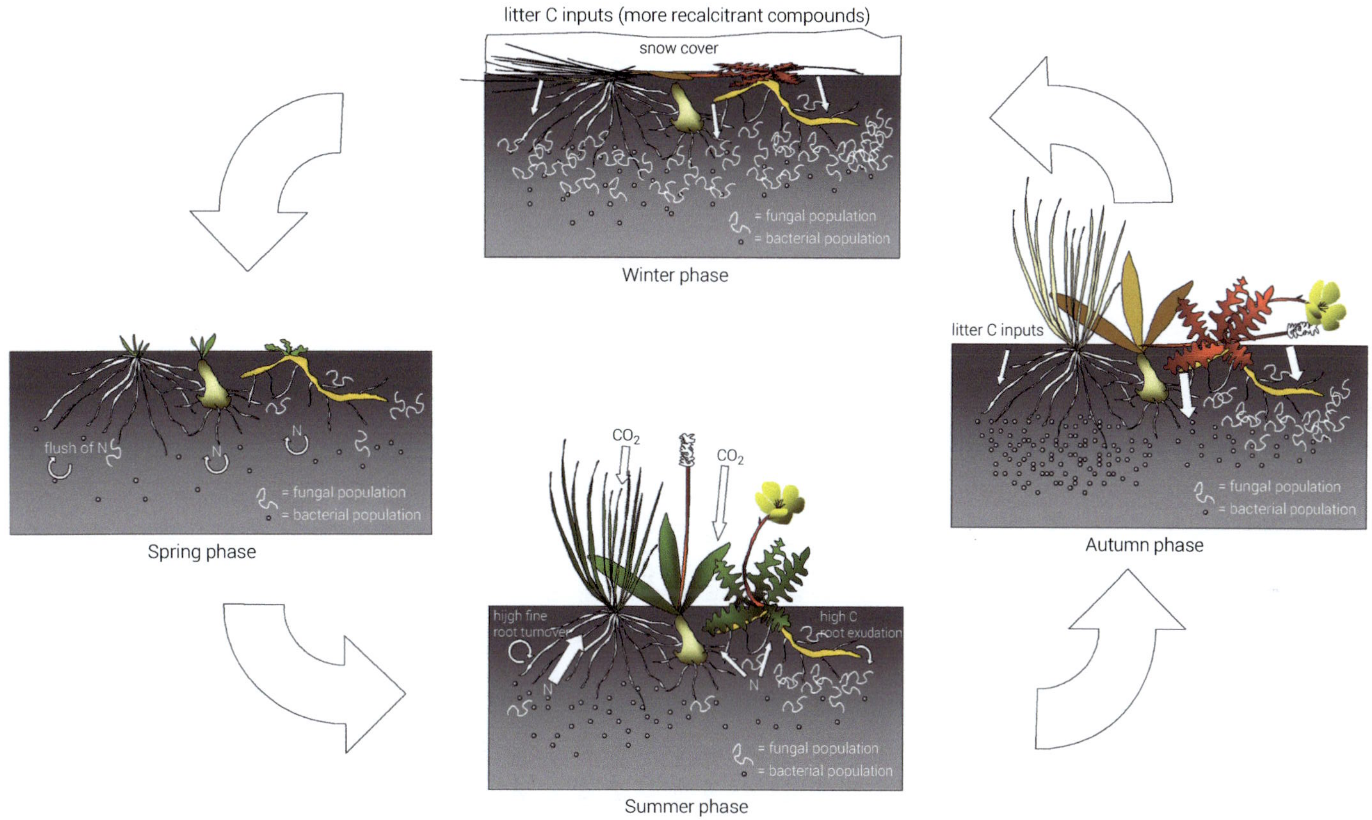

Figure 4.18 Schematic of seasonal trends in plant and microbial resource demands in strongly N-limited alpine ecosystems. In autumn, senescing plants provide a pulse of labile C to support microbial growth, and in winter the microbial biomass, and especially fungi which degrade recalcitrant polyphenolic compounds, continues to increase as C and N in plant litter are consumed and mineralised by the microbial community. In spring, rapid changes in microclimate and the exhaustion of labile C compounds lead to enhanced turnover of microbial community, with concomitant release of labile N for plant uptake, and in summer plant uptake of N continues to meet growth demands, followed by a period of C sequestration and loss to soil microbes. From Bardgett et al. (2005) by permission of Elsevier.

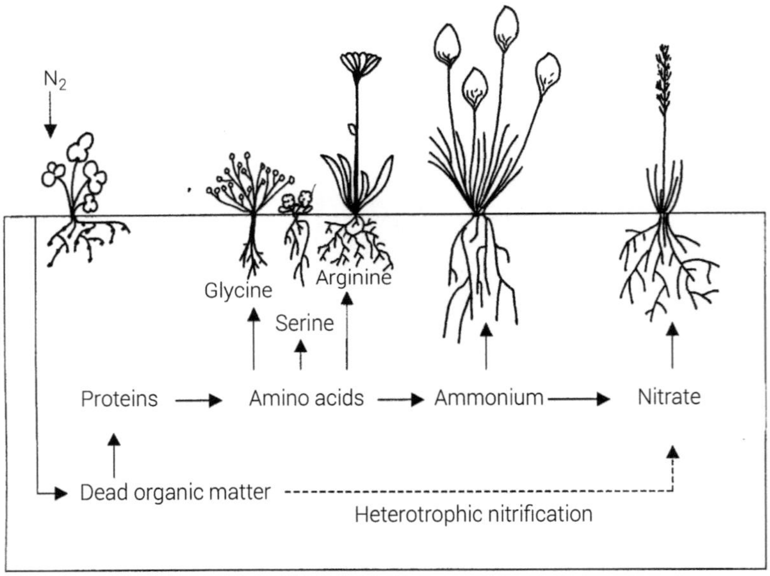

Figure 4.19 Individual plant species access different chemical forms of N, both inorganic and organic, enabling species to coexist in terrestrial ecosystems. From Bardgett and Wardle (2010) by permission of Oxford University Press.

have experimentally tested this idea. One such study used P isotope labelling to show that the relative abundance of European grassland plant species was linked to the acquisition of specific P forms, both inorganic and organic, thereby supporting the existence of P resource partitioning (Ceulemans et al. 2017). Also in grasslands, isotope labelling of organic and inorganic P sources was used to show that co-occurring plants acquire P from a range of chemical forms of P in different proportions, and that the differences in P source use were consistent with the species' root adaptations for P acquisition (Phoenix et al. 2020). Furthermore, the net benefit arising from niche plasticity, defined as the gain in P uptake for a species in a mixed community compared to monoculture, correlated with species abundance in the field, suggesting that niche plasticity for P is an important driver of community structure (Phoenix et al. 2020). ECM and AM trees have also been shown to partition the soil P pool in tropical and subtropical forests, which suggests that P resource partitioning may play an important role in promoting species-rich forested ecosystems (Liu et al. 2018b). Together, these studies lend support to the idea that P resource partitioning and niche plasticity might explain the high levels of plant diversity often found in P-limited ecosystems.

It has been proposed that the importance of P resource partitioning likely increases as ecosystems age and P becomes progressively more limiting (Laliberté et al. 2013). The idea behind this is that as ecosystems develop in the long term (i.e., tens of thousands or millions of years), they typically enter a retrogressive phase, during which there is a significant decline in ecosystem productivity and standing plant biomass (Wardle et al. 2004b, Peltzer et al. 2010). Such declines in primary productivity have been linked to leaching and occlusion of P in older, strongly weathered soils, which leads to P limitation of plant growth (Walker and Syers 1976, Crews et al. 1995, Peltzer et al. 2010), as well as reductions in litter P and decomposer activity, which further intensifies P limitation (Wardle et al. 2004b). However, while the total amount soil P typically declines during long-term ecosystem development, organic P accumulates, and the diversity of plant nutrient-acquisition strategies typically increases (Turner 2008, Laliberté et al. 2013). Supporting this, a marked increase in the diversity of plant nutrient-acquisition strategies was detected with declining soil P availability across a > 2-million-year chronosequence on sand dunes in southwest Australia (Zemunik et al. 2015). Moreover, almost all plant nutrient-acquisition strategies currently known, including plant species that can 'mine' P from organic P pools, co-occurred on the oldest, most P-impoverished soils (Zemunik et al. 2015). It was proposed that such high functional diversity in nutrient-acquisition strategies in old, P-impoverished soils could favour resource partitioning for soil P among co-occurring plant species, thus promoting high species diversity (Zemunik et al. 2015).

Conclusions

There is now an abundance of evidence that the growth of individual plant species and shifts in vegetation diversity and composition can have considerable impacts on soil biota, often with major consequences for ecosystem functioning. These effects are largely attributed to the functional traits of the dominant plant species, including those of their roots, although mechanisms also exist whereby plant diversity can positively impact soil biota and their functioning, for instance via increased: (i) supply and heterogeneity of plant-derived resources entering soil; (ii) probability of including plants with specific functional traits in more diverse mixtures, such as specific root exudation profiles or root morphological traits; and (iii) niche space for plant species-specific microbial associations, especially symbionts.

It is also evident that plant species effects on soil biota can induce negative and positive feedback that shape plant population and community dynamics, with consequences for the course of succession and species coexistence and invasions. Several mechanisms operate simultaneously to regulate plant–soil feedback, including changes in nutrient availability and the accumulation of both symbionts and natural enemies. Positive feedback typically occurs when a given plant species encourages the accumulation of beneficial microbes, especially symbionts that increase resource acquisition, thereby increasing its relative performance. In contrast, negative feedback occurs when a given plant species performs worse in its own soil due to the accumulation of host-specific pathogens, thereby preventing any single species becoming dominant and enabling more species to coexist. Moreover, the 'escape' from host-specific pathogens and negative feedback can contribute to species invasions and climate change-induced range shifts, thereby further impacting plant community dynamics and ecosystem functioning in new territories.

Plant–soil feedback is an important driver of ecosystem development over longer timescales of hundreds to thousands of years, given that negative feedback can cause species replacements, thereby altering the course of succession. Also, long-term changes in nutrient limitation can set in motion interactions between decomposer organisms and plant litter, which contribute to retrogressive succession and associated declines in ecosystem productivity and standing plant biomass. Whilst the significance of interactions between plants and soil biota is becoming increasingly apparent to ecologists, much is still to be learned about the mechanisms that drive them and their importance relative to abiotic forces that structure communities and ecosystems, especially at larger spatial and temporal scales. Moreover, major gaps in knowledge exist over the role of root traits, especially changes in the quantity and quality of root exudates, including shifts in phenolic profiles, in modifying soil biota, their feedback to plants, and ecosystem functioning.

CHAPTER 5

Above-ground consumers, soil biota, and ecosystem function

Key messages

- Herbivores influence soil biota and their functioning mainly through altering the quantity and quality of resources entering soil and via soil disturbance.
- Herbivore effects on soil biota are highly variable and depend on a range of factors including soil properties, climate, and the size, type, and diversity of herbivores.
- Predators modify the influence of herbivores on soil biota and their functioning through altering their behaviour and the spatial distribution of animal wastes.
- Herbivores have important effects on the amount and persistence of C in soil, with implications for climate change mitigation.

As discussed in the previous chapter (Chapter 4), feedbacks between plants and soil organisms play a key role in regulating plant community dynamics and ecosystem functioning. But these feedbacks between plants and soil biota are also modulated by other trophic groups, especially herbivores that consume plant material. Herbivores have long been known to be powerful drivers of plant communities and ecosystems (Crawley 1983, McNaughton et al. 1989), which is evident from the dramatic impact of past megafaunal extinctions on ecosystem and Earth-system processes (Gill et al. 2009, Gill 2014, Rule et al. 2012, Malhi et al. 2016). Indeed, herbivores can consume considerable amounts of primary productivity, although their abundance and diversity, and hence the amount of plant material they consume, digest, and excrete, varies enormously across ecosystems, depending on factors such as climate, soil fertility, and herbivore type (McNaughton et al. 1989, Olff et al. 2002, Hempson et al. 2015). Herbivore impacts on ecosystems are also modified by their predators, which alter herbivore abundance and cause shifts in foraging behaviour in response to predation risk (Schmitz et al. 2010, Le Roux et al. 2018, Atkins et al. 2019). For this reason, herbivore impacts on soil biota and ecosystem functioning need to be considered from a multitrophic context, whereby herbivores and their predators directly and indirectly interact with plants and soil biota to shape terrestrial ecosystems (Schmitz et al. 2010, Schmitz and Leroux 2020). The need for improved understanding and awareness of how herbivores shape ecosystems is heightened by recent evidence that they are the trophic group at highest risk of extinction from anthropogenic activities (Atwood et al. 2020).

This chapter outlines how above-ground consumers, including mammals, insects, and their predators, impact soil biota and the processes they drive, and how consumer-induced modification of below-ground communities can induce feedback effects on nutrient and C cycling, plant productivity, and forage supply. First, the mechanisms by which herbivores impact soil biota and ecosystem functioning within and across ecosystems are considered. Then, factors that modify how herbivores impact soil biota and ecosystem functioning will be considered, including how climate and top predators modulate herbivore effects on soils and ecosystems. Finally, the wider implications of

The Ecology of Soil. Richard D. Bardgett, Oxford University Press. © Richard D. Bardgett (2025). DOI: 10.1093/9780191858710.003.0005

above-ground consumers for soil processes that underpin ecosystem services, especially soil C storage, which is of key importance for climate mitigation, will be considered. The goal of the chapter is to illustrate how an understanding of ecosystem structure and function requires considering the multitrophic interactions between above-ground consumers, plants, and soil biota.

Mechanisms and pathways

Herbivores influence soil biota and ecosystem functioning via a wide variety of mechanisms. These can be direct, for example through trampling, and the return of animal excreta and frass (i.e., insect excreta) to soil, or indirect, via herbivore-mediated feedbacks between plants and soil biota that alter rates of decomposition, nutrient cycling, and plant productivity (Bardgett and Wardle 2010). Moreover, these mechanisms vary in their importance across ecosystems, and they operate at a variety of temporal and spatial scales, ranging from short-term effects of herbivory on plant C allocation and root exudation at an individual plant level to longer-term effects of herbivore-induced shifts in vegetation composition on soil biogeochemical cycles at the ecosystem and landscape scale (Figure 5.1). Given the variety of mechanisms at play, herbivore effects on soil biota and ecosystem functioning are highly variable and context dependent (Bardgett and Wardle 2003, 2010). Nevertheless, in order to provide a general framework to understand how herbivores shape terrestrial ecosystems, the various mechanisms involved are considered in the context of two broad alternative pathways: the *accelerating* and *decelerating effect* (*sensu* Ritchie et al. 1998),

(a) (b) (c) (d)

Figure 5.1 Fenced grazing exclosures set up across contrasting ecosystems demonstrate profound impacts of herbivores on vegetation and soils: (a) exclusion of sheep (*Ovis aries*) grazing from hill grasslands in Llyn Llydaw, Snowdonia, Wales, has resulted in a dramatic shift from grass-dominated to dwarf-shrub (*Calluna vulgaris*)-dominated vegetation, with regeneration of trees; (b) exclusion of free-ranging sheep, cows, and horses from grasslands in the Western Pyrenees, Spain, leading to shrub encroachment by gorse (*Ulex gallii*); (c) exclusion of sheep grazing accelerates succession on glacier foreland, Rotmoosferner glacier, Austria; and (d) exclusion of European red deer (*Cervus elaphus*) leading to regeneration of birch (*Betula pubescens*) forest, Scottish Highlands. Images by Richard Bardgett.

which are respectively associated with increased and reduced rates of nutrient cycling and positive and negative impacts on plant production and herbivore carrying capacity (Figure 5.2).

Accelerating effect

The accelerating effect operates in situations where above-ground herbivores favour soil decomposers and increase rates of nutrient cycling, leading to dominance of fast-growing, N-rich plant species that can tolerate herbivory and associated increases in plant productivity and forage supply (Figure 5.3a). This pathway is most common in fertile, productive ecosystems, especially grasslands where optimisation of above-ground plant productivity (termed compensatory growth) is frequently reported at intermediate levels of grazing by large mammalian herbivores (McNaughton 1985, De Mazancourt et al. 1999, Bardgett et al. 2001b, Frank et al. 2016). Moreover, under such situations, grazing typically increases the diversity of most

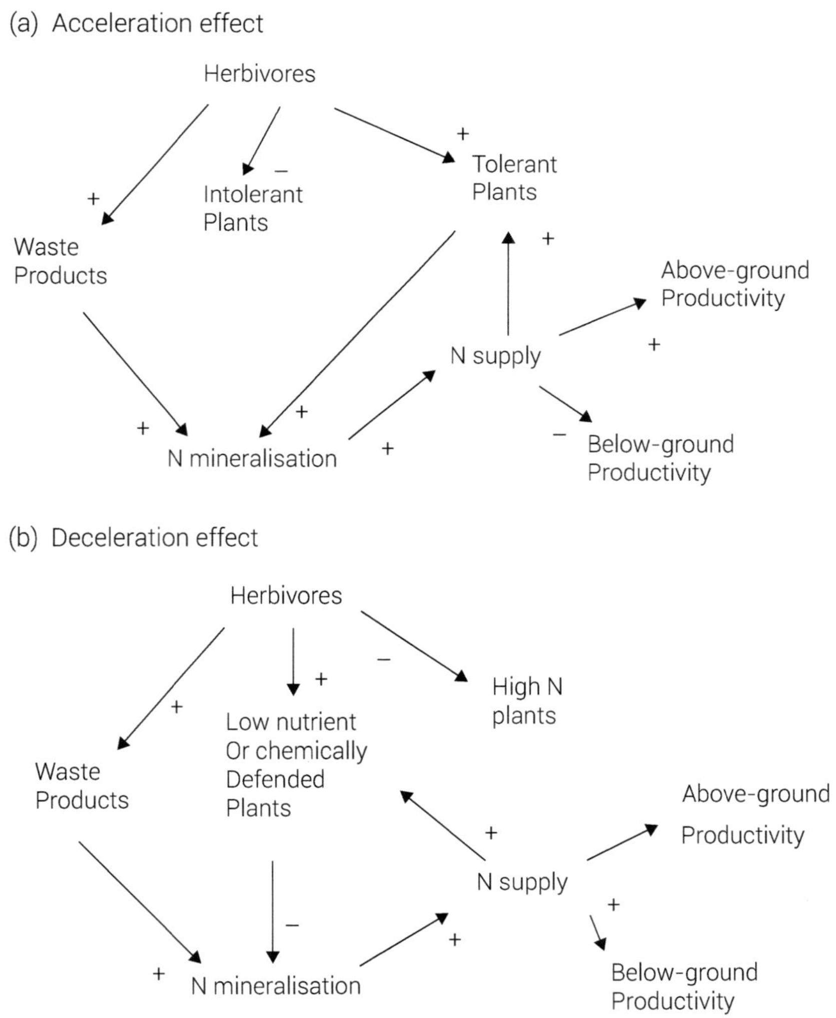

Figure 5.2 Feedback loops illustrating (a) the acceleration effect and (b) the deceleration effect of herbivores on feedback between plant species and nutrient cycling. Arrows indicate net indirect effect of herbivores on the abundance of plants or the rate of the process. From Ritchie et al. (1998) by permission of John Wiley and Sons.

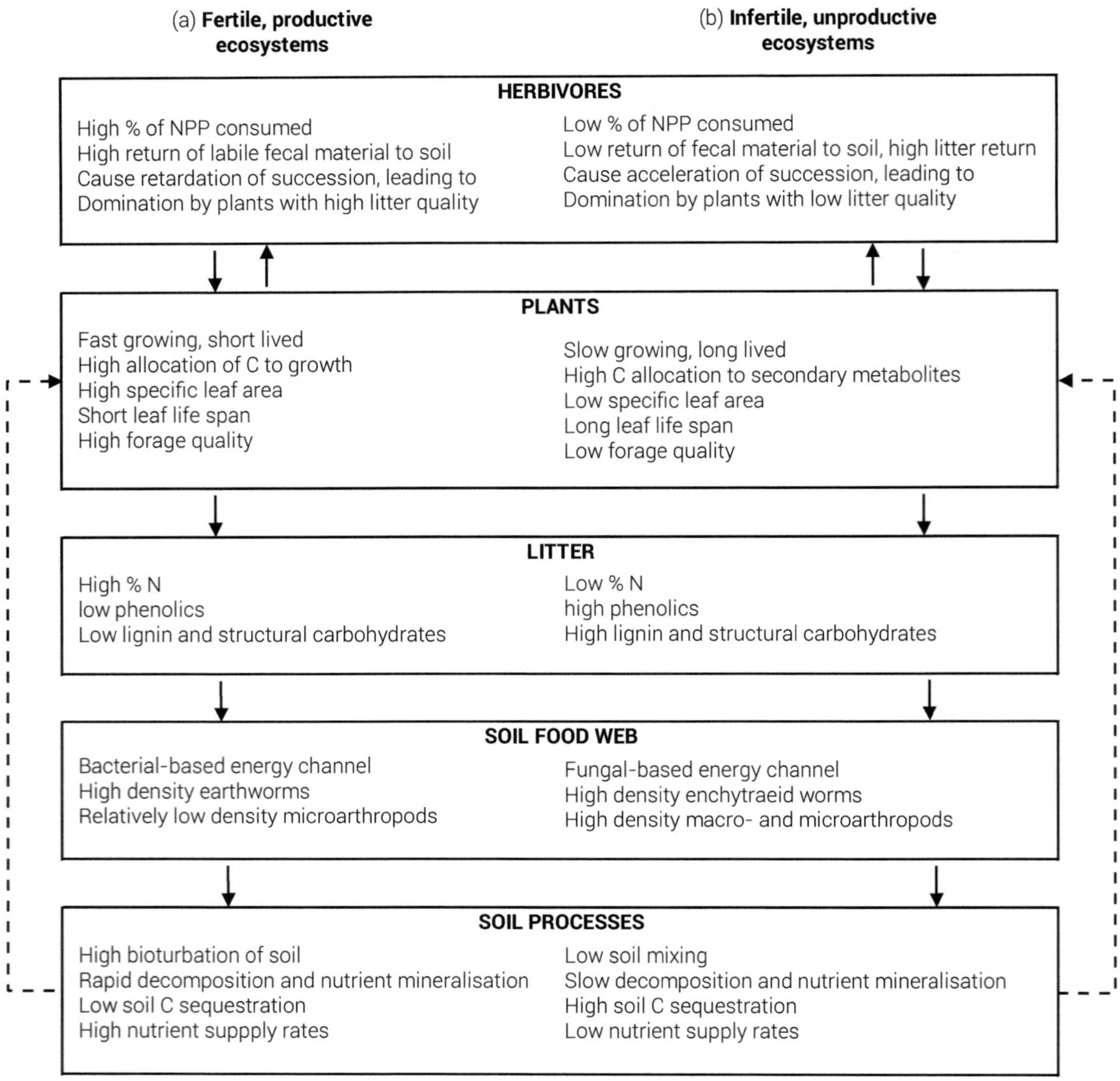

Figure 5.3 Contrasting pathways by which herbivores impact above-ground and below-ground ecological processes and feedbacks in (a) fertile habitats that support high herbivory and (b) infertile habitats which support low herbivory. These linkages between below-ground and above-ground systems feedback (dotted line) to the plant community positively in fertile conditions (a) and negatively in infertile ecosystems (b). From Wardle et al. (2004a).

groups of soil microbes and fauna, at least compared to when grazers have been removed (Schrama et al. 2023) (Figure 5.4). However, herbivore-induced acceleration of soil nutrient cycling has also been reported in arctic tundra (Van der Wal et al. 2004, Sitters et al. 2017, Tuomi et al. 2019), high alpine ecosystems (Reider and Schmidt 2021), and boreal forest (Stark et al. 2000), suggesting that it is not limited to nutrient-rich, productive ecosystems. Indeed, in nutrient-poor arctic systems, heavy grazing by reindeer, and associated trampling and dunging, causes a shift in vegetation from a moss-dominated to graminoid-dominated state, which is typically associated with higher litter quality

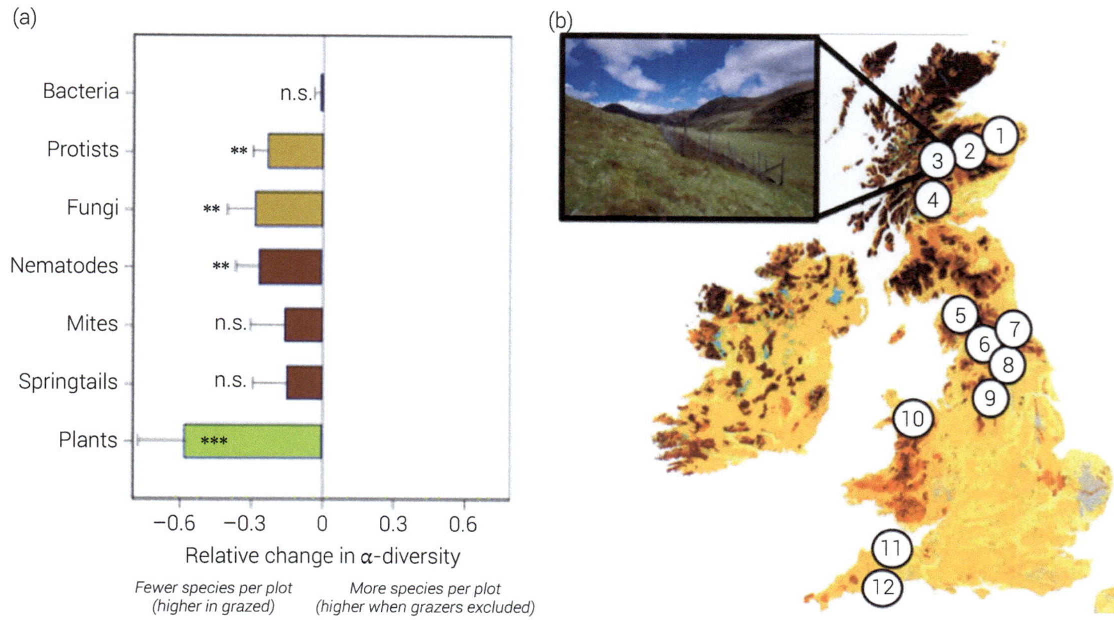

Figure 5.4 Effects of cessation of grazing on (a) α-diversity (species richness) of different groups of soil microbes and fauna across (b) 12 grassland sites in the UK, each with paired grazed and ungrazed sites. For (a), data are expressed as response ratios ($\pm$ SD), with negative values indicating a decrease in diversity. Light-brown bars indicate soil microbes, dark-brown bars indicate soil fauna, green bars indicate plants, and asterisks indicate significant differences (***P < 0.001; **P < 0.01; n.s., not significant). For (b), numbers correspond to different sites, with the background map depicting SOM concentrations, with darker colours indicating higher C stocks. Inset picture shows a typical pattern in response to grazing, i.e., higher grass cover at the grazed side of the fence, with dominance of ericaceous shrubs (e.g., *Calluna vulgaris*) on the ungrazed side. From Schrama et al. (2023) with inset image by Maarten Schrama. Cessation of grazing causes biodiversity loss and homogenization of soil food webs, Volume: 290, Issue: 2011, DOI: (10.1098/rspb.2023.1345).

and rates of nutrient cycling (Van der Wal 2006). Similarly, high vole densities in low productivity tundra were found to accelerate soil microbial activity and rates of nutrient cycling, which was explained by an associated decline in the abundance of dwarf shrubs and increase in forbs, which produce N-rich litter that decomposes rapidly (Tuomi et al. 2019). The evolutionary basis for the accelerating effect remains unresolved, but it has been suggested that grazers modify ecosystem processes in such a way as to elevate the carrying capacity of an ecosystem, hence benefiting the herbivore population (McNaughton et al. 1997).

A primary mechanism underlying the accelerating effect is the deposition of animal wastes that promote the activities of soil organisms and rates of nutrient cycling, as well as the supply of nutrients to fast-growing plants, which compensates for tissue losses resulting from herbivory (Figure 5.5). Herbivorous mammals return large quantities of undigested and non-assimilated nutrients to the soil as dung and urine, and this effectively shortcuts the litter decomposition pathway, providing soil organisms with a highly decomposable resource that is rich in labile substrates for soil organisms. This, in turn, stimulates the activity of soil microbes and animals, leading to enhanced mineralisation of N and increased availability of nutrients for plant uptake (Bardgett and Wardle 2010). Such acceleration of nutrient cycling and plant productivity via animal wastes has been widely reported. For example, a positive association between herbivore dung deposition and plant production was detected in grasslands of Yellowstone National Park (Frank and McNaughton 1992, Tracy and Frank 1998), whereas in the Serengeti, the return of faeces and

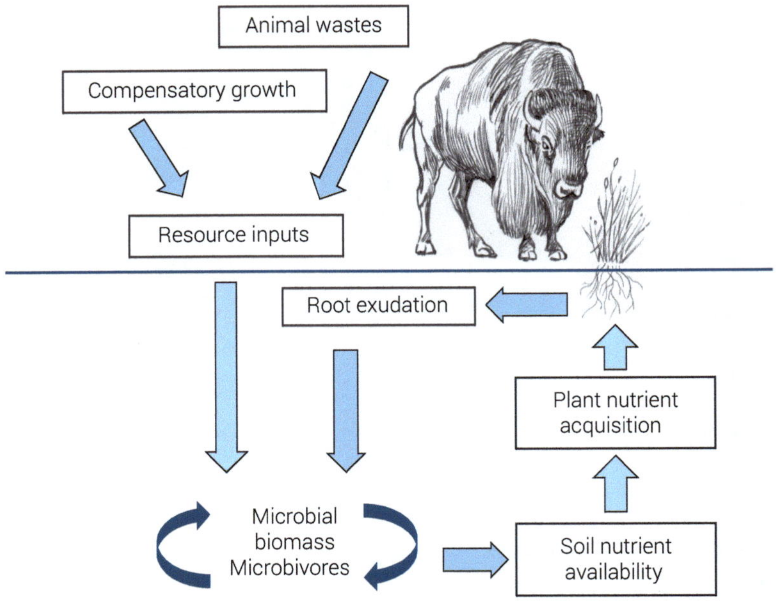

Figure 5.5 Schematic of grazer effects on below-ground resource supply, soil biota, and nutrient cycling in fertile grasslands. These mechanisms are common in productive grasslands, where dominant plant species benefit from herbivory through positive feedback between herbivores, plants, and soil biota, and through preventing colonisation by later successional plants that produce poorer litter quality. From Bardgett and Wardle (2003) by permission of John Wiley and Sons. Image of bison by Derek Colquhoun.

urine of free-ranging mammalian grazers was found to accelerate soil N cycling, contributing to greater plant productivity in grazed areas (McNaughton et al. 1997). Long-term livestock grazing across grasslands in South America was also found to stimulate the soil N cycle through faster N returns to the soil via urine and dung (Piñeiro et al. 2006), and in arctic tundra, dunging by reindeer (*Rangifer tarandus*) stimulated soil microbial biomass and grass growth, thereby increasing food supply for the reindeer (Van der Wal et al. 2004).

Invertebrate herbivores can also stimulate soil microbial activity, nutrient supply, and plant production via generating frass (Lovett and Rueseink 1995, Frost and Hunter 2007, 2008), which is enriched in macro- and micronutrients (Chen and Forschler 2016, Houben et al. 2020), or via preferential feeding on slow-growing plants, which selects for plants with litter of higher N content that decomposes rapidly and liberates nutrients for plant use (Belovsky and Slade 2000). While responses are variable, a recent meta-analysis revealed that the net effect of insect herbivory in forests on the abundance of soil fauna and processes of C and N cycling is positive (Kristensen et al. 2020). However, such effects are typically localised, as they also are for mammalian herbivores that deposit their dung in the places where they congregate, and/or in latrines (Willot et al. 2000, Reider and Schmidt 2021); as such, they generally influence nutrient cycles in small patches of an ecosystem and/or during insect outbreaks (Kristensen et al. 2020), thereby contributing to spatial heterogeneity in below-ground communities and ecosystem functioning.

An additional, albeit less studied mechanism underlying the accelerating effect involves changes in plant physiology in response to defoliation that triggers a positive feedback loop to plant growth. Several studies show that, for some grassland plant species, defoliation causes a short-term pulse in root exudation, which stimulates the abundance and activity of microbes and their consumers (Mawdsley and Bardgett 1997, Guitian and Bardgett 2000, Hokka et al. 2004, Mikola et al. 2009). In turn, this increases rates of N mineralisation in the rhizosphere (Paterson and Sim 2000, Hamilton et al. 2008, Medina-Roldán and Bardgett 2011), ultimately benefiting the regrowth of defoliated plants via enhanced nutrient supply (Hamilton and Frank 2001, Sørensen et al. 2008) (Figure 5.6). It should be noted that there is much variation among plant species, plant growth stages, and groups of soil organisms in their response to defoliation (Guitian and Bardgett 2000, Ilmarinen et al. 2005, Barto and Rillig 2010), and responses vary depending on the timing, duration, and frequency of defoliation events (Bardgett and Wardle 2010). Nevertheless, grassland field experiments demonstrate that defoliation can be a dominant factor explaining grazer effects on soil nutrient cycling, the proposed mechanisms being changes in rhizosphere C supply (Mikola et al. 2009, Liu et al. 2015a). Much less is known about how leaf herbivory impacts rhizosphere C fluxes in forest ecosystems, but pot experiments indicate that pulses of root exudation and associated increases in rates of soil N mineralisation could play a role in mediating tree seedling responses to herbivory, potentially helping them compensate for loss of tissues through herbivory (Ayres et al. 2004). However, a meta-analysis of studies on below-ground impacts of above-ground insect herbivory in forest ecosystems found that while, on average, rates of C and N cycling increase in response to insect herbivory, below-ground C allocation to roots and root exudation is reduced (Kristensen et al. 2020).

A final mechanism by which grazers can accelerate soil nutrient cycles is by affecting soil temperature. This is of particular importance in cold ecosystems, such as arctic tundra, where low soil temperatures hamper soil microbial activity and nutrient cycling. As an example, positive effects of reindeer grazing on soil N availability and plant productivity in arctic tundra have been put down, in part, to increased soil temperature, which results from a reduced depth of the moss layer (Brooker and van der Wal 2003, Van der Wal et al. 2004). Positive effects of grazers on soil temperature have also been reported in upland steppe within Yellowstone National Park (Coughenour 1991), high country grassland in New Zealand (McIntosh et al. 1997), and oak savanna in Minnesota (Ritchie et al. 1998), which likely contributes to acceleration of N cycling in these ecosystems. However, effects of grazing on soil temperature (and moisture) can be dynamic, being sensitive to factors such as grazing intensity and seasonality. For example, a study on meadow steppe in Inner Mongolia found that soil temperature increased with increasing grazing

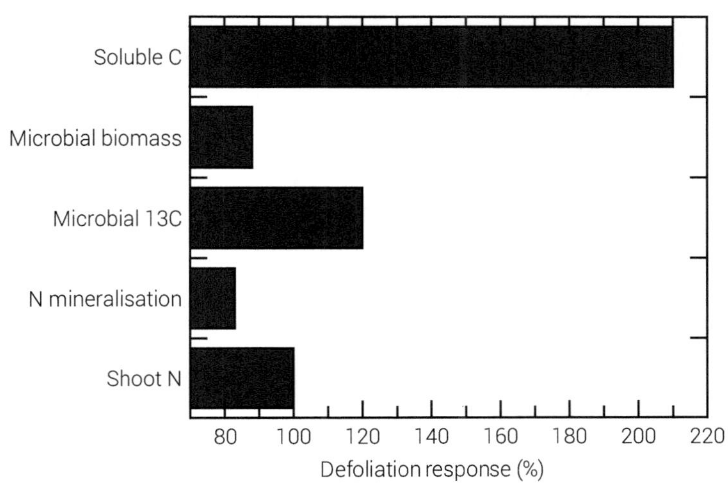

Figure 5.6 Positive effects of defoliation of the grass *Poa pratensis* on the transfer of recently photosynthesised ^{13}C to rhizosphere microbes, leading to stimulation of microbial biomass, net N mineralisation, and plant N acquisition. Data are expressed as a percentage increase relative to undefoliated control. These data indicate that grazed plants stimulate rhizosphere microbes and their own N supply. From Bardgett (2005) using data from Hamilton and Frank (2001) by permission of Oxford University Press.

pressure during the warm season due to the removal of above-ground biomass, but it decreased with increasing grazing pressure during the winter due to decreases in above-ground biomass and snow accumulation (Yan et al. 2018). In northern latitudes, grazers can also indirectly impact soil temperature by compacting snow during the winter, which decreases the thermal insulation efficiency of snow, leading to deeper winter freezing and colder mean annual soil temperatures (Beer et al. 2020). This observation has led to the suggestion that the increasing herbivore density in tundra ecosystems could increase snow compaction and help prevent thawing of permafrost soils, thereby serving as an important climate change mitigation strategy (Beer et al. 2020). Such grazer-induced changes in soil temperature, and associated changes in soil moisture, clearly have potential to impact soil organisms and the processes they drive; hence, they should be considered as important drivers of herbivore effects on ecosystem processes.

Decelerating effect

The decelerating effect typically operates when above-ground herbivores depress soil microbial activity and rates of nutrient cycling, ultimately reducing primary productivity and the carrying capacity of an ecosystem. This decelerating effect is especially prevalent in unproductive, low fertility ecosystems where low herbivore population densities favour selective feeding on the most nutritious plants of high litter quality and decomposability, leading to the dominance of well-defended plants that produce recalcitrant litter that decomposes slowly (Ritchie et al. 1998, Bardgett and Wardle 2010). Given that nutrients are mostly returned to soil as litter in such situations, rather than being recycled via animal dung, the dominance of low quality, slowly decomposing litter has the net effect of suppressing soil biological activity, nutrient mineralisation, and rates of plant nutrient supply (Figure 5.3b). Moreover, this suppression in nutrient availability gives a competitive advantage to slow-growing plant species with poor-quality litter that thrive under such conditions, thereby strengthening the decelerating effect of herbivores on ecosystem biogeochemical cycling.

A classic example of the decelerating effect is a study by Pastor et al. (1993) of the ecosystem impacts of moose (*Alces alces*) browsing in boreal forests of the Isle Royale National Park, Michigan, USA (Figure 5.7). Here, selective browsing by moose on deciduous hardwood trees with

(a)

(b)

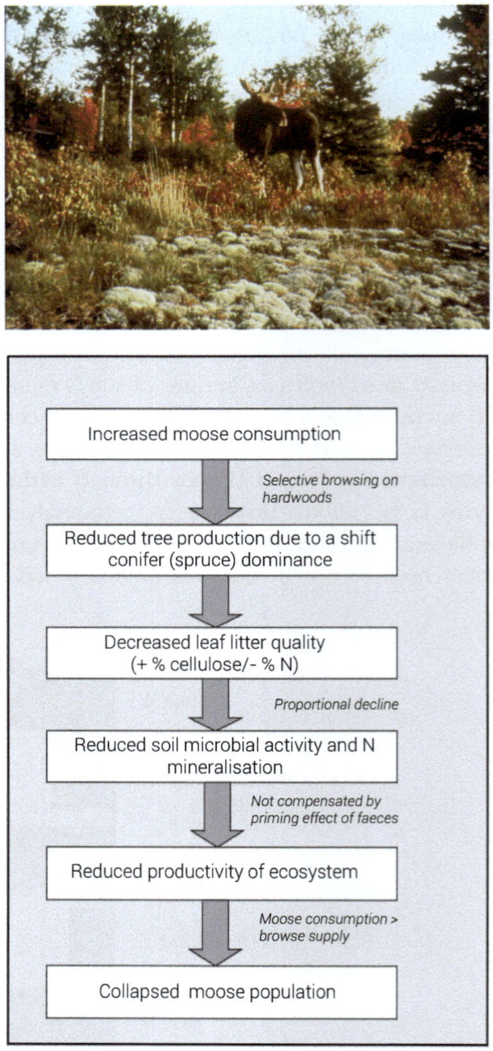

Figure 5.7 (a) Image of a moose (*Alces alces*) with heavily browsed deciduous species and unbrowsed conifers. (b) Schematic of the cascading effects of moose browsing on ecosystem properties in the Isle Royale National Park, Michigan. From Bardgett and Wardle (2010) using data from Pastor et al. (1993) by permission of Oxford University Press. Image by John Pastor.

N-rich foliage led, over time, to the dominance of less palatable slow-growing spruce trees that produce litter of low N content and decomposability. This, in turn, caused a build-up of slowly decomposing litter, reduced rates of soil N mineralisation and primary productivity, and ultimately reduced the carrying capacity of the ecosystem. Other examples include a study by Ritchie et al. (1998), who found that selective foraging by deer, rabbits, and insects in oak savanna depressed rates of nutrient cycling; this was attributed to the removal of nutrient-rich woody plants and legumes, and subsequent dominance of less palatable grasses that produce nutrient-poor, slowly decomposing litter. Selective grazing by geese, hares, and rabbits on salt marsh has also been shown to depress rates of soil nutrient cycling due to an increase in the abundance of unpalatable plants, such as common sea-lavender (*Limonium vulgare*), which produces nutrient-poor, tannin-rich litter of low decomposability (Van Wijnen et al. 1999).

Another mechanism contributing to the decelerating effect is the production of secondary metabolites by plants, such as polyphenols, that act as defence compounds against herbivory. While these compounds defend plants against herbivore attack, they also reduce the decomposability of plant litter, thereby slowing rates of decomposition and nutrient cycling. It is well known that some plants produce high concentrations of secondary metabolites that deter herbivores, and when these plants are avoided by herbivores, they become dominant in a plant community. Moreover, certain plants can increase the production of secondary metabolites when under herbivore attack, especially by insects, and these compounds serve as defence compounds in subsequently produced foliage (Rhoades 1985, Agrawal et al. 1999, Nykanen and Koricheva 2004). While it remains unclear whether this response is 'induced' for plant defence against herbivory or whether it is a physiological response of plants recovering from the nutritional deficiency associated with tissue loss, the net result is the production of less decomposable leaf litter that inhibits soil microbial activity and nutrient supply to plants (Chomel et al. 2016). Experimental evidence for this mechanism of deceleration is scant, although in a now classic study, cellular damage to seedlings of *Populus deltoides* by spider mites was found to increase the concentration of polyphenols in foliage, resulting in a 50% reduction in decomposition rate of subsequently produced leaf litter (Findlay et al. 1996). Also, herbivory by a leaf-galling aphid (*Pemphigus betae*) on riparian trees (*Populus* spp.) was found to induce the production of secondary compounds in foliage, which decelerated rates of nutrient cycling through their 'afterlife' effects on leaf litter quality (Schweitzer et al. 2005). While questions remain about the role of herbivore-induced secondary metabolite production for ecosystem processes, these studies demonstrate that it has potential to alter the quality of litter inputs to soil, their decomposability, and ecosystem-level rates of nutrient cycling.

Herbivores can also negatively impact soil organisms and nutrient cycling via physical disturbance of soil. This is at its most dramatic in situations where high livestock densities lead to soil compaction, erosion, and poaching of the soil surface, causing severe grassland degradation with negative consequences for soil biota, plant, and livestock production (Bardgett et al. 2021). In some parts of the African Sahel, for example, overgrazing by cattle has caused a catastrophic shift to a degraded ecosystem state that can no longer sustain livestock production (Rietkerk et al. 2004). Also, in China, an estimated 30–50% of grasslands on the Qinghai-Tibetan plateau are severely degraded, largely due to overgrazing by livestock, but also climate change (Harris 2010). Studies of grasslands also show that some groups of soil organisms, especially collembolans and mites, are highly sensitive to trampling from heavy livestock grazing, which increases soil compaction and reduces habitable pore space (King and Hutchinson 1976, King et al. 1976, Cole et al. 2008).

Herbivore-driven physical disturbance of soil is common in natural ecosystems, often with major consequences for ecosystem properties. For example, increasing numbers of lesser snow geese (*Chen caerulescens*) on the coastal salt marshes of the Hudson Bay, caused by high quality agricultural food subsidies in their wintering grounds and along migration routes, has led to increased geese foraging (termed grubbing), vegetation loss, and soil physical degradation, with likely negative effects on soil biota (Handa et al. 2002, Jefferies and Rockwell 2002). Similarly, in Svalbard, climate-change-related increases in population size of pink-footed geese (*Anser brachyrhynchus*) have led to increased grubbing for below-ground rhizomes, roots, and tubers, causing major soil disturbance and localised declines in plant cover, C uptake, and soil C storage (Figure 5.8) (Van der Wal et al. 2007). High levels of grazing by reindeer (*R. tarandus*) in boreal forests of Fennoscandia have been shown to cause reductions in lichen cover, exposing soil organisms to a less favourable microclimate (Stark et al. 2000), and soil compaction (Stark et al. 2003), which negatively impacts soil fauna, especially collembolans and enchytraeids (Sørensen et al. 2009).

It should be noted that negative effects of heavy grazing on soil physical properties and soil biota are mostly localised, for example, occurring around supplementary feeding or watering areas and gates, and at times of excessive soil wetness. Moreover, they vary depending on the physical properties of soils themselves, especially soil texture. For example, in a long-term study of salt marsh in the Netherlands, cattle grazing was found to decrease rates of N mineralisation in clay soil, but not in sandy soil (Schrama et al. 2013). This was because on fine-textured clay soils, which are more susceptible to compaction, trampling by large herbivores increased water-filled soil porosity, induced more negative soil redox potentials, and reduced the abundance of soil macrofauna and decomposition activity. In contrast, on sandy soil, which is more freely drained and resistant to compaction, no changes in these variables in response to grazing were detected (Schrama et al. 2013). In a subsequent study on clay soil, negative effects of soil compaction on soil organisms were most severe when soils were waterlogged, leading to a four-fold decrease in microarthropods abundance (Van Klink et al. 2015). These are just a couple of studies, but they clearly demonstrate that to understand how herbivores impact soil biota and ecosystem functioning requires consideration of impacts on soil physical properties (Tuomi et al. 2021).

As a final point on this topic, it has been suggested that there is an interplay between the biotic processes of herbivore-driven soil compaction and bioturbation by soil macrofauna (e.g., earthworms, termites, and ants), which contributes to the formation of vegetation mosaics in grazed ecosystems (Howison et al. 2017, Andriuzzi and Wall 2018b). The general idea here is that soil compaction caused by high herbivore densities leads to patches of

Figure 5.8 Spring foraging by pink-footed geese (*Anser brachyrhynchus*) (a) for below-ground rhizomes, roots, and tubers (termed grubbing) is associated with (b) significant disturbance to vegetation and soil, causing localised soil C loss in the high arctic. Data show that the amount of C stored in the surface organic horizon of the soil was reduced by goose grubbing. From Van der Wal et al. (2007) by permission of John Wiley and Sons. Image by Christiane Hübner.

drier soils, due to low rates of infiltration and increased bare soil evaporation. In such situations, only 'lawn grasses' of small stature and prostrate growth can cope with both the abiotic stress and high levels of grazing. In contrast, the foraging, tunnelling, and nesting activities of soil fauna loosen up the soil, increasing its porosity and the formation of aggregates, which positively affects oxygen and water availability, as well as rates of nutrient mineralisation, leading to high productivity patches dominated by tall, more light-competitive plant species. Hence, the interplay of these biotic processes contributes to mosaics of vegetation within grazed ecosystems, and switches from a *compacted state* dominated by 'lawn grasses' to a *bioturbated state* with a more productive tall plant community, thereby contributing to patchiness of vegetation in grazed ecosystems (Figure 5.9).

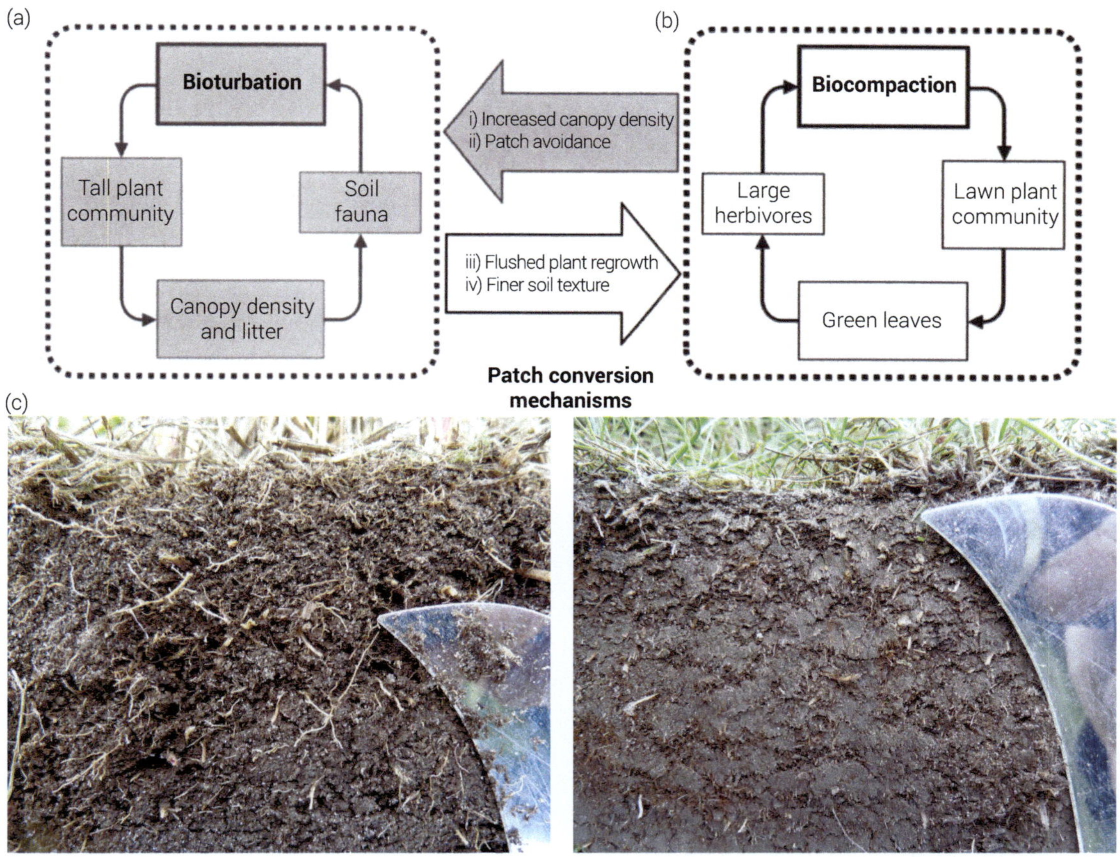

Figure 5.9 Schematic of the mechanisms promoting vegetation mosaics in grasslands. (a) Tall plant communities are promoted by bioturbation activities of soil fauna, whereas (b) lawn plant communities form when large grazers cause abiotic stress via trampling and soil compaction. (c) Patch conversion mechanisms from lawn to tall involve: (i) herbivore aggregation or seasonal migration, which releases defoliation pressure and increases plant biomass, providing shade and increased litter, thereby favouring soil fauna; and (ii) patch avoidance through plant defence (chemical or physical) and the presence of carcasses or dung. Conversion from tall to lawn mechanisms include flushed regrowth being highly attractive to large grazers (iii), and bioturbating soil fauna (ants, termites, earthworms) bringing fine particulate subsoil (clay, loam) preferentially to the surface over coarser particles (sand, gravel), with the resulting finer texture promoting compaction (iv). The images show differences in soil structure resulting from bioturbation, which leads to well-structured soil created by the bioturbating activities of soil fauna (left), and biocompaction, which leads to compacted soil, compressed plant roots, and reduced soil pore space (right) caused by large grazers. From Howison et al. (2017) by permission of John Wiley and Sons. Images by Han Olff.

Herbivores and ecosystem stoichiometry

The studies above, discussing herbivore effects on soil nutrient cycling, focussed on single nutrients and predominately on N. However, mammalian herbivores can also modify the stoichiometry of nutrients in terrestrial ecosystems, especially of N and P. Few studies have explored this topic, but those that have reveal contrasting responses, with grazers promoting both plant P and N limitation due to differential effects on the cycling and availability of P and N. Studies in arctic tundra and Yellowstone grasslands, for example, show that grazing by large mammals can promote plant P limitation by enhancing the availability of N more than P (Frank 2008, Sitters et al. 2017, 2019), whereas browsing by red deer has conversely been shown to promote plant N limitation in regenerating *Betula pubescens* woodland by reducing the availability of N relative to P (Carline et al. 2005). Contrasting responses of N:P stoichiometry to grazing were also detected across grasslands in Inner Mongolia: grazing promoted plant P limitation in typical and desert steppe, suggesting that grazers accelerate N more than P cycling, but not in meadow steppe where grazing increased the availability of P relative to N (Bai et al. 2012).

These contrasting responses of N:P stoichiometry to herbivores point to there being several mechanisms at play that differ in their relative importance across ecosystems. The classic mechanism for promotion of P limitation in grazed ecosystems is 'consumer-driven recycling', which is based on the notion that mammalian herbivores have a high P demand because of their high bone mass, so they recycle more N than P through their wastes; as such, their wastes are enriched in N relative to P (Sterner and Elser 2002, Sitters and Venterink 2018). However, while 'consumer-driven recycling' provides a plausible explanation as to why herbivores will accelerate N cycling more than P cycling, thereby promoting plant P limitation, there is limited evidence that this mechanism explains herbivore effects on nutrient stoichiometry in field settings. Rather, studies point to two alternative mechanisms: the first involves herbivore-driven shifts in vegetation composition and the stoichiometry of litter inputs to soil, which can have differential effects on rates of N and P mineralisation and loss pathways in soil; and the second involves the export of nutrients in animal wastes that are deposited away from feeding grounds (Sitters and Venterink 2018, Daufresne 2021).

Both of the above two mechanisms were proposed to explain the promotion of P limitation with heavy reindeer grazing across different vegetation types in arctic tundra: grazing in heath vegetation encouraged N-rich graminoids with highly decomposable litter, which stimulated soil N availability without a concurrent increase in the availability of P, whereas in meadow vegetation, the promotion of plant P limitation was attributed to the export of large amounts of P from the system in reindeer antlers, which are shed in autumn and winter along their migration routes (Figure 5.10) (Sitters et al. 2017, 2019). Although a different response, the promotion of plant N limitation by red deer browsing in regenerating forest in Scotland was also attributed to a shift in vegetation composition and the quality of litter entering soil, which reduced soil N availability relative to P (Carline et al. 2005). Browsing by red deer reduced the growth of the dominant tree species, *B. pubescens*, thereby reducing the beneficial effects of its N-rich leaf litter on rates of N mineralisation, and hence the availability and plant uptake of N relative to P. In contrast, the removal of browsing by fencing led to more vigorous growth of *B. pubescens*, which greatly stimulated rates of N mineralisation, but had no impact on P mineralisation, thereby triggering a shift from N to P limitation (Carline et al. 2005).

As noted above, the topic of how herbivores impact the stoichiometry of nutrient availability in terrestrial ecosystems has not received much attention, so the mechanisms involved remain poorly understood. However, the studies discussed here demonstrate that herbivores can be powerful drivers of ecosystem nutrient stoichiometry and nutrient limitation. As such, they suggest that future studies of herbivore effects on below-ground communities and ecosystem functioning should consider the cycling of both N and P.

Figure 5.10 Schematic of changes in plant and soil compartments because of heavy, long-term reindeer grazing on arctic heath and meadow vegetation. The plant compartments show changes in dominant plant functional type and nutrient limitation from lightly to (→) heavily grazed sites. Total plant biomass decreases (↓) or remains the same (=) due to heavy grazing in heath and meadow vegetation, respectively. The soil compartments show increases (↑), decreases (↓), or no changes (≈) in nutrient availability due to heavy grazing. Adapted from Sitters et al. (2019) by permission of John Wiley and Sons. Image of Svalbard reindeer by Richard Bardgett and inset images of plants and soils by Jill Colquhoun Bardgett.

Herbivore type, size, and diversity

The direction, magnitude, and spatial scale of herbivore impacts on soil biota and ecosystem functioning also depend on the type and body size of herbivores, which ranges from a few grams to more than a tonne (Fritz and Loison 2006). The importance of herbivore body size is abundantly clear from studies of the Late Pleistocene megafaunal extinctions, which triggered transformational biome shifts: mixed open woodlands were transformed to more uniform, closed forests (Gill et al. 2009, Rule et al. 2012, Bakker et al. 2016), and the extensive mammoth steppe, which prevailed in Eurasia and North America, underwent a shift from productive graminoid-dominated steppe to low productivity moss-dominated tundra and boreal forest (Zimov et al. 1995, 2012). However, recent studies indicate that herbivore type and body size is also an important contemporary ecosystem driver, with consequences for soil biota and ecosystem functioning.

Herbivore type and body size can influence soil biota and ecosystem functioning via several routes, the most obvious being the modification of vegetation composition and productivity, and hence the amount and quality of plant material entering soil. As already discussed, large herbivores selectively feed on more palatable tree species, leading over time to the dominance of chemically defended trees that produce leaf litter of low decomposability, thereby slowing rates of nutrient cycling. However, megaherbivores (> 1,000 kg) also impact vegetation composition through their physical strength and height. In African savannas, for example, elephants not only pull shrubs out of the ground and push trees over, but also debark trees and feed on saplings and adult shrubs and trees, thereby keeping tree growth and abundance in check (Bakker et al. 2016). These effects can be substantial: upward of 80% of all woody plant loss in savannas has been attributed to elephants (Asner and Levick 2012) and their exclusion can increase tree densities by as much as 42% (Kimuyu et al. 2014). Moreover, the density of bull elephants has been identified as one of the main drivers of changes in above-ground woody vegetation C stocks in Kruger National Park, South Africa, with reductions in C stocks occurring when bull densities exceed 0.5 bulls km^{-2} (Davies and Asner 2019). While the consequences of such dramatic shifts in tree abundance and above-ground C stocks for below-ground communities and their activities are largely unexplored, they are likely to be substantial, given their scale and strong influence on the amount and quality of organic material entering soil.

Herbivore impacts on soil biota via the physical disturbance of soil are also type and size dependent: herbivorous insects and small herbivores generally create relatively few soil disturbances, unless at peak abundance, whereas large mammalian herbivores have a unique capacity to physically modify and engineer the soil, especially when at high densities (Olff and Ritchie 1998). Trampling by high densities of reindeer in boreal forests, for example, has been shown to increase the amount of barren ground and cause soil compaction (Väre et al. 1996, Stark et al. 2003), which negatively impacts soil fauna (Sørensen et al. 2009). Also, grizzly bears digging for forage (roots, insects, and small mammals) in subalpine meadows create large patches of disturbed soil (> 15 m^2 in size and cover ~ 12% of land area) with elevated soil nutrient availability (Tardiff and Stanford 1998), and soil disturbance associated with foraging by feral pigs (*Sus scrofa*) in New Zealand forests has been shown to stimulate soil respiration and cause shifts in microbial community structure (Parkes et al. 2015). As already mentioned, large herbivores and soil macrofauna can also co-engineer mosaics of vegetation, with the bioturbation activities of soil macrofauna loosening the soil, which promotes the formation of tall vegetation patches, while grazing and trampling by large herbivores causes compaction and the formation of lawn patches (Howison et al. 2017).

Few studies have examined how herbivore type and body size affect soil nutrient dynamics, but those that have indicate that effects are body size dependent. For example, the selective exclusion of different components of the herbivore community in a Dutch grassland revealed contrasting effects of small and large herbivore species on soil nutrient cycling: exclusion of large herbivores (i.e., cattle) led to a 1.5-fold increase in rates of soil N mineralisation due to a build-up of plant litter on the soil surface, whereas the combined exclusion of cattle and rabbits, but not meadow voles, altered the timing of N cycling, leading to an autumn peak in soil N mineralisation after it had already declined under cattle grazing (Bakker et al. 2004). This autumn peak in soil N mineralisation coincided with a peak in vole density and high levels of N input to soil from vole faeces (Bakker et al. 2004). The removal of herbivores based on body size was also found to influence soil N cycling across a series of subalpine grasslands in Switzerland, where progressive size-selective exclusion of large, medium, and small vertebrates and invertebrates—simulating a realistic scenario of human-induced defaunation—led to higher rates of soil N mineralisation regardless of vegetation type (Risch et al. 2015). Moreover, in a related study, the removal of all vertebrates increased the strength of relationships between above- and below-ground communities and their abiotic environment (i.e., ecosystem coupling), while the additional loss of invertebrates led to poor ecosystem coupling, highlighting their importance for maintaining ecosystem functioning in an increasingly defaunated world (Risch et al. 2018).

Evidence also exists of facilitative interactions between animal species of different size that have consequences for below-ground communities and functions. Notably, a study of semi-arid grasslands in China found that grazing by domestic cattle (*Bos taurus*) increased the abundance of ants (*Lasius* spp.) by reducing plant litter accumulation and allowing more light to reach the soil surface, whereas ants facilitated increased foraging of cattle by increasing soil N availability and the N content and biomass of forage (Li et al. 2018b). In a related experiment, ants and cattle together were found to stimulate soil microbial biomass and rates of decomposition and nutrient cycling much more than when each of these animals was alone (Li et al. 2021a). These studies not only demonstrate the potential for positive interactions between different animal taxa to shape below-ground communities and their functioning, but also have implications for land management, given the benefits of interactions between different herbivores for forage production and quality.

Herbivore body size has also been shown to influence the redistribution of nutrients across ecosystems and the ratios of nutrients supplied to plants. Notably, a study of grazing lawns in South African savanna with diverse mammal herbivore communities spanning a 10-fold gradient in body size, from impala (*Aepyceros melampus*) to white rhino (*Ceratotherium simum*), found that dung from larger-bodied herbivores contained lower quantities of P per unit mass and had a higher N:P ratio (Figure 5.11) (Le Roux et al. 2020). As such, lawns on which mainly larger herbivores defecated received faecal material of higher N:P ratio, potentially driving plants towards P limitation, whereas those with mostly smaller herbivores received faeces of low N:P ratio, potentially creating conditions of N limitation (Le Roux et al. 2020). The consequences of the observed size-dependent effects of herbivores on nutrient stoichiometry for decomposer organisms and their activities were not tested, but they could potentially be far reaching, given the known importance of resource quality as a driver of below-ground processes and ecosystem functioning.

Herbivore diversity can also act as an important driver of below-ground communities and their functioning. Notably, a 5-year grazing experiment on semi-natural meadow steppe in Inner Mongolia, China, which included treatments grazed by either single herbivore species (i.e., cattle or sheep individually) or both (i.e., cattle and sheep together), demonstrated that mixed grazing promoted the diversity of a wide range of above- and below-ground organisms and ecosystem multifunctionality, including a range of soil functions related to nutrient and C cycling, and plant–microbe symbiosis (Wang et al. 2019). Furthermore, this positive effect of mixed grazing on ecosystem multifunctionality was indirect, being driven predominately by a concurrent increase in above- and below-ground diversity; this positive relationship between multitrophic diversity and multifunctionality was stronger when multiple groups of above-ground and below-ground organisms were considered simultaneously, rather than when considered individually. This finding suggests that the relationship between biodiversity and ecosystem function in grazed grasslands is more positive and significant when more trophic levels, both above-

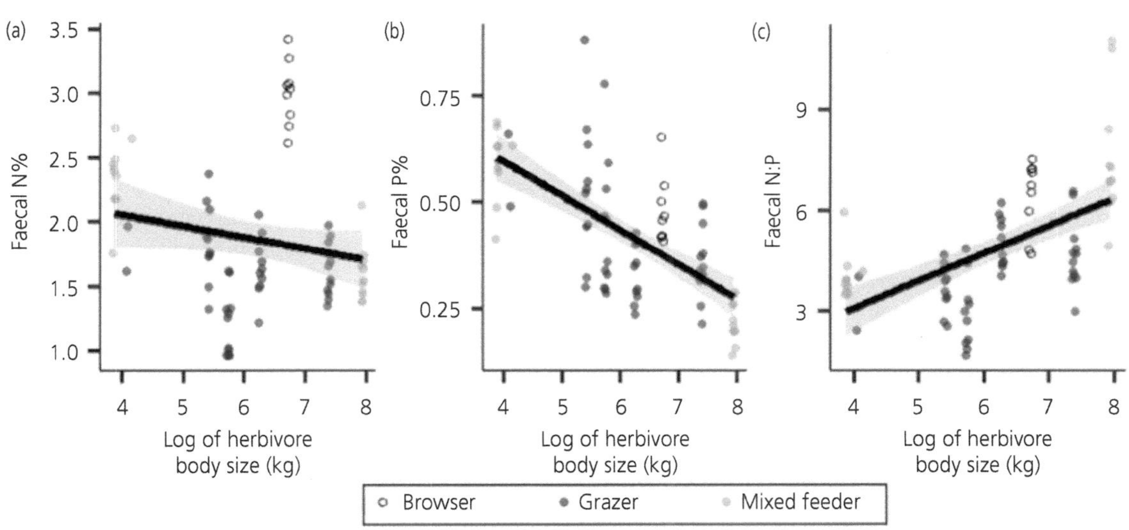

Figure 5.11 N (a), P (b), and N:P ratio (c) of dung in relation to the log of herbivore body size and categorised according to feeding type in South African savanna. From le Roux et al. (2020).

and below-ground, are considered (Wang et al. 2019). Moreover, they suggest that diversifying livestock could be a useful strategy to maintain multifunctionality in managed ecosystems by promoting multitrophic diversity of above- and below-ground organisms (Wang et al. 2019).

Herbivore movement and nutrient redistribution

Herbivores are not static: they move from place to place. As herbivores move, they transport nutrients from one place to another, thereby influencing spatial patterns of soil biota, nutrient cycling, and vegetation. The movement of herbivores is driven by many factors, such as the search for food, water, or better climatic conditions, and the avoidance of predators. These movements can cover considerable distances, for instance when large herbivores migrate seasonally between home ranges (Fryxell and Sinclair 1988); or they can be over short distances to select feeding sites where forage is more abundant or nutritious, to carry out non-foraging activities such as resting and ruminating, or to avoid predators (Bailey and Provenza 2008). Also, in livestock systems, herbivores are actively moved from one place to another by farmers or herders to maximise forage use or to allow them to cope with seasonal changes in forage supply (Boone et al. 2008). In the Mongolian steppe, for example, herders move their livestock hundreds of kilometres between the mountains and lowlands to make best use of forage supply, and in mountainous regions of Europe, herders typically move their livestock from lower valleys to higher elevations during the summer months to make use of snow-free, high elevation mountain meadows (Boone et al. 2008).

Animal-mediated movement of nutrients is well documented in livestock systems, where the use of different parts of a paddock or a rangeland by livestock for feeding and excretion leads to the patchy return of nutrients and hot spots of soil nutrient cycling and plant production. Cattle, for example, deposit most of their dung in areas where they rest and ruminate, leading to high soil concentrations of nutrients in the small patches where the animals spend much of their time (Afzal and Adams 1992, Kohler et al. 2006). The consequences of this for spatial patterns of soil nutrient availability can be significant. In mountain pastures in Switzerland, for example, cattle were found to redistribute substantial amounts of P by feeding and defecating in different places (Jewell et al. 2007); this led to substantial enrichment of P (> 50 kg P ha^{-1} $year^{-1}$) in small patches where cattle rest and ruminate and drop most of their dung; conversely, there was depletion of P in the larger areas of pasture where cattle graze, with an overall negative impact on pasture productivity (Jewell et al. 2007). High stocking rates of cattle in Brazilian pastures have similarly been shown to cause declines in pasture productivity due to soil N enrichment in areas where cattle defecate and urinate, with associated increases in gaseous and leaching losses of N from soil, and reduced soil N availability in the areas where cattle graze (Boddey et al. 2004). Also, cattle grazing in African savannas has been shown to cause a net export of C and N from paddocks due to an offtake of nutrients via grass consumption than return of nutrients in dung (Sitters et al. 2020a). This effect, however, was negated by the presence of wild megaherbivores (mainly elephants), which import C and N to soil through their dung and the breaking, killing, and toppling of trees (Sitters et al. 2020a).

Wild mammalian herbivores also transport significant quantities of nutrients in their dung across ecosystems, with implications for spatial patterns of soil nutrient cycling. Studies in Europe and North America, for example, show that deer transport in their dung significant quantities of N and P from fertilised croplands to neighbouring forests, thereby subsidising forest nutrient cycles and altering their biogeochemistry (Seagle 2003, Abbas et al. 2012). Free-ranging grazers in the Serengeti, including Thomson's (*Gazella thomsoni*) and Grant's (*Gazella granti*) gazelles, topi (*Damaliscus korrigum*), and kongoni (*Alcelaphus buselaphus*), have been shown to stimulate soil N mineralisation and Na availability in areas where they preferentially graze and defecate, thereby contributing to the spatial heterogeneity of soil fertility at a landscape scale (McNaughton et al. 1997). Similarly, native migratory ungulates of grasslands in Yellowstone National Park, USA, including elk (*Cervus elaphus*), bison (*Bison bison*), and pronghorn (*Antilocapra americana*), have been shown to affect the spatial distribution of soil

N and rates of N mineralisation, largely due to landscape-level variations in grazing intensity and site selection for resting and dunging (Augustine and Frank 2001). The impacts of such herbivore-driven nutrient redistribution on spatial patterns of below-ground communities have not been studied, but given documented impacts on nutrient cycling, they are likely of importance at both an ecosystem and landscape scale.

Mammalian herbivores also redistribute nutrients via their carcasses, which create hot spots of biological activity and nutrient availability in soil. A growing number of studies demonstrate this across a range of ecosystems. For instance, a study in Yellowstone National Park showed that carcasses of wolf-killed bison and elk increased soil respiration and vegetation nutrient content and left a strong fingerprint on the composition of bacterial and fungal communities in soil (Risch et al. 2020). Similarly, ungulate carcasses have been shown to create long-lasting hot spots of soil fertility and plant production in a range of ecosystems (Towne 2000, Bump et al. 2009b, Parmenter and MacMahon 2009), and in boreal forests, the spatial distribution of moose carcasses and their impact on soil nutrients and microbial communities is modulated by their predators, namely wolves (Bump et al. 2009a). The impacts of carcasses on the spatial patterning of below-ground communities and functions are still poorly understood. But these examples collectively suggest that herbivore carcasses can contribute to the heterogeneity of plant and below-ground communities at both local and ecosystem scales.

Herbivorous invertebrates can contribute to patchiness in below-ground communities and functions, especially during outbreaks or periodic emergences. An excellent example of this is resource pulses from periodic deposition of carcasses of cicadas of the genus *Magicicada* in North American deciduous forests. These insects spend most of their time underground feeding on roots, but every 13–17 years they emerge from the ground in vast numbers, mate, lay eggs, and die; this creates a major resource pulse that has been shown to stimulate biomass and activity of soil microbes, soil inorganic N availability, and the N content and seed mass of understorey plants (Yang 2004). Similarly, large-scale defoliation events by insects can have cascading effects on below-ground communities and biogeochemical cycles. For example, a large-scale defoliation event in subarctic birch forests (*B. pubescens*) caused by the caterpillars *Epirrita autumnata* and *Operophtera brumata* was found to reduce hyphal production of EM fungi and caused a shift in EM fungal community composition (Parker et al. 2017). Moreover, these changes in mycorrhizal growth and community structure were associated with a reduction in soil respiration and build-up of inorganic N in soil, suggesting a slowing down of soil processes (Parker et al. 2017). These are just a couple of examples, but they suggest that outbreaks of above-ground herbivorous insects represent an important, but often neglected, driver of spatial and temporal heterogeneity of below-ground communities and their functioning across ecosystems.

Predator control of below-ground processes

Above-ground predators eat herbivores, thereby influencing their densities and foraging behaviour. Given the high potential for herbivores to modify below-ground communities and processes, it therefore holds that any effect of predators on herbivore density and/or foraging behaviour will potentially have cascading effects on vegetation and soils. While a relatively understudied topic, studies are beginning to emerge demonstrating the potential for predators to shape below-ground communities and processes and their spatial patterning. These predator effects operate via a variety of mechanisms, which can be broadly divided into (i) consumptive predator effects, which operate by altering carcass distribution and herbivore density, and (ii) non-consumptive predator effects, which influence foraging behaviour (Schmitz et al. 2010). These two broad mechanisms are considered in turn below, but the interested reader should see Schmitz et al. (2010) and Monk and Schmitz (2022).

One of the ways consumptive predator effects influence below-ground communities and processes is through the concentration of prey carcasses at kill sites, often in areas of high predator activity and hunting success (Monk and Schmitz

2022). As discussed above, animal carcasses can create long-lasting hot spots of soil microbial activity, nutrient cycling, and plant productivity and lead to shifts in microbial community structure; as such, they contribute to spatial heterogeneity in below-ground communities and processes within ecosystems. However, consumptive predator–prey interactions also impact below-ground communities and processes through controlling herbivore densities, hence their impact on vegetation composition and productivity, and amount and quality of organic matter entering soil. In North American forests, for example, predators, such as cougars and wolves have been shown to reverse the negative effects of herbivores, especially elk and moose, on the growth of palatable deciduous plant species and resulting shifts in forest vegetation (Ripple and Beschta 2004, Creel and Christianson 2008), presumably changing the quality of litter entering the soil (e.g., Pastor et al. 1993). Also, in a classic study of top-down control of ecosystems, the exclusion of predators (i.e., snakes, raptors, jaguar, and puma) from small islands created by hydroelectric impoundment at Lake Guri, Venezuela, caused an explosion in plant-eating rodents and herbivores (howler monkeys, iguanas, and leaf-cutter ants), leading to overexploitation of palatable canopy trees and a shift in vegetation towards less preferred plant species that produce poorer quality litter (Terborgh et al. 2001). While the below-ground impacts of these changes in vegetation were not studied, they were likely profound, given the known potential for herbivore-driven shifts in vegetation to have cascading effects on soil biota and processes, as discussed earlier in this chapter.

In contrast, non-consumptive predator effects can impact below-ground communities and processes through predation risk, or by creating 'landscapes of fear', where herbivores change their behaviour and spatial foraging patterns to avoid risky habitats (Schmitz et al. 2010, Monk and Schmitz 2022). In doing so, they can have profound below-ground effects either by concentrating herbivory in safe sites, which generates hot spots of soil fertility, or by redistributing nutrients in faeces between risky and safe sites, thereby contributing to spatial heterogeneity in below-ground communities and processes. While it is well known that herbivores concentrate in safe areas to avoid predation risk, below-ground consequences are rarely documented. However, examples do exist from studies done in African savannas, where grazing lawns, or patches of high soil fertility and plant production, are maintained by herbivores that concentrate in high visibility areas as a collective antipredator strategy (McNaughton 1983, McNaughton et al. 1989). As discussed earlier in this chapter, this predator-driven concentration of herbivores accelerates rates of soil nutrient cycling through localised dung deposition and increases the dominance of fast-growing, N-rich plants that tolerate herbivory, with associated increased forage supply (McNaughton et al. 1997). It has also been proposed that these hot spots could create a positive feedback loop, whereby the congregation of herbivores attracts predators, which increases the density of carcasses, further enhancing rates of decomposition and nutrient cycling, and the production of nutrient-rich vegetation, which provides forage for herbivores (Monk and Schmitz 2022) (Figure 5.12).

While the accumulation of fearful prey can create hot spots of soil fertility and plant production, evidence suggests that megaherbivores, which are largely invulnerable to predators, counteract this by redistributing nutrients away from these fear-driven fertility hot spots. This was found to be the case in a study of African savanna by Le Roux et al. (2018); mesoherbivores (wildebeests, impalas, and warthogs) concentrated faecal nutrients in more open habitat with low predation risk, but megaherbivores (elephants, white rhino, and giraffes) partly counteract this fear-driven accumulation of faecal nutrients by feeding within the open habitat yet defecating more evenly across the risk gradient. Recent evidence also suggests that 'landscapes of fear' can be converted into 'landscapes of fearlessness' through the removal of predators, again with consequences for vegetation and nutrient redistribution within ecosystems. As an example, the extirpation of predators (leopards and African wild dogs) from Mozambique's Gorongosa National Park, which happened during the Mozambican Civil War, led to the expansion of forest-dwelling antelopes (*Tragelaphus sylvaticus*) into treeless floodplains with more nutritious

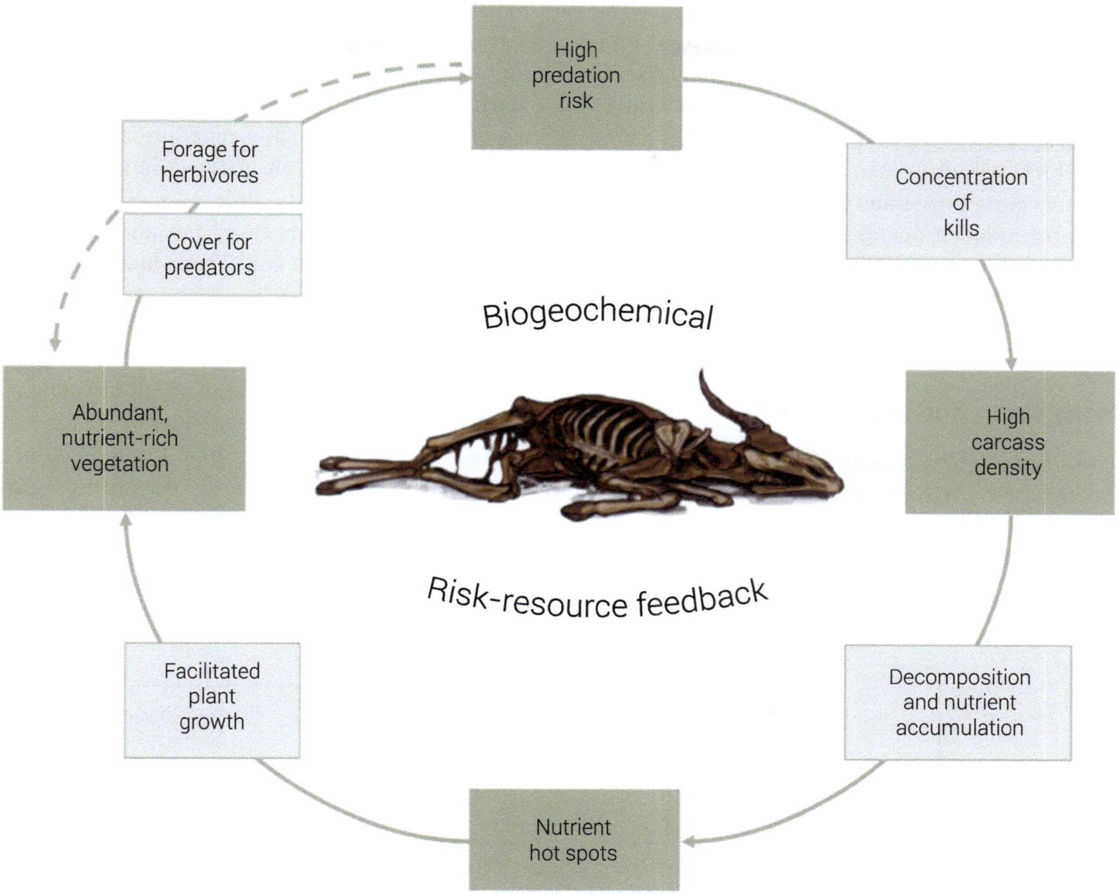

Figure 5.12 Positive feedback loop between predation risk, soil nutrient availability, and forage supply in the biogeochemical landscape of fear. Hot spots of soil fertility and forage supply in grasslands lead to the congregation of herbivores, which attracts predators, increases the density of carcasses, and further enhances rates of decomposition and nutrient cycling and the production of nutrient-rich vegetation, which provides forage for herbivores. From Monk and Schmitz (2022) by permission of John Wiley and Sons.

vegetation, which they previously avoided due to predation risk; this in turn has led to the suppression of preferred food plants and shifts in vegetation composition (Atkins et al. 2019). Below-ground effects were not examined, but they were likely significant, given the known importance of herbivore-driven vegetation shifts as a driver of below-ground biota and processes (Bardgett and Wardle 2010).

Another way that predators can indirectly influence below-ground processes is via herbivore stress responses to predator risk. The general idea here is that physiological stress responses to predation risk cause elemental changes in the composition of herbivore prey and the quality of herbivore biomass entering soil as detritus, which in turn modifies below-ground community function. Specifically, the risk of predation is known to increase prey metabolism, which increases demand for digestible carbohydrate-C relative to protein-N, which ultimately increases body C:N and the stoichiometry of detrital inputs to soil from herbivores (Hawlena and Schmitz 2010). This idea was tested by Hawlena et al. (2012), who, in an elegant combination of field and laboratory studies, demonstrated that grasshopper herbivores stressed by spider predators had a higher body C:N ratio than did grasshoppers raised without spiders, and that this change in elemental composition of herbivore carcasses slowed rates of decomposition of plant litter through affecting the functioning of the soil community. In doing so, they identified an additional mechanism by which predators can indirectly affect belowground communities and their functioning through predator-induced changes in prey chemical composition, which is even more important, given the high rates at which predators are being lost from ecosystems worldwide (Estes et al. 2011).

Herbivores and soil C cycling

The amount of C stored in soil depends on the balance of C input to soil via plants (dead leaves, roots, and exudates) and its loss via decomposition, and associated C release by heterotrophic respiration and as dissolved organic carbon (DOC) in leachates. The predominant controls on soil C storage are abiotic, especially moisture, temperature, and soil mineralogy, but above-ground herbivores can strongly modify the processes of C input and loss via a variety of mechanisms, and, hence, the amount of C stored in soil. These mechanisms are broadly similar to those by which herbivores influence rates of soil nutrient cycling (Figure 5.13), as already discussed in this chapter, and are related to changes in the amount and quality of organic matter entering soil as plant litter (shoot and roots) and as root exudates and the rate of decomposition of plant-derived organic matter by heterotrophic soil

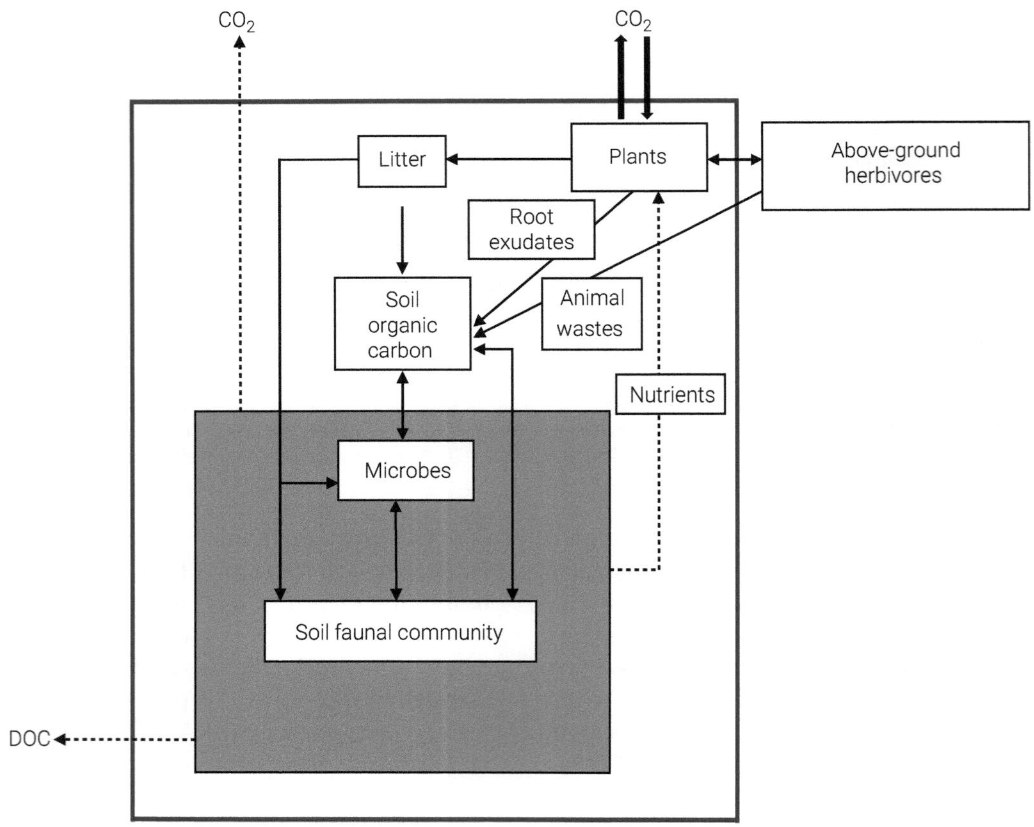

Figure 5.13 Schematic of the various routes by which above-ground consumers influence C fluxes in terrestrial ecosystems. Solid lines represent inputs of C to the soil system, for instance from plant litter, root exudates, and animal wastes. Coarse dashed lines represent outputs of C in the form of DOC and CO_2. Fine dashed line represents a feedback to plant production and C input via altered nutrient availability. From Bardgett and Wardle (2010) by permission of Oxford University Press.

organisms. But also, herbivores can impact the persistence, or stabilisation, of soil C, by promoting the formation of mineral-associated organic matter (MAOM), which is protected from decomposition via chemical sorption to mineral particles and/or in microaggregates. This section considers the role of above-ground herbivores in regulating soil C storage in both managed and natural ecosystems, which is key to understanding how they might contribute to climate mitigation.

Grazers and soil C storage

Studies exploring how grazers influence soil C storage paint a mixed picture. An extensive review of the available literature by Milchunas and Lauenroth (1993) reported both positive and negative effects of grazing by large mammals on soil C, and, similarly, individual studies report positive (e.g., Schuman et al. 1999, Reeder and Schuman 2002, Hewins et al. 2018), negative (e.g., Han et al. 2008, Golluscio et al. 2009, He et al. 2011, Bai et al. 2012), and neutral (e.g., Shrestha and Stahl 2008, Bai et al. 2012, Medina-Roldán et al. 2012, Roy and Bagchi 2022) effects, sometimes even in the same study (Piñeiro et al. 2009, Bai et al. 2012). These variable responses have been put down to several factors, including climate, soil type, grazing intensity, and nutrient availability. In an extensive review of 83 studies of extensive grazing, covering 164 sites across different countries and climates, the response of soil C to grazing was found to be climate-dependent: in warm, moist climates, grazing increased soil C by 7.6% on average, whereas in cool, moist climates it reduced soil C by 19% on average (Abdalla et al. 2018). However, in another meta-analysis, grazer effects on soil C in grasslands were found to depend on an interplay between climate and soil type: on fine-textured soils of high clay content, grazing was found to negatively influence soil C stocks at higher precipitation, whereas on coarse-textured sandy soils, grazing had the opposite effect, increasing soil C at higher precipitation (McSherry and Ritchie 2013) (Figure 5.14). These authors also found a strong interaction between grassland type and grazing intensity: for grassland dominated by C3 grasses, grazing positively affected soil C under light grazing intensities, but negatively affected soil C at moderate to heavy intensities; whereas for grasslands dominated by C4

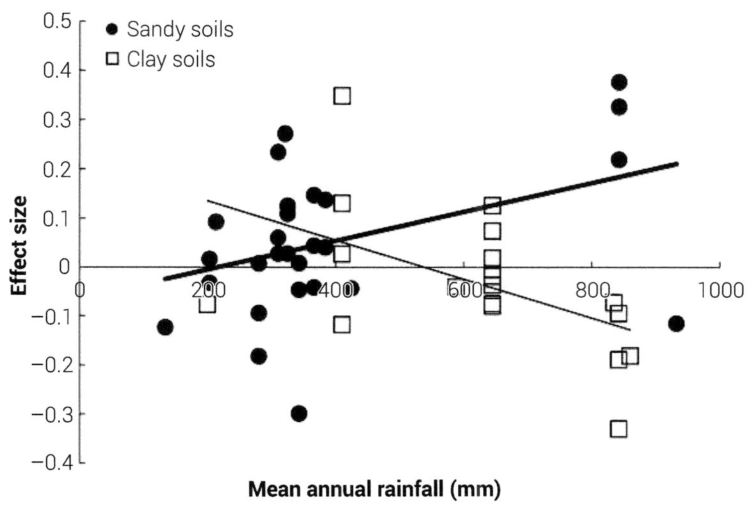

Figure 5.14 Significant regressions of effect size of grazing on SOC associated with mean annual precipitation for sandy soils (solid circles; df = 1, 26, $P = 0.033$, $R^2 = 0.162$) and finer-textured soils (open squares; df = 1, 17, $P = 0.024$, $R^2 = 0.264$). From McSherry and Ritchie (2013) by permission of John Wiley and Sons.
Global Change Biology, Volume: 19, Issue: 5, Pages: 1347–1357. First published: 20 January 2013, DOI: (10.1111/gcb.12144).

grasses, grazer effects shifted from slightly negative at light grazing intensities to positive for moderate and heavy intensities. It should be noted that no data from tropical grasslands and savannas were included in this meta-analysis. Nevertheless, the results indicate that grazer effects on soil C storage are highly context-specific, and hence that grazers need to be managed differently in different regions to enhance or protect soil C.

Grazing intensity has been identified as a key factor explaining the reported contradictory responses of soil C to grazing across grassland ecosystems. A meta-analysis of 115 published studies examining how livestock grazing influences soil C and N pools in grasslands found that, on average, light grazing contributed to soil C sequestration, whereas moderate to heavy grazing increased soil C and N losses (Zhou et al. 2017). The mechanisms explaining these responses remain elusive, but the positive response of soil C to light grazing was attributed to a greater above-ground biomass and litter accumulation and increased below-ground C allocation to roots and root exudates, which increases C input to soil. In contrast, increased soil C loss under moderate and heavy grazing was explained by a decrease in C input to soil due to the removal of plant material by livestock, reduced litter mass, and reductions in root mass, which is commonly associated with heavy grazing (Zhou et al. 2017, Dai et al. 2019). Reductions in soil C with heavy grazing may also be related to soil physical properties, in that moderate and heavy grazing is typically associated with reduced soil moisture due to greater evapotranspiration and lower rates of water infiltration due to soil compaction (Zhou et al. 2017).

Declines in soil C under heavy grazing may be related to vegetation shifts, especially an increase in dominance of palatable, nutrient-rich plants that readily supply labile substrates to soil as nutrient-rich litter and root exudates. Consistent with the acceleration hypothesis, an increase in the supply of labile substrates, along with greater input of dung and urine, should promote the activities of decomposer organisms and bacterial-based energy channels, enhancing C mineralisation and loss from soil (Bardgett and Wardle 2003). This idea was broadly tested by Klumpp et al. (2007, 2009), who studied monoliths from grasslands that had been under constant high and low grazing pressure for 14 years; they found that monoliths exposed to low grazing pressure were dominated by slow-growing, high stature plant species, fungal-based energy channels, and greater soil C content. In contrast, grassland monoliths exposed to high grazing were dominated by small-stature, fast-growing plant species, bacterial-based energy channels, higher above-ground plant productivity, and lower soil C storage (Klumpp et al. 2007, 2009). Further exploration of the underpinning mechanisms revealed that heavy grazing caused a decline in root biomass and a proliferation of soil bacteria, which was associated with faster organic C mineralisation and increased C loss from soil via heterotrophic respiration (Klumpp et al. 2009) (Figure 5.15). These findings, and those of other studies (e.g., Schuman et al. 1999, Piñeiro et al. 2009, Wilson et al. 2018), suggest that high grazing intensities trigger soil C loss by promoting 'fast' cycling, bacterial-dominated soil microbial communities that accelerate rates of organic C mineralisation, and/or by reducing the input of C to soil via roots. In contrast, light grazing promotes soil C accumulation by selecting for unpalatable, slow-growing plants with low quality litter that decomposes slowly, and/or by promoting below-ground allocation to roots.

Another factor that plays an important role in explaining variable responses of soil C to grazing in grasslands is soil nutrient availability. Grasslands commonly receive nutrient inputs from the addition of fertilisers to land, and it has been proposed that this might enhance the positive impact of herbivores on plant production and C supply to soil, as well as its transfer to more stable C pools, including microbial necromass, forming a large and persistent part of the soil C pool (Sitters et al. 2020b). This idea was tested by measuring soil C pools across 22 grasslands distributed across the globe, each with experimental plots grazed or not by mammalian herbivores and with ambient and elevated nutrient supply (Sitters et al. 2020b). Across sites, the impact of herbivore exclusion on soil C storage was found to be dependent on fertilisation, being greater in grazed plots receiving fertiliser and

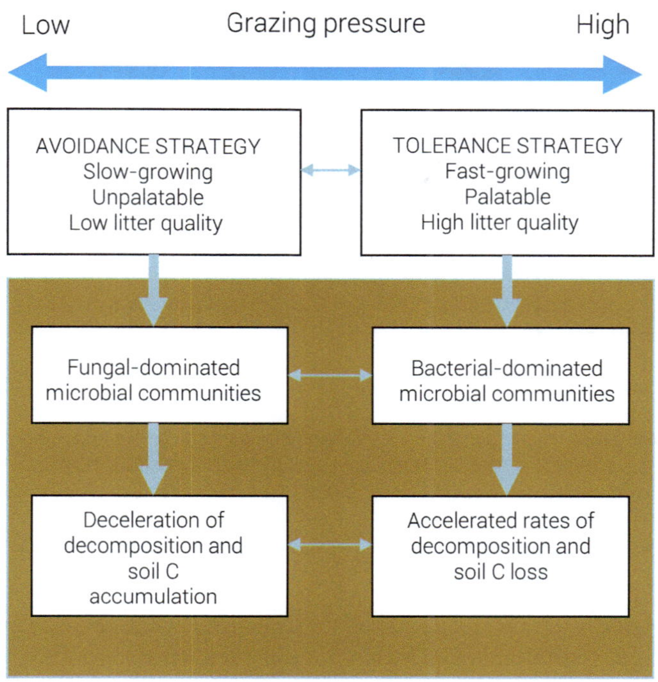

Figure 5.15 Schematic showing how shifts in vegetation and soil microbial community composition influence SOM decomposition and C storage in temperate grassland. In this model, heavy grazing promotes the dominance of small-stature, fast-growing plant species, bacterial-dominated soil communities, higher above-ground plant productivity, faster rates of organic matter decomposition (especially old particulate organic matter), and lower soil C storage. In contrast, low grazing pressure promotes the dominance of slow-growing, high stature plant species, fungal-dominated soil communities, reduced rates of organic matter decomposition, and greater soil C content. From Bardgett and Wardle (2010) based on Klumpp et al. (2009) by permission of Oxford University Press.

lower in plots where grazers had been excluded, but with elevated nutrient supply; under ambient, unfertilised conditions, no effect of grazing on soil C was detected (Figure 5.16) (Sitters et al. 2020b). While the underlying mechanisms remain uncertain, the higher soil C content of grazed and fertilised plots was associated with greater above-ground plant biomass and microbial activity, and it was proposed that this increased C storage in microbial necromass (Sitters et al. 2020b). It is also likely that herbivores will have been attracted to fertilised plots in response to higher forage availability, thereby increasing C input to soil via dung and its incorporation into the soil C pool via the activities of invertebrates (e.g., dung beetles and/or termites) and heterotrophic microbes. Taken together, these findings suggest that grassland SOC storage increases when grazing and nutrient enrichment are combined, especially when the removal of grazers decreases plant biomass and/or microbial activity in soil.

As touched on earlier, the type of herbivore is also of high relevance for soil C. This is especially clear from studies that have considered the replacement of wild herbivores by domestic grazers, which typically has substantial negative impacts on biodiversity and ecosystem functioning, especially at high cattle stocking densities (Du Toit and Cumming 1999, Hempson et al. 2017), and it can also reduce soil C storage. Indeed, overgrazing by livestock is recognised as a major driver of grassland degradation worldwide, resulting in major declines in the ecosystem services supplied by grasslands, including the storage of C in soil (Bardgett et al. 2021). Even when at comparable grazing intensities, the replacement of wild herbivores by livestock can

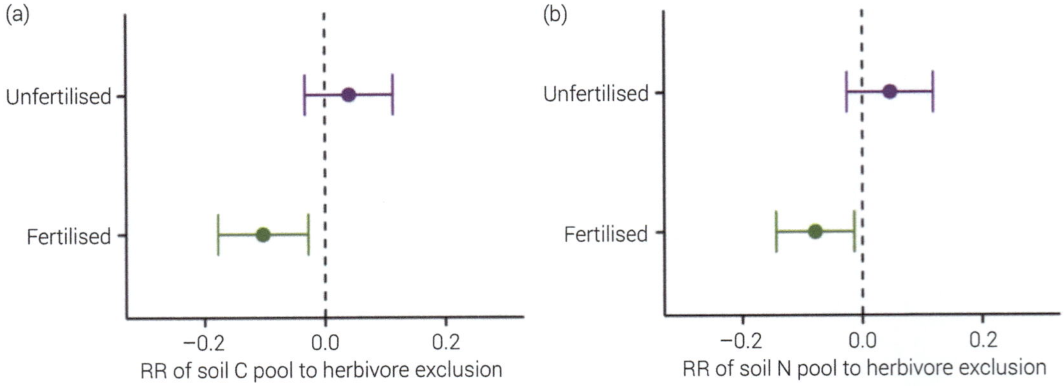

Figure 5.16 Data from 22 globally distributed grassland sites showing that effects of herbivore exclusion on soil C (a) and N (b) storage depend on fertilisation: herbivore exclusion reduced soil C and N storage in fertilised grasslands but had no effect on soil C and N pools in unfertilised grasslands. Data are log response ratios (RR) for soil C pool (a) and N pool (b) to herbivore exclusion calculated as RR = ln(fenced/unfenced) for unfertilised (purple) and fertilised (NPK) (green) plots. RR = 0 if herbivore exclusion had no effect on the variable; RR < 0 if herbivore exclusion decreased the variable, and RR > 0 if herbivore exclusion increased the variable. From Sitters et al. (2020b).

diminish the potential soil C storage capacity of grasslands due to shifts in vegetation composition and reductions in plant production, and, hence, C input to soil (Bagchi and Ritchie 2010). There is also evidence that veterinary antibiotics contained in the dung and urine of livestock can alter the composition of soil microbial communities and contribute to reductions in microbial C use efficiency (CUE), which in turn restricts C sequestration in grassland soils (Roy et al. 2023).

Despite the above, it has been proposed that the integration of moderate numbers of cattle and wild herbivores could offer an effective 'wildlife friendly' management strategy for biodiversity conservation (Keesing et al. 2018), with potential benefits for soil C storage. This idea was tested by measuring soil C in experimental plots of the Kenya Long-term Exclosure Experiment, established in 1995 to test how different combinations of wildlife and cattle influence savanna ecosystem properties (Young et al. 1997). It was found that after 20 years of cattle grazing at moderate densities, soil C and N content was decreased, whereas the presence of megaherbivores, mainly elephants, had the opposite effect on soil C and N and reversed the negative effects of cattle (Sitters et al. 2020a). The mechanisms explaining this contrasting response to cattle grazing and the presence of megaherbivores remain unclear. But, the depletion of soil C and N in cattle-grazed plots was attributed to a net export of C and N outside the plots, due to high grass consumption within plots and the deposition of dung and urine elsewhere, namely in sites where they are corralled over-night (i.e., bomas). In contrast, the positive effect of megaherbivores was explained by the import of C and N from their dunging, which partially compensated for losses of C and N from cattle, but also an increased input of C to soil from the breaking up, toppling, and killing of woody vegetation. These results are of importance because they demonstrate the contrasting impact of wild and domestic herbivores on soil C storage. But also, they have implications for the sustainable management of savannas, given that they suggest that a mix of cattle at moderate densities and wild herbivores can be sustainable and beneficial for climate mitigation (Sitters et al. 2020a).

Herbivores and C persistence in soil

SOC is highly complex and its persistence in soil depends on many factors, including the formation of stable MAOM, which is protected from microbial decomposition and constitutes the bulk of SOM in most soils (Jackson et al. 2017). The SOM pool has two major components: (i) particulate organic matter (POM), which is made up of lightweight fragments of relatively undecomposed plant litter; and (ii) MAOM, which is formed by mineral adsorption and protection of relatively low molecular weight compounds produced by the depolymerisation of plant litter, or from the microbial transformation of organic material and formation of persistent microbial necromass (Lehmann and Kleber 2015, Lavallee et al. 2020). The formation of MAOM is a topic of considerable recent interest, largely because of its persistence and resistance to disturbance, and hence its importance for soil C storage and climate mitigation. As such, there is much interest in understanding what controls the formation of MAOM, including the potential role of herbivores in enhancing the persistence of C in soil.

While very little is known about how herbivores influence MAOM formation, there are three potential pathways involved (Kristensen et al. 2022) (Figure 5.17). The first pathway concerns the role of herbivores in promoting the flux and turnover of organic matter through the plant–soil system, which can enhance microbial metabolism and the formation of microbial necromass and its subsequent mineral protection. As highlighted earlier in relation to the accelerating effect, there are many situations where herbivores, especially at low to moderate densities and in productive grasslands, promote C input to soil through increasing plant litter, root turnover and exudation, and animal excreta, which stimulates microbial biomass production and microbial C use efficiency (CUE), a key trait controlling microbial necromass production and MAOM formation (Buckeridge et al. 2022, Tao et al. 2023).

The second pathway concerns the promotion of fast-growing, productive grasses in grazed grasslands, which allocate large amounts of C to roots and typically invest more C in fine roots to increase nutrient capture (Eissenstat 1992, Gordon and Jackson 2000). This high investment in roots, especially fine roots, means that much of the soil matrix in grazed grasslands is in intimate contact with roots. This in turn ensures a rich supply of root exudates to soil, which stimulates microbial metabolism and the production of polysaccharide gums, thereby contributing formation of microaggregates (Tisdall and Oades 1982, Totsche et al. 2018) that physically protect C from microbial decomposition (Lavallee et al. 2020). Fine roots physically enmesh soil particles, which also promotes aggregation and hence the formation of MAOM (Gould et al. 2016). It should be noted, however, that this potential pathway of MAOM formation is likely of most significance in lightly/moderately grazed grasslands, given that, as discussed above, heavy grazing typically reduces root biomass and causes a proliferation of soil bacteria, which accelerates organic C mineralisation and soil C loss via heterotopic respiration (Klumpp et al. 2009).

The third and final pathway by which herbivores can potentially influence MAOM formation in soil involves the exposure of SOM to organo-mineral interactions through physical soil mixing or bioturbation. This mixing of organic material into mineral soil can be caused by bioturbation of the herbivores themselves, for example through the grubbing activities of wild boar, which has been shown to increase the incorporation of litter-derived C into the underlying mineral soil and MAOM formation (Don et al. 2019), or the digging, trampling, and mixing of elephant dung into soil in areas where they congregate (Haynes 2012). Alternatively, it can be driven by the activities of associated soil engineers, especially earthworms, which are abundant in grazed grasslands, especially beneath herbivore dung pats (Bacher et al. 2018), and play a key role in MAOM formation through the promotion of soil aggregates (Bossuyt et al. 2005, Lubbers et al. 2017) and conversion of labile plant-derived compounds into microbial necromass (Angst et al. 2019). These are just a few examples, but they highlight the potential, but poorly understood, role of bioturbation by large herbivores and associated soil engineers as drivers of MAOM formation in grassland soils.

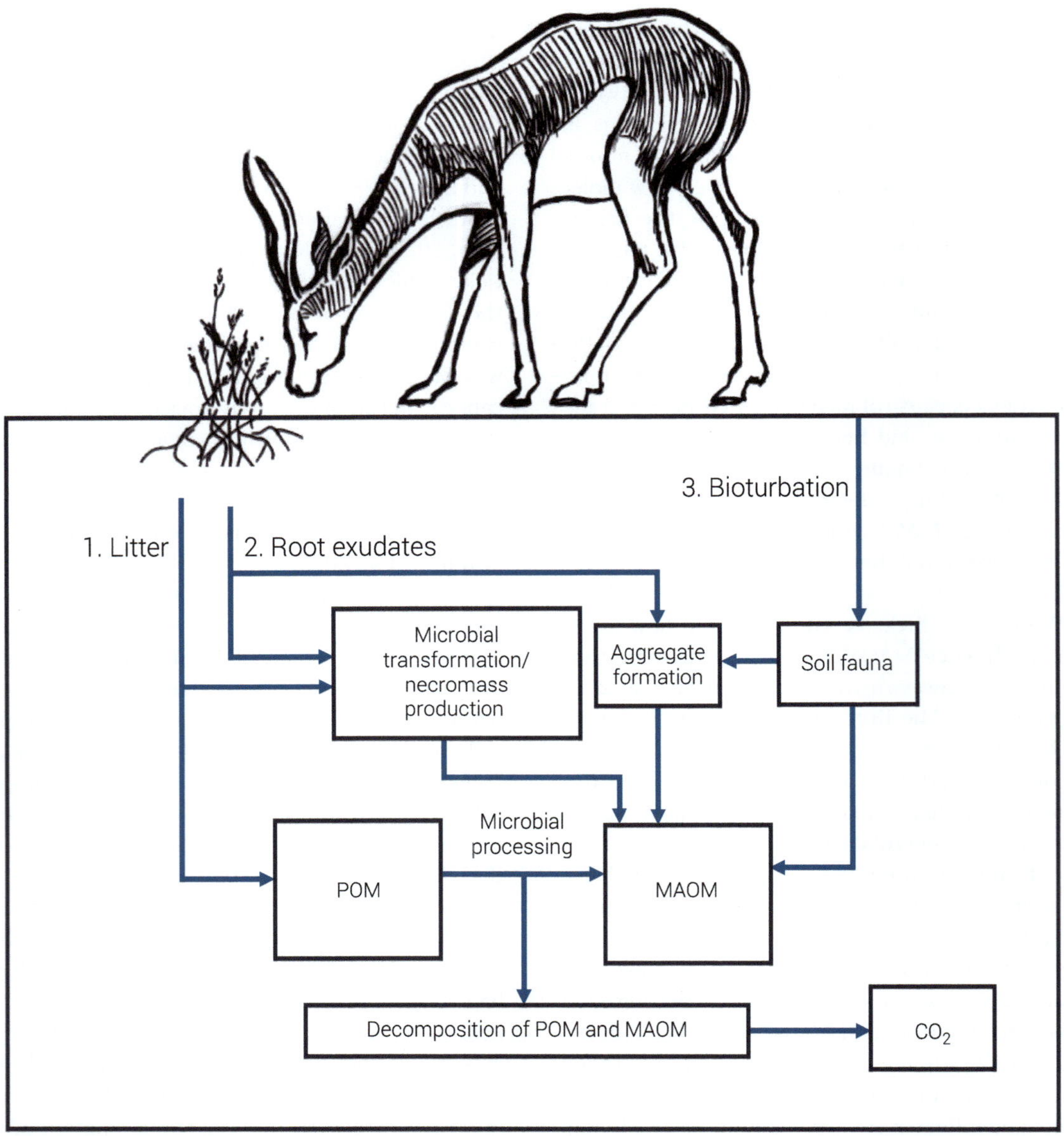

Figure 5.17 Hypothesised pathways by which the presence of large herbivores enhances the persistence of organic C, based on lightly grazed grasslands. Pathway 1 involves the promotion of organic matter inputs to soil via plant litter, root exudates, and animal excreta, which enhances POM and microbial C use efficiency, a key trait controlling the formation of microbial necromass and formation of MAOM. Pathway 2 involves the promotion of fast-growing, productive grasses that allocate large amounts of C to roots, especially fine roots that ensure a rich supply, or root exudates and promote soil aggregation by enmeshing soil particles, thereby enhancing MAOM formation. Pathway 3 involves enhanced mixing of organic matter into mineral soil by bioturbation, for example through the digging, trampling, grubbing, and soil-mixing activities of herbivores, and via associated promotion of soil macrofauna. Adapted from Kristensen et al. (2022). Image by Derek Colquhoun.

Conclusions

Understanding the roles of soil biota in ecosystems requires that they are studied in the context of interactions between above-ground and below-ground communities. As discussed in this chapter, below-ground communities and their functioning are strongly modified by organisms that feed on plants, with ecosystem-level consequences for biogeochemical cycles and C storage. This chapter has identified a range of mechanisms by which mammalian and invertebrate herbivores can directly and indirectly affect soil organisms and their functioning. These mechanisms operate either in the short term, involving physiological responses of individual plants to herbivore attack, or in the long term, involving alteration of plant productivity and vegetation composition, and subsequent changes in the quantity and quality of litter inputs to soil, or through physical disturbance of the soil. Through their movements and the transport of nutrients in animal wastes, herbivores play an important role in creating spatial heterogeneity in plant and soil communities, often creating hot spots of biological activity in soil. However, due to the variety of possible mechanisms at play, the effects of herbivores on soil biota and ecosystem functioning are highly variable, with positive, negative, or neutral effects of herbivory being possible depending upon the balance of these different mechanisms as well as abiotic factors, such as climate and chemical and physical properties of soil.

An important conclusion to emerge from the literature is that the direction and magnitude of herbivore effects on soil biota and their functioning, and consequences for ecosystem function, is highly context dependent. While frameworks exist to explain the contrasting responses of nutrient and C cycling to herbivory, based largely on soil fertility and the proportion of primary productivity that is consumed by herbivores, it is evident that many other factors, both biotic and abiotic, play a role. For instance, there is growing evidence that herbivore type, size, and diversity can determine how herbivores affect soil biota and their functioning, and that predators can indirectly affect below-ground communities by creating landscapes of fear. There is also evidence that soil abiotic properties and climate can explain variability in the response of soil C cycling to herbivores, and that grazer intensity plays an important role, with positive effects of grazing on soil C being detected at light to moderate intensities of grazing in managed systems. Finally, although poorly understood, emerging evidence suggests that wild and domestic herbivores can have contrasting impacts on soil C cycling, and presumably the soil organisms involved, with implications for the management of ecosystems for ecosystem services, especially climate mitigation.

CHAPTER 6

Soil biota and climate change

Key messages

- Climate change impacts below-ground communities and the biogeochemical cycles they regulate via a variety of direct and indirect mechanisms.
- Direct effects relate to changing temperature and precipitation and climate extremes, whereas indirect effects result from changes in plant growth and community composition.
- Soil organism groups and taxa differ considerably in their response to climate change, with high potential to disrupt ecological interactions between plants and soil organisms.
- Interactive effects of global change factors on below-ground communities and the biogeochemical cycles they regulate are common and often additive.

Human activities have a substantial and ever-growing influence on ecosystems to the extent that much of the Earth's land surface and underlying soil has been transformed by a suite of global change phenomena. The most obvious influence of humans on Earth is land use change, especially the transformation of natural ecosystems for agriculture and forestry, which currently represents the primary cause of biodiversity loss worldwide (IPBES 2019) and is considered in Chapter 2. However, terrestrial ecosystems are also under increasing pressure from climate change, and unless there are rapid and large-scale reductions in greenhouse gas emissions, further rises in average temperatures and a greater occurrence of extreme weather events are to be expected (IPCC 2021). Indeed, trends in heatwave frequency, duration, and intensity are accelerating (Perkins-Kirkpatrick and Lewis 2020), as are those for associated droughts (Chiang et al. 2021), and under high greenhouse gas emissions scenarios, record-breaking heatwaves are predicted to be 2.7 times more probable compared to recent decades (Fischer et al. 2021), posing a considerable threat to biodiversity (Murali et al. 2023). The risk of floods is also increasing in many parts of the world owing to climate change (Kreibich et al. 2022), and economic damage from floods is expected to double globally with a mean temperature rise of 2°C (Dottori et al. 2018).

Whilst the significance of climate change for terrestrial ecosystems is widely recognised, we lack a deep understanding of the mechanisms that drive ecosystem responses. However, what is clear is that to fully understand how climate change impacts terrestrial ecosystems requires explicit consideration of interactions between plants and below-ground communities, which drive processes of decomposition, nutrient recycling, and C cycle feedback to climate change (Bardgett et al. 2013, Classen et al. 2015, Pugnaire et al. 2019). This chapter explores this issue, considering how climate change impacts below-ground communities and their activities, and how these effects in turn influence biogeochemical cycles, especially land–atmosphere C exchange and soil C storage. Earth's ecosystems are subject to multiple, simultaneous assaults of global change phenomena; as such, this chapter considers how multiple global change factors can amplify, but also suppress or neutralise, climate change-driven impacts on soil organisms and their functioning, and potentially C cycle feedbacks.

The Ecology of Soil. Richard D. Bardgett, Oxford University Press. © Richard D. Bardgett (2025). DOI: 10.1093/9780191858710.003.0006

There are many ways that climate change impacts soil organisms, their activities, and the biogeochemical cycles they regulate, which can be broadly divided into direct and indirect effects (Bardgett et al. 2008). Direct effects include the influence on soil organisms of temperature, changing precipitation, and extreme climatic events, whereas indirect effects are plant-mediated, resulting from climate-driven changes in plant growth and community composition, which alter the supply of C to soil via litter and root exudates, and the structure and function of soil organisms involved in decomposition processes and C release from soil (Figure 6.1). This chapter uses this broad framework of direct and indirect effects to illustrate the main pathways by which climate change impacts soil organisms and their activities, as well as the consequences for biogeochemical cycles, soil C storage, and C cycle feedback to climate change.

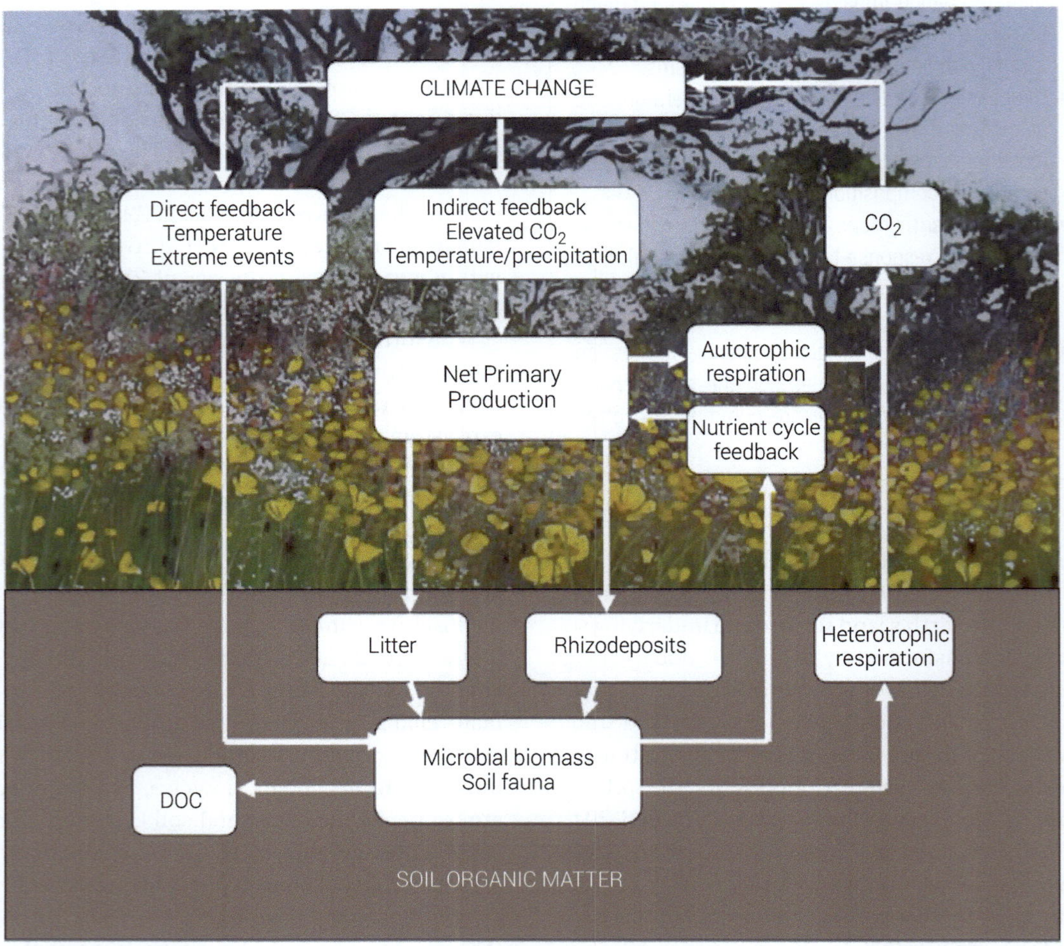

Figure 6.1 Direct and indirect effects of climate change on below-ground communities and routes of feedback to global warming through CO_2 production. Direct effects include the influence on soil microbes and greenhouse gas production of temperature, changing precipitation, and extreme climatic events, whereas indirect effects result from climate-driven changes in plant productivity and vegetation structure which alter soil physicochemical conditions, the supply of C to soil and the structure and activity of microbial communities involved in decomposition processes and C release from soil. From Bardgett et al. (2008) by permission of Oxford University Press.

Direct pathways

The direct effects of climate change on below-ground communities, and the processes of nutrient and C cycling they perform, are largely driven by responses to temperature and extreme climatic events, including droughts, heatwaves, and floods. This section considers the direct effects of warming and climate extremes on below-ground communities, SOM decomposition, and feedback consequences for soil C storage and land–atmosphere C exchanges.

Warming effects on soil biota

Warming effects on soil biota have been a major focus of research in recent decades, largely due to concerns that warming will enhance soil microbial activity and rates of SOM decomposition, thereby increasing C loss from soil via respiration and the release of DOC in leachates (Figure 6.2) (Jenkinson et al. 1991, Davidson and Janssens 2006). Moreover, because rates of C release to the atmosphere from soil in response to warming are expected to outweigh soil C gains from increased primary production, models predict that this will stimulate positive feedback on climate change: the so-called C cycle feedback (Cox et al. 2000, Friedlingstein 2015). However, much uncertainty exists over the nature of the relationship between temperature, soil microbial activity, and organic matter decomposition, which contributes to large uncertainties in predicting C cycle feedback to climate change (Davidson and Janssens 2006, Kirschbaum 2006, Bradford et al. 2016). Syntheses of data from field experiments provide some insights into the response of soil C to warming, for instance showing that the effects of warming are contingent on the size of the initial soil C stock, with considerable

(a) (b) (c) (d)

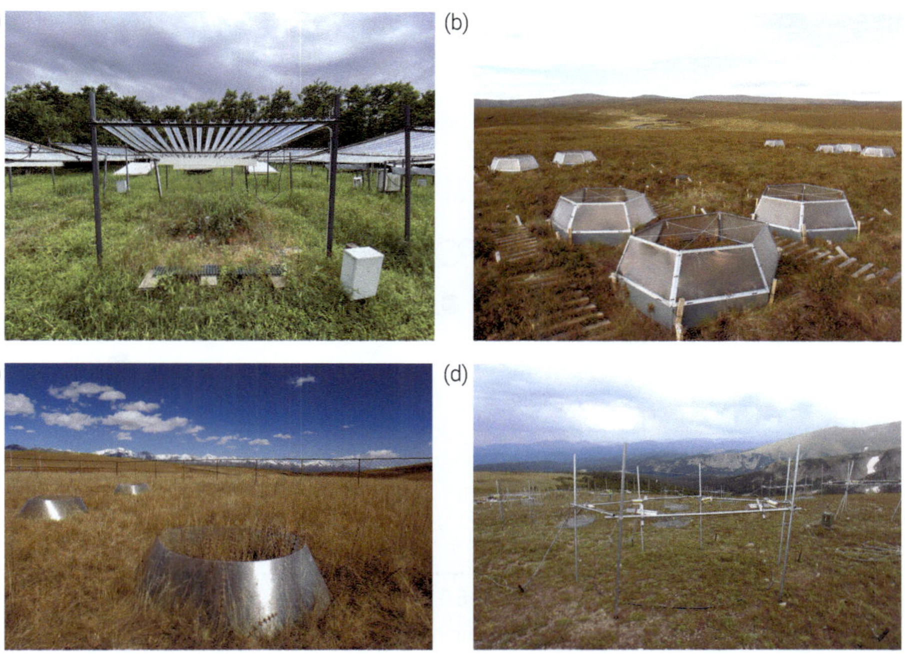

Figure 6.2 Warming experiments on: (a) tallgrass prairie, Great Plains of Central Oklahoma, USA, with warming as the main factor and clipping as a nested subfactor, and warmed plots being heated using infrared heaters suspended above the plot; (b) ombrotrophic blanket bog within the Moor House National Nature Reserve in northern England, studied using open-top chambers (OTCs); (c) alpine meadow at the Haibei Research Station, northeast region of the Qinghai-Tibetan plateau, using OTCs; and (d) alpine tundra, Niwott Ridge, in the Front Range of the Colorado Rocky Mountains, USA, using infrared heaters. Images by Richard Bardgett.

losses occurring in high latitude areas (Figure 6.3) (Crowther et al. 2016). However, global predictions of soil C loss remain challenging because of a lack of studies in the southern hemisphere and at high latitudes in the northern hemisphere, and a paucity of long-term data (Van Gestel et al. 2018). It also remains unclear the extent to which documented soil C losses in response to warming are offset by C gains from increases in plant production, and hence increased input of C to soil from plant litter and exudates (Zhou et al. 2012, Lu et al. 2013, Van Gestel et al. 2018). Nevertheless, a growing number of long-term studies demonstrate significant soil C loss in response to warming across ecosystems (Melillo et al. 2017, Walker et al. 2018), sometimes to considerable soil depth (Soong et al. 2021, Chen et al. 2024), and in response to permafrost thaw despite parallel increases in plant production (Xue et al. 2016, Schuur et al. 2021). Also, global soil respiration data indicate a rising trend in recent decades, a clear signal of soil C loss across the globe (Bond-Lamberty et al. 2018).

Historically, SOM quality was thought to be the primary determinant of the temperature sensitivity of SOM decomposition. The general idea here, which is consistent with enzyme kinetic theory, is that microbial decomposition of low quality, recalcitrant substrates (i.e., the slow C pool) requires high activation energy, and hence is more sensitive to temperature than is higher quality substrate (i.e., the active C pool). Several incubation studies support this theory, showing that the temperature sensitivity of relatively more resistant organic matter is greater than that of relatively more labile substrates

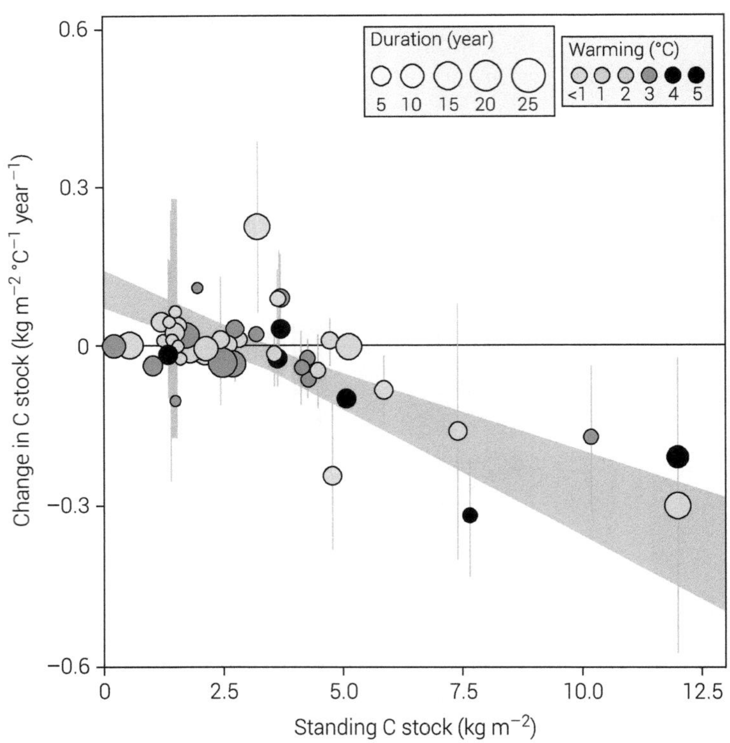

Figure 6.3 The effect of warming on soil C losses depends on the initial standing soil C stock across 49 climate change experiments covering six biomes, ranging from arctic permafrost to dry Mediterranean forests. Data show that the interaction between warming (degree-years) and standing soil C stock is a key determinant of the final warmed soil C stock in the surface 10 cm of soil. Each point represents the difference (mean ± SE) between soil C stocks in warmed and ambient plots within an individual experiment; the size of each point represents the length of an individual study, and the colour indicates the amount of warming. From Crowther et al. (2016) by permission of Springer Nature.

(Knorr et al. 2005, Conant et al. 2008, Yan et al. 2017). However, others paint a different picture, with the temperature sensitivity of more recalcitrant substrates being similar to (Fang et al. 2005, Conen et al. 2006, Tang et al. 2017), or even less than (Luo et al. 2001, Melillo et al. 2002), more labile substrates. This uncertainty has led to the suggestion that factors that limit the availability of C to soil microbes, such as the binding of SOM with minerals and physical protection of SOM within aggregates, are more important than SOM quality in determining the temperature sensitivity of decomposition to warming (Conant et al. 2011, Schmidt et al. 2011). Consistent with this, a recent analysis of a global dataset of > 9,000 soil profiles revealed that while soil C storage declines with increasing annual temperature, the reduction in soil C storage was three times greater in coarse-textured soils with limited capacities for stabilising organic matter than in fine-textured soils with greater stabilising capacities (Figure 6.4) (Hartley et al. 2021). Similarly, the protection of SOM in microaggregates, which spatially separates substrates from microbes and enzymes and limits microbial activity by restricting oxygen diffusion, was found to dampen the temperature response of SOM decomposition in cropland soils (Qin et al. 2019). Together, these studies suggest that mechanisms that stabilise soil C, such as aggregation and mineral binding of organic matter, can dampen warming effects on SOM decomposition, while unprotected pools remain vulnerable (Qin et al. 2019, Hartley et al. 2021).

Another major source of uncertainty in C cycle feedback projections concerns the way soil microbial communities respond to warming, given that they can adjust their response over time by changing their metabolism and/or community structure. Reflecting this, several warming experiments report an initial short-term stimulation of soil microbial activity and respiration, followed by a dampening of this warming effect on soil C loss in the longer term (i.e., ~5 years or more) (Luo et al. 2001, Zhou et al. 2012, Melillo et al. 2017, Verbrighe et al. 2022). This decrease in the temperature

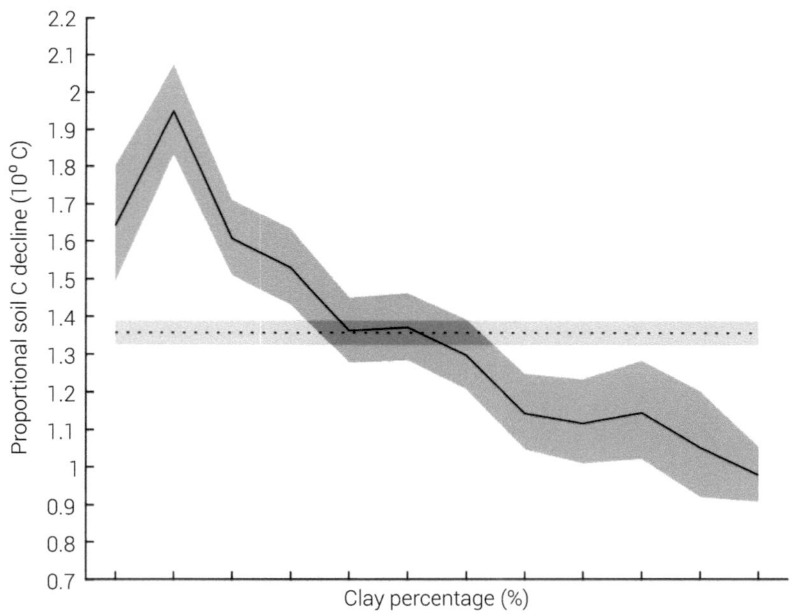

Figure 6.4 The effect of soil texture (clay content) on the relationships between C storage in the surface 50 cm of mineral soil and mean annual temperature, based on analysis of > 9,000 soil profiles. The y-axis refers to the proportional reduction in C storage for each 10°C increase in mean annual temperature, with higher values indicating greater reductions in soil C with temperature, i.e., a value of 1 indicates no change in soil C storage with temperature, values < 1 indicate soil C stocks increase with temperature, and values > 1 indicate C stocks decline with temperature, e.g., a value of 2 indicates that C stocks halve for every 10°C increase in temperature. From Hartley et al. (2021).

sensitivity of microbial activity over time, termed 'thermal adaptation' of soil respiration (Bradford et al. 2019), has been attributed to various mechanisms. These include decreases in labile organic C availability (i.e., substrate limitation) (Hartley et al. 2008, Carey et al. 2016, Walker et al. 2018) and acclimatisation of microbial activity due to shifts in microbial community composition and/or physiological acclimation of soil microbes to warming (Luo et al. 2001, Crowther and Bradford 2013, Guo et al. 2020). Longer-term studies, however, suggest that microbial responses are dynamic and go through distinct phases. This was shown to be the case over 26 years of warming in a temperate forest in New England, USA. Following an initial phase of decreasing warming-induced respiration and declining soil C loss, the soil microbial community entered an impoverished phase characterised by minimal warming-induced respiration and C loss. This was followed by a third phase of rising respiration and soil C loss as a reorganised microbial community utilised more recalcitrant C pools, such as lignin, and a fourth phase of gradually declining respiration (Figure 6.5) (Melillo et al. 2017). These findings are in line with those of other studies showing that, in the long term (i.e., > 5 years), warming causes a shift in C-degrading enzymes from cellulase to ligninase activities that catalyse the degradation of more recalcitrant C pools, thereby potentially exacerbating C cycle feedback to climate change (Chen et al. 2018a).

Other long-term studies show that the response of soil C to warming can be mediated by microbial feedback that helps to maintain plant production and below-ground C supply, as well as soil C storage. This was found to be the case in a long-term study on tallgrass prairie in Oklahoma, USA, where no effect of 2°C warming on total soil C content was detected after 8 years, despite large shifts in microbial community composition and increases in microbial respiration (Zhou et al. 2012) and accelerated soil priming (Tao et al. 2024). This absence of a warming effect on total soil C was attributed to a series of mechanisms involving differential stimulation of microbial populations and functional genes involved in the decomposition of different organic matter pools and processes of N cycling that sustain plant growth (Zhou et al. 2012). Specifically, warming stimulated microbial populations and genes involved in the rapid cycling of labile organic matter and processes of N cycling, which enhanced soil N availability, thereby sustaining an

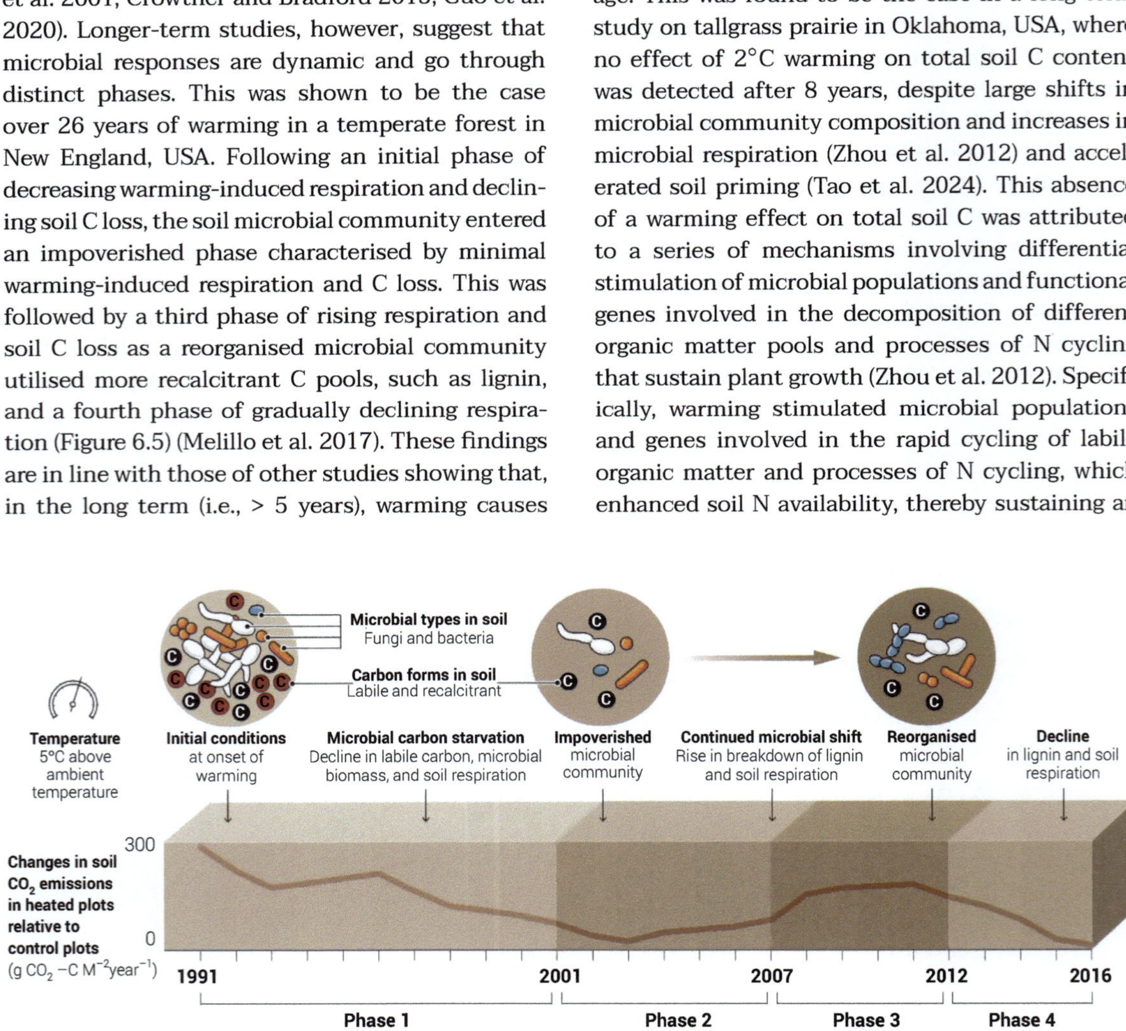

Figure 6.5 Reorganisation of soil microbial communities under long-term warming in temperate forest. After an initial phase of declining warming-induced respiration and reduced soil C loss, the soil microbial community enters an impoverished phase characterised by minimal warming-induced respiration and C loss. This is followed by a third phase of increasing respiration and soil C loss as a reorganised microbial community was better able to use more recalcitrant C pools, such as lignin, and a fourth phase of gradually declining respiration. From Metcalfe (2017) based on data from Melillo et al. (2017) by permission of the American Association for the Advancement of Science.

increase in plant growth and C supply to soil. However, warming had no effect on populations and/or genes involved in the breakdown of recalcitrant C, thereby helping to maintain the soil C store. Taken together, these findings indicate that mechanisms involving feedback between soil microbes and plants, and differential responses of microbial groups with distinct functions, can mediate effects of climate warming on the amount of C stored in soil.

Relatively few studies have explored how soil animals respond to warming, and the traditional view has been that small changes in temperature will have relatively little effect on them due to their wide temperature optima; this view is broadly supported by classic literature synthesis that reports only weak relationships between mean temperature and densities of major faunal groups (Petersen and Luxton 1982, Wardle 2002). However, there is evidence that some soil faunal groups are highly responsive to warming and that warming-induced changes in their abundance and distribution can have significant consequences for decomposition processes and, potentially, C cycle feedback to climate change. For example, research in Iceland revealed that geothermal warming by 6°C for more than 50 years has induced a 'trophic downgrading' of the soil food web, with soil fauna decreasing in abundance and predatory bacteria and viruses becoming relatively more abundant, with important implications for soil C sequestration and nutrient dynamics (Dahl et al. 2023). In other studies, population densities of the dominant fauna of northern peatlands, the enchytraeid worm, were shown to increase with warming, which in turn increased rates of decomposition and the release of DOC into drainage waters (Briones et al. 1998, Cole et al. 2002). A global study of wood decay by termites found that both the discovery and consumption of deadwood was more sensitive to temperature than was microbial decay (Zanne et al. 2022). This finding led the authors to suggest that warming and associated tropicalisation will lead to the expansion of termites into subtropical regions, enhancing their contribution to global wood decay and C cycle feedback to climate change (Zanne et al. 2022). Similarly, the global distribution of earthworms has been shown to be strongly related to climate variables, especially precipitation and temperature, suggesting that future changes in climate will alter earthworm communities globally, with implications for ecosystem functioning (Phillips et al. 2019). It should be noted, however, that warming effects do vary among groups of soil fauna (Thakur et al. 2023): some studies report that soil fauna are resistant to long-term warming (e.g., Holmstrup et al. 2018, Peng et al. 2022), whereas others report negative effects, for instance on the nematode *Scottnema lindsayae*, the dominant soil animal of the McMurdo Dry Valleys, Antarctica (Simmons et al. 2009), and on collembolans on Signy Island (Bokhorst et al. 2008) and in subarctic grassland (Thakur et al. 2023).

Drought

Most research on climate change impacts on soil biota has focused on gradual changes in climate. But extreme climatic events, such as droughts, floods, heatwaves, freezing events, and fires, are becoming more frequent and severe with climate change, posing a significant threat to soil biota and the processes soil organisms drive (Reichstein et al. 2013). Most of the research on this topic has focussed on resistance (i.e., capacity to resist) and resilience (i.e., capacity to recover) of soil microbial communities to drought (Figure 6.6), which is expected to become more frequent and intense due to anthropogenic climate change (Chiang et al. 2021). There is growing evidence that drought has high potential for significant and long-lasting impacts on the composition and functioning of soil microbial communities (Bardgett and Caruso 2020). For example, studies reveal that rewetting of droughted soil is immediately followed by highly dynamic patterns of microbial growth and mortality (Blazwicz et al. 2020), but also that drought impacts on soil microbial communities and their functioning can persist long after rewetting (De Vries et al. 2018). Recent studies also show that soil microbial communities and their functions display non-linear responses to increasing drought intensity (Oram et al. 2023) and that high intensity drought can trigger abrupt and persistent shifts in the soil microbial community, characterised by reduced complexity and functionality (Figure 6.7) (Cordero et al. 2023). Such drought-induced shifts

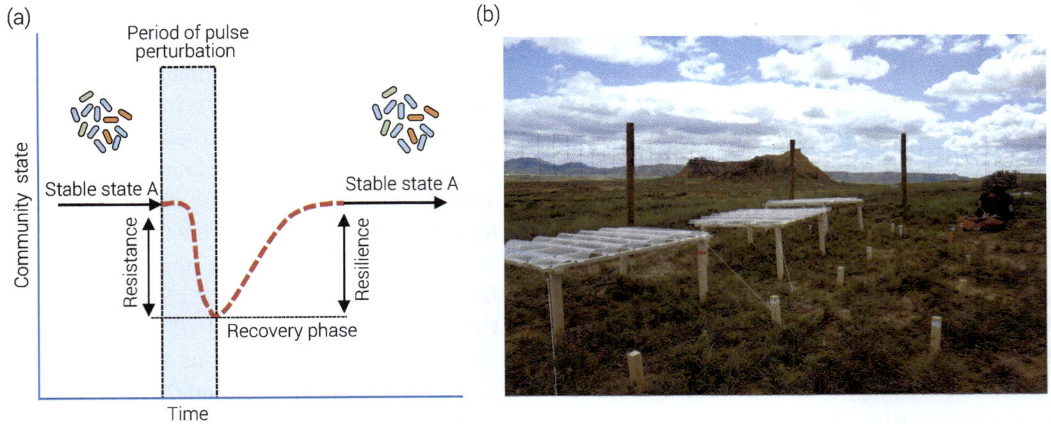

Figure 6.6 Conceptual diagram illustrating the resistance and resilience of a microbial community to an abrupt, intense 'pulse' perturbation, such as a drought event (a). The scale of impact on the microbial community or microbial-mediated functions is termed resistance, whereas a resilient community returns to its normal compositional and functional state. The image (b) is of rainfall shelters in the Bardenas Reales, Navarre, Spain, set up to test for the resistance and resilience of below-ground communities and processes to a single drought event. Image by Richard Bardgett.

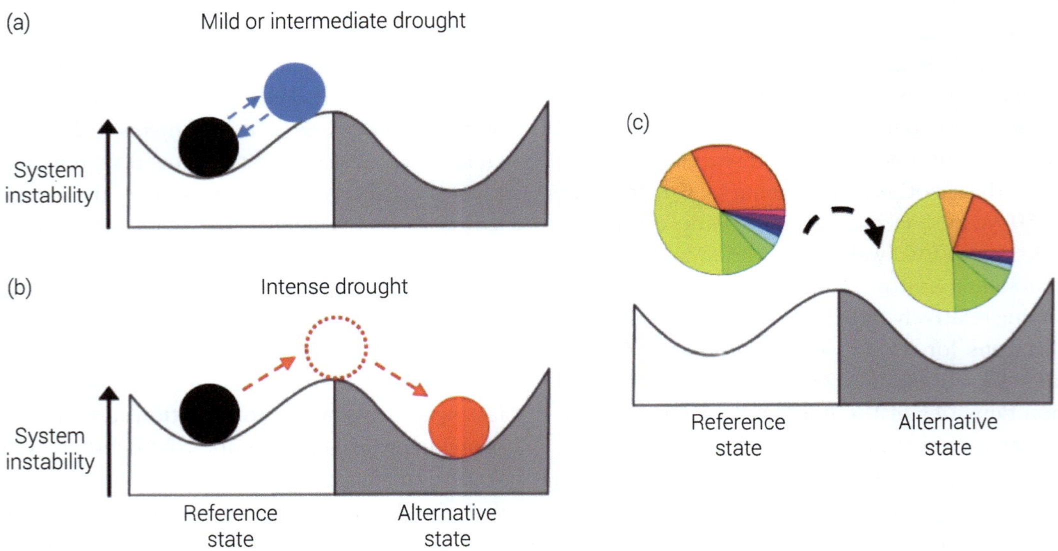

Figure 6.7 Schematic of the effects of intense drought on the soil microbial system, represented as a ball moving up and out of the stability basin. Mild or intermediate drought does not push the ball far enough, bouncing back to the reference state (a), while high intensity drought pushes the ball out of the reference state into a different stability basin, an alternative state (b), with distinct functionality and microbial community composition, the later represented by differences in the relative abundance of microbial families within the pie charts (c). From Cordero et al. (2023).

in soil microbial communities can have significant legacy effects that modify responses to subsequent droughts (Müller and Bahn 2022). For example, soil bacterial communities with a history of exposure to low precipitation were found to display high resistance and resilience to drought (Tang et al. 2023b), and recurring droughts have been shown to create a 'memory' that buffers microbial communities

Box 6.1 Mechanisms underpinning below-ground resistance and resilience

The concepts of resistance and resilience, or the ability of an ecological system to resist (i.e., resistance) and recover (i.e., resilience) from a perturbation and maintain its structure and function, are a major focus of ecological research. One of the most frequently discussed ideas linking community diversity to the resistance and resilience is the so-called insurance hypothesis, i.e., that high diversity insures against declines in ecosystem function because many species provide greater guarantees that some will maintain their function if others do not (Yachi and Loreau 1999). The mechanisms underpinning this relationship, however, remain unclear, especially for highly complex and species-rich below-ground communities. This is in part because many different facets of diversity, including taxonomic and functional diversity (i.e., the range and variance of traits), as well as the traits of dominant species, influence resistance and resilience to perturbations, such as climate extremes. Despite a general lack of empirical evidence, two main mechanisms are likely to operate in stabilising the effects of below-ground biodiversity and food web composition on ecosystem functioning in the face of climate extremes, namely: (i) asynchrony of species responses, based on microbial traits related to the tolerance and speed of recovery; and (ii) changes in the strength of food web interactions and the asymmetry of soil food web energy channels.

Asynchrony of species responses: The idea here is that species differ in their fundamental niches, leading to dissimilar or divergent responses among species to environmental fluctuations that stabilise ecosystem functions. In other words, an increase in the abundance of some species in response to environmental fluctuations compensates for declines in others, ultimately yielding more stable aggregate ecosystem properties (Loreau and De Mazancourt 2013). Species also differ in the speed at which they respond to and recover from perturbations, which generates asynchronous population responses that promote more stable ecosystem properties (Loreau and De Mazancourt 2013). Divergent species responses to climate extremes, such as drought, are well documented for below-ground communities and have been linked to broad life-history strategies and specific traits associated with resistance and resilience (Bardgett and Caruso 2021). For example, microbial traits associated with heterotrophic growth strategy and dormancy potential have been linked to an enhanced ability of microbial-mediated functions to buffer perturbations (Nemergut et al. 2016, Kearns and Shade 2018). As such, asynchrony of species responses and difference in the speed that species respond to and recover from perturbations based on their functional traits are likely to be key mechanisms contributing to below-ground resistance and resilience to climate extremes.

Asymmetry of food web energy channels: An approach to understand how soil food webs are structured to confer stability is to consider them as coupled fast and slow energy channels that broadly relate to rates of energy and nutrient flux in the soil system. The general idea here is that soil food webs are broadly structured into the 'fast' energy channel, which is formed of bacteria and their consumers, cycles nutrients rapidly, and recovers quickly from perturbations, thereby increasing resilience, and the 'slow' energy channel that comprises fungi and their consumers cycles nutrients more slowly and dampens responses to perturbations, thereby increasing resistance (Moore and Hunt 1988, de Ruiter et al. 1995). These complementary functions of the 'fast' and 'slow' energy channels should produce a rapid, yet stable recovery of the soil food web from a perturbation (Rooney et al. 2006). Moreover, any factor that destroys this complementarity or changes the asymmetry of the two channels, such as increased resource supply or removal of organisms, should destabilise the food web and its functioning relative to an unperturbed state (Rooney et al. 2006). Some theoretical and empirical evidence supports this idea, for instance showing that intensive agriculture can shift the distribution of biomass and energy flow to the bacterial rather than the fungal channel, rendering it less resistant to drought (De Vries et al. 2012b).

against future drought (Canarini et al. 2021). Similarly, high intensity drought can have strong legacy effects on soil bacterial communities, inducing an enhanced growth rate following wet-up after a subsequent drought (Cordero et al. 2023).

Drought impacts vary considerably among microbial groups and taxa due to differences in their life history strategies (Box 6.1). At a broad taxonomic level, fungi are generally thought to be more resistant to (i.e., able to withstand) drought than are bacteria, due to their thick cell walls, capacity to produce osmolytes and invest in storage compounds, and conservative growth strategy (Hawkes et al. 2011, Barnard et al. 2013,

De Vries et al. 2018, Canarini et al. 2024). It has been proposed that oligotrophic bacteria, which thrive under conditions of low nutrient availability, are more resistant to drought due to their low growth rate and high resource use efficiency; in contrast, copiotrophs, which prosper under conditions of high resource availability, are more resilient (i.e., able to recover) post-drought due to their high growth rate and low resource use efficiency (Fierer et al. 2007a). Similarly, the resistance and the resilience of soil microbial communities to drought has been explained based on the relative abundance of r- and K-strategists, as well as environmental factors like nutrient availability. Specifically, it has been proposed that microbial communities dominated by K-strategists (i.e., typically larger, with slower growth rates) in nutrient-poor environments have high resistance, whereas a dominance of r-strategists (i.e., smaller, fast growing) in nutrient-rich environments confers high resilience, or recovery post-drought (Figure 6.8) (De Vries and Shade 2013). It was also recently discovered that the rapid growth, or resuscitation, of bacteria following the 'wet-up' of droughted soils is associated with a large increase in the biomass of viruses that lyse bacterial cells, thereby acting as a major driver of microbial death during soil wet-up and the release of microbial debris into the soil environment (Nicolas et al. 2023). Root-associated AM fungi are also known to ameliorate the negative effects of drought on plant performance (Tang et al. 2023a) and can influence the drought tolerance of bacteria in the hyphosphere, i.e., the region surrounding fungal hyphae. Notably, inoculation of plants with AM fungi was found to facilitate greater growth potential, growth efficiency, and diversity of hyphosphere bacteria in droughted soil, thereby enhancing the resistance of bacteria to drought (Hestrin et al. 2022).

Contrasting phylum-level responses of soil bacteria to desiccation and rewetting have been linked to their life history strategies. For example, members of the *Actinobateria* phylum, which are common in dryland soils (Maestre et al. 2015), are highly resistant to soil drying due to their high tolerance to desiccation, whereas members of the *Acidobacteria* phylum are less resistant, but more resilient, given their remarkable recovery due to their fast growth strategy (Barnard et al. 2013). Individual bacterial taxa have been classified as having tolerant, opportunistic, or sensitive drought-response strategies, which vary in their relative abundance depending on environmental factors, such as precipitation history. Soil bacterial communities shaped by a history of moisture pulses, for example, have been shown to have a greater proportion of taxa exhibiting a drought-tolerant strategy than those subject

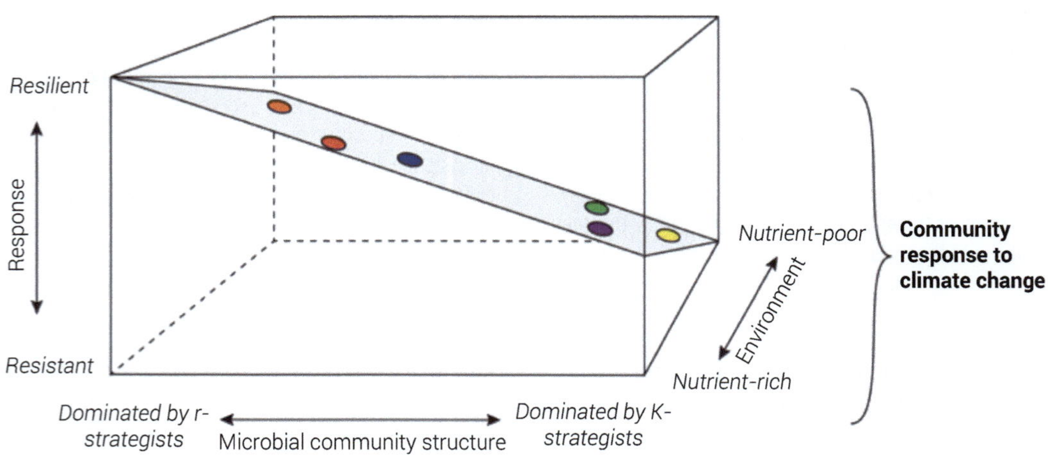

Figure 6.8 Framework for predicting microbial community responses to pulse perturbations, such as drought, as determined by the relative abundance of r- and K-strategists and extrinsic factors, such as nutrient availability. A K-strategist-dominated microbial community in a nutrient-poor environment likely has high resistance, whereas an r-strategist-dominated community in a nutrient-rich environment likely has high resilience. The exact shape of the surface will vary depending on specific circumstances. From De Vries and Shade (2013).

to ambient conditions, which have a greater abundance of drought-sensitive taxa (Evans and Wallenstein 2014). Similarly, soil microbial community resistance to drought has been shown to be greater in warmer and drier sites than in colder and wetter sites due to a greater abundance of drought-tolerant phyla (Ochoa-Hueso et al. 2018, Knight et al. 2024), and to be increased by recurring droughts due to community reorganisation in favour of drought-tolerant microbial taxa (Evans and Wallenstein 2012). These findings are consistent with modelling studies showing that microbial assemblages from more constant historical environments are more sensitive to environmental change than are those from fluctuating environments (Hawkes and Keitt 2015), which is likely explained by both community reorganisation and acclimatisation. They also emphasise the point that extrinsic factors, in this case climate history, but also vegetation and soil nutrient availability, shape the intrinsic attributes and traits of microbial communities that confer resistance and/or resilience (Figure 6.9).

Drought-induced shifts in soil microbial communities have legacy effects that influence subsequent responses of plants to drought, with implications for plant community dynamics. The adaptation of soil microbial communities to recurring droughts, for example, has been shown to improve the fitness and ability of plants to withstand subsequent droughts (Lau and Lennon 2012, De Vries et al. 2023). Drought has also been shown to favour plant associations with mycorrhizal fungi and mutualistic soil bacteria, thereby enhancing plant drought tolerance via improved access to water and hormonal signalling (Mariotte et al. 2017, Rubin et al. 2017). Studies indicate that legacy effects of drought on soil microbial communities can alter the strength and/or direction of plant–soil feedback, with consequences for plant competitive interactions. For example, legacy effects of drought on soil microbial communities were found to switch the direction of plant–soil feedback from positive and/or neutral to negative for certain grassland plant species, presumably due to an increase in pathogens (Kaisermann et al. 2017), and to neutralise plant–soil feedback for others (Fry et al. 2018a), with potential consequences for plant competitive interactions. Similarly, drought legacy effects on soil microbial communities can impact plant invasiveness, promoting exotics over natives (Meisner et al. 2013), especially in more nutrient-rich soils with bacterial-dominated communities (Schrama and Bardgett 2016). This latter finding is because drought-induced dry–wet cycles cause a pulse in soil inorganic N availability, which is especially pronounced in nutrient-rich soils with bacterial-dominated communities; in turn, this benefits N acquisition and growth of exotic plants over natives (Schrama and Bardgett 2016). Research on these topics is still in its infancy, but evidence is growing that drought legacy effects on soil microbial communities can have significant consequences for plant community dynamics and ecosystem functioning, as well as future plant and microbial responses to drought.

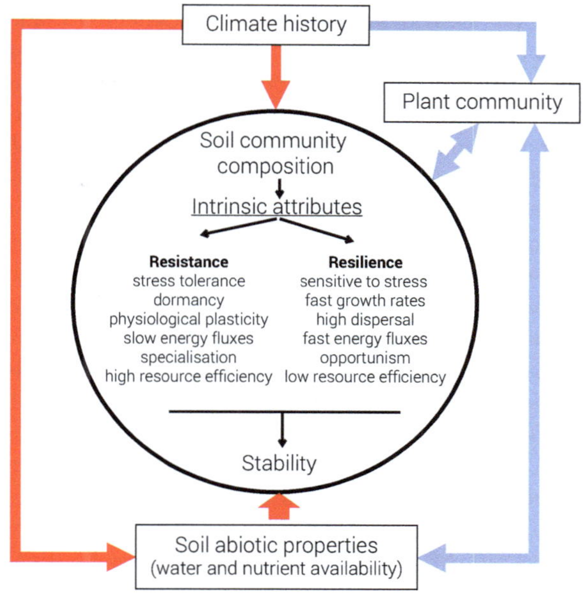

Figure 6.9 Schematic of the pathways by which extrinsic factors can change the attributes of soil microbial communities that confer resistance and resilience both directly (red pathways), via climate history and inherent variation in soil properties, and indirectly (blue pathways), via changes in the plant communities which alter soil resource availability. From Bardgett and Caruso (2020) by permission of Royal Society (U.K.).

Soil-dwelling animals are sensitive to drought, especially those that live in water films or water-filled pore spaces, such as nematodes. However, larger bodied animals, such as Collembola and mites, are also highly sensitive to reductions in moisture, which directly interferes with their behaviour and physiology, leading to higher mortality and/or lower fecundity (Holmstrup et al. 2001, Gundersen et al. 2014, Aupic-Samain et al. 2021). Reflecting this, many studies document marked declines in the abundance and diversity of larger bodied faunal groups in response to drought, including enchytraeids, mites, and Collembola (Lindberg et al. 2002, Tsiafouli et al. 2005, Santonja et al. 2017. Martin et al. 2024), as well as reduced invertebrate feeding activity (Siebert et al. 2019) and changes in predator–prey relationships (Lang et al. 2014, Aupic-Samain et al. 2021). Simulated drought has also been shown to simplify the trophic structure of nematode communities, favouring fungal-feeding nematodes over bacteria-feeders (Cesarz et al. 2015), and the abundance of root-feeding (Franco et al. 2019a) and opportunistic (Siebert et al. 2019) nematodes. Moreover, drought reduces the below-ground transfer of recent photosynthetic C from plants to soil fauna (Chomel et al. 2019), thereby disrupting a critical pathway of energy flow for soil animal communities (Pollierer et al. 2007), although drought could favour smaller bodied fauna over larger bodied ones as they can survive in soil microhabitats, including deeper soil layers (Aupic-Samain et al. 2021). Given these far-ranging responses of soil fauna to drought, it is likely that future increases in drought frequency and intensity will cause major shifts in soil food webs, with potential consequences for ecosystem functions and their capacity to buffer subsequent climate extremes (Bardgett and Caruso 2021).

Floods, heatwaves, freezing, and fire

Another form of climate extreme that is becoming more common is flooding, resulting from heavy precipitation events (Bavacqua et al. 2019), which causes waterlogging of soils and anoxic conditions. Very few studies have explored how flooding events and associated waterlogging impact soil biota, but available evidence suggests they can be profound, supressing or killing populations that require well oxygenated soil conditions and benefitting others that thrive in anaerobic soil. A recent pot experiment, for example, found that flooding dramatically restructured the soil bacterial communities associated with wheat roots, increasing the abundance of bacterial taxa with anaerobic respiratory capabilities and of potential detrimental taxa, but decreasing aerobic bacteria and several plant-beneficial bacterial taxa (Francioli et al. 2021). Flooding events can also modify the composition of mycorrhizal fungal communities, causing reductions in the richness and relative abundance of ECM fungi (Barnes et al. 2018), and increases in AM fungi that thrive in waterlogged and anoxic conditions (Camy et al. 2003, Yang et al. 2016). Other organisms that benefit from flooding events are methanogens, which produce the potent greenhouse gas methane (CH_4) (Conrad 2007), and denitrifying bacteria, which are facultative anaerobes that reduce NO_3^- to gaseous N_2O (Tomasek et al. 2019). In addition to increasing the abundance of denitrifiers, flooding has been shown to promote AOA, which play a key role in the process of nitrification and can be tolerant to low O_2 conditions (Barneze et al. 2023). However, flooding effects on methanogens and denitrifying bacteria are generally temporary and reversed on soil drainage (Ratering and Conrad 1998, Tomasek et al. 2019).

Heatwaves, although interrelated with drought, are becoming more frequent and intense (Perkins-Kirkpatrick and Lewis 2020), and record-shattering temperatures are expected to become more common in coming decades (Fischer et al. 2021). As with drought, extreme heat generally causes abrupt declines in population sizes of soil microorganisms, given that most microbes can only tolerate temperatures under 40°C (Bérard et al. 2015). However, as with drought, resistance and resilience of soil microbiomes to heat has been shown to be greater in warmer and drier climates than in colder and wetter climates (Knight et al. 2024), although there is considerable variation in the heat tolerance among soil trophic groups, suggesting that heatwaves may cause shifts in soil food web architecture. For instance, rhizosphere bacterial communities have been shown to be more vulnerable to extreme heat than their protist predators, suggesting that lower trophic groups could be more vulnerable to heatwaves than their predators (Thakur et al. 2021). The reason for this is unclear, but it might be explained by predator strategies that minimise thermal stress, such as remaining inactive and/or decreasing body size, which gives them an advantage over their prey, unless the prey community collapses (Thakur et al. 2021). There also appears to be much variation in heat tolerance across trophic levels of soil arthropods, with prey species, such as the abundant Collembola *Isotoma riparia*, being shown to be more sensitive to high temperature than their predators (i.e., various spiders) (Franken et al. 2018). This again suggests that heatwaves could directly alter soil food web dynamics and ecosystem function due to differences in the heat tolerance across trophic groups.

Another consequence of climate change is increased freeze–thaw activity due to reduced snow cover in alpine and arctic regions. In the European Alps, for example, snow cover is predicted to decline by around 50% above 2,500 m elevation by the end of the century, with spring snow melt occurring 50–60 days earlier (Beniston 2012). Arctic snow cover duration is decreasing at a rate of ~ 3–5 days per decade, due to earlier spring snow melt and later onset of snow cover (Derksen et al. 2015). Snow insulates soil against freezing temperatures, enabling many soil organisms to remain active over winter (Schimel et al. 2004, Buckeridge and Grogan 2008, Gavazov et al. 2017) As such, reductions in snow cover, or the complete removal of snow due to unseasonably warm winter weather, diminish this insulating effect, resulting in more frequent and severe freeze–thaw cycles that damage, or even kill, soil organisms (Bokhorst et al. 2016, Ji et al. 2022). Snow removal in boreal forest, for example, can greatly reduce the abundance of soil microarthropods, due to their intolerance to soil freezing (Bokhorst et al. 2013), and a global meta-analysis revealed that freeze–thaw activity generally reduces microbial biomass and causes shifts in microbial community structure, with bacteria being generally more sensitive than fungi (Ji et al. 2022). Reductions in snow depth and earlier snow melt in alpine grasslands have been shown to advance the transition from winter to summer microbial communities, with an associated decline in bacterial and fungal biomass on thawing (Broadbent et al. 2021). Given that soil microbes retain large amounts of N during the winter months, this abrupt decline in microbial biomass at snow melt causes a large pulse in soil N availability. In turn, this can increase N_2O emissions and N loss in drainage waters, especially if the N pulse is out of synch with plant N requirements and/or if plants are damaged by increased freeze–thaw activity (Darrouzet-Nardi et al. 2019, Ji et al. 2022, Broadbent et al. 2024). Indeed, it was recently shown that the combination of reduced snow cover and shrub expansion disrupted this seasonal coupling of plant and soil microbial N-cycling processes in alpine grasslands, potentially diminishing the capacity of these ecosystems to retain N and support plant productivity under future climate change (Broadbent et al. 2024). However, significant questions remain concerning how variation in the timing and frequency of freeze–thaw events impact below-ground communities and their functioning, and how responses are moderated by extrinsic factors, such as vegetation type and soil nutrient availability (Broadbent et al. 2022, 2024).

The threat of fire has also increased over a large portion of the Earth's land surface with climate change (Jolly et al. 2015). While relatively little is known about how fire impacts below-ground communities, available evidence suggests that different components of the soil food web respond differently to fires: while most organism groups respond negatively, including protists (Qin et al. 2017), fungi (Day et al. 2019, Pressler et al. 2019), and microfauna (Gongalsky et al. 2021), others, such as bacteria and bacterial-feeding nematodes, generally respond positively (Butenko et al. 2017, Whitman et al. 2019, Gongalsky et al. 2021). Moreover, while the scale of fire impacts on soil organisms varies with the severity and frequency of fire (Malmström 2010, Whitman et al. 2019), they can be sizable. A recent meta-analysis, for example, revealed that, across a wide range of ecosystems, fire decreases fungal abundance by 47.6% and total microbial abundance by 33.2%, with effects being especially pronounced in high intensity wildfires (Pellegrini et al. 2022), and other meta-analyses show that fire typically reduces the abundance of fungi relative to bacteria (Dove and Hart 2017, Pressler et al. 2019). Moreover, negative effects of fire on soil microbes can be long-lasting (Waldrop and Harden 2008), as can impacts on surface-dwelling animals that are unprotected from fire and/or have limited dispersal ability (Zaitsev et al. 2016, Malmström 2010, Gongalsky and Persson 2013).

Fire-induced shifts in below-ground communities have implications for processes of soil C cycling. In particular, fire-induced reductions in microbial and fungal abundance have been linked to declines in rates of decomposition (Dooley and Treseder 2012, Wang et al. 2012, Holden et al. 2013), and repeated fire has been shown to reduce the activity of hydrolytic enzymes involved in SOM decomposition across a range of ecosystems (Köster et al. 2014, Pellegrini et al. 2020). Across a broad range of forests in European Russia, fire-induced shifts in soil food webs reduced the flow of C through the soil food web, especially in northern taiga forests, which have a thick humus layer that is highly susceptible to fire (Gongalsky et al. 2021). However, fire can also enhance rates of decomposition and soil C loss via the production of fire-derived charcoal, which has been shown to promote the decomposition of humus in boreal forests due to its stimulatory effect on microbial activity, potentially offsetting the benefits of charcoal as a long-term soil C sink (Wardle et al. 2008). This suggests that fire-induced shifts in below-ground communities can both decelerate and accelerate SOC loss (Figure 6.10), although the environmental and biological factors that regulate the balance of effects, and consequences for soil C storage, remain unclear (Pellegrini et al. 2022).

Indirect pathways

Climate change has marked indirect effects on below-ground communities via impacts on plant growth and community composition, which influence both the amount and the quality of C inputs to soil from plant litter (shoot and root) and root exudates (Figure 6.1). This section considers how elevated CO_2 indirectly alters below-ground processes via changes in C supply to soil via roots, and how, over longer timescales, shifts in vegetation resulting from climate change can modify below-ground communities, with consequences for biogeochemical cycles and soil C storage.

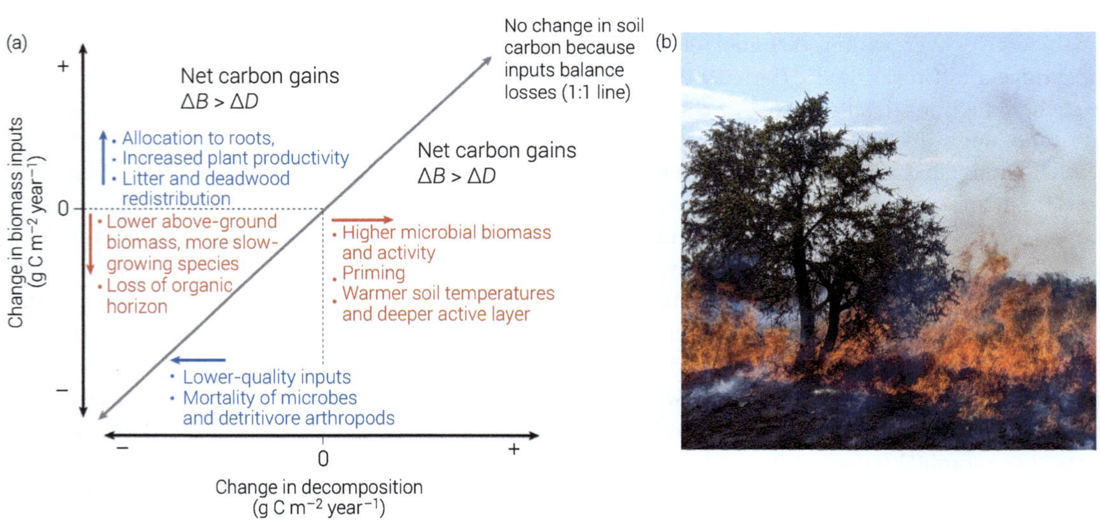

Figure 6.10 The balance between decomposition-based soil C losses versus plant biomass-based C inputs influences how fire affects soil C stocks (a). Fire effects on biomass-based C inputs (*B*) and decomposition-based C losses (*D*) can promote soil C gains (blue), including higher root biomass, leaching of ash into soil, reduction in detritivore biomass, and formation of aggregates. Fire effects can also promote soil C losses (red) via declines in plant biomass inputs, erosion and leaching of organic matter, priming of microbial activity, and warmer soil temperatures. Coloured arrows represent the vector direction caused by the processes adjacent to the arrow. From Pellegrini et al. (2022) by permission of Springer Nature. Image of a controlled fire on mopane savanna in Kruger National Park (b) by Jacob Wieczorkowski.

Below-ground processes and elevated CO_2

A major focus of research on plant-mediated indirect effects of climate change on below-ground communities has been rising atmospheric CO_2, which typically stimulates plant photosynthesis and growth, especially under nutrient-rich conditions (Curtis and Wang 1998, De Graff et al. 2006, Poorter et al. 2022). This stimulation in plant growth is generally associated with an increased flux of C below-ground, to roots and their symbionts, and exudation of easily degradable sugars and other soluble organic compounds into the rhizosphere, thereby stimulating the growth and metabolic activity of free-living microbes and the abundance of their consumers (Bardgett et al. 2013). The generality of these responses is supported by meta-analyses that have used data from a wide range of studies to show that elevated atmospheric CO_2 typically leads to an expansion of the root system, increasing root length, diameter, and biomass, and root exudation (Figure 6.11) (Nie et al. 2013). Elevated CO_2 is known to lead to substantial increases in mycorrhizal abundance (Treseder 2004) and the biomass of host plants, especially those associated with AM fungi (Dong et al. 2018). It also typically reduces plant tissue N (Cotrufo et al. 1998, Curtis and Wang 1998, Poorter et al. 2022), thereby increasing the C:N ratio of litter inputs to soil, which reduces their decomposability, potentially leading to SOM accumulation.

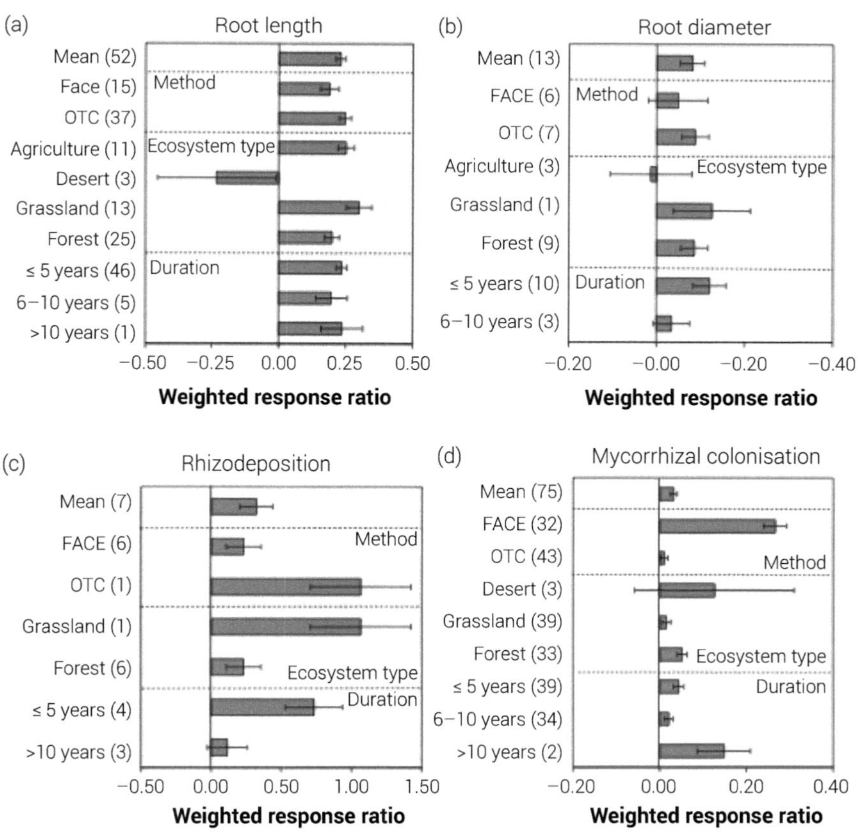

Figure 6.11 Responses to elevated CO_2 of root length (a), root diameter (b), rhizodeposition (c), and mycorrhizal colonisation (d). Bars represent the weighted response ratio ± 95% confidence interval and values near each variable are the number of cases synthesised. FACE refers to experiments done using free-air CO_2 enrichment, whereas OTC refers to experiments using open top chambers. From Nie et al. (2013) by permission of John Wiley and Sons.

It might be expected that increases in below-ground C flux to roots will increase the biomass of soil microbes, but evidence for this is mixed, with positive, negative, and neutral responses being reported (Zak et al. 2000, Blankinship et al. 2011). Effects of elevated atmospheric CO_2 on the composition of microbial communities are also mixed, with studies reporting negligible responses among different microbial groups (Zak et al. 1996, Kampichler et al. 1998, Gutknecht et al. 2012), increases in the relative abundance of fungi (Carney et al. 2007, Zhong et al. 2018) and bacteria (He et al. 2010), or differential responses among specific bacterial and fungal groups (Drigo et al. 2009, He et al. 2010). This discrepancy in soil microbial community responses to elevated CO_2 is due to several factors, including differences in experimental and environmental conditions (e.g., the duration of experiments, CO_2 concentrations used, and soil abiotic properties) (Blankinship et al. 2011, Sun et al. 2021b) and the degree to which components of the soil microbial community are regulated by predation; any potential increase in microbial biomass, or the abundance of a particular microbial group, could be dampened by increased consumption by microbial-feeding soil animals (Bardgett and Wardle 2010). Added to this, responses of soil fauna to elevated CO_2 are highly variable and differ for different trophic groups: a meta-analysis of responses of soil biota to elevated CO_2 showed that while detritivores tend to respond positively, other trophic groups, including herbivores, bacterivores, fungivores, and predators, show little response (Blankinship et al. 2011). Similarly, some animal groups, such as mites and collembolans, have been shown to respond positively to elevated CO_2 (Eisenhauer et al. 2012), while others, such as nematodes, show more variable responses (Mueller et al. 2016). This variability makes it difficult to draw firm conclusions on how the composition of soil food webs responds to increases in atmospheric CO_2.

Irrespective of these divergent microbial and faunal responses, many studies indicate that elevated CO_2 can increase root C transfer, stimulating microbial metabolism, with important consequences for biogeochemical cycles and soil C storage. Increased plant investment of C to roots and root exudates under elevated CO_2 can stimulate microbial activities and the breakdown of SOM by priming, i.e., an increase in SOM decomposition rate after fresh organic matter input to soil resulting from an increase in microbial activity (Fontaine et al. 2003, Dijkstra et al. 2013). This priming effect appears to be widespread under elevated CO_2 and has important consequences for soil C storage. In a study of a wide range of European tree species, for example, elevated CO_2 stimulated the transfer of root-derived C to soil, enhancing microbial activity and SOM decomposition and leading to a decline in soil C sequestration (Heath et al. 2005). A meta-analysis of CO_2 enrichment studies showed that this response is widespread: across a wide range of studies, CO_2 enrichment increased both the input and turnover of C in soil, which ultimately limited soil C accumulation due to priming of native SOM (Van Groenigen et al. 2014b).

As well as affecting the SOM turnover, priming has important feedback consequences for plant growth under elevated CO_2, because the breakdown of SOM also releases N for plant uptake. In turn, this can set in motion a positive feedback mechanism that increases rates of plant production. An excellent example of this comes from a study by Drake et al. (2011), who found that elevated CO_2 increased C flow through the plant root–microbial–soil system of a low fertility pine forest in North Carolina, USA, which stimulated microbial activity and the breakdown of SOM, but also the release of N and its uptake by trees. This set in motion a positive feedback mechanism that increased tree production, and hence C storage in tree biomass but not in soil (Figure 6.12). A companion study by Phillips et al. (2011) explored the mechanisms behind this C cascade: over 3 years, elevated CO_2 increased the rate of exudation from tree roots by 55%, leading to a 50% annual increase in soluble organic inputs to soil; in turn, this stimulated microbial activity, organic matter decomposition, and the release of microbial extracellular enzymes involved in organic N breakdown, thereby enhancing SOM loss and N availability for plant use. A subsequent study revealed that the accelerated turnover of root exudates under elevated CO_2—which also releases N—was of sufficient magnitude to offset an increase in below-ground inputs to soil, thereby explaining the

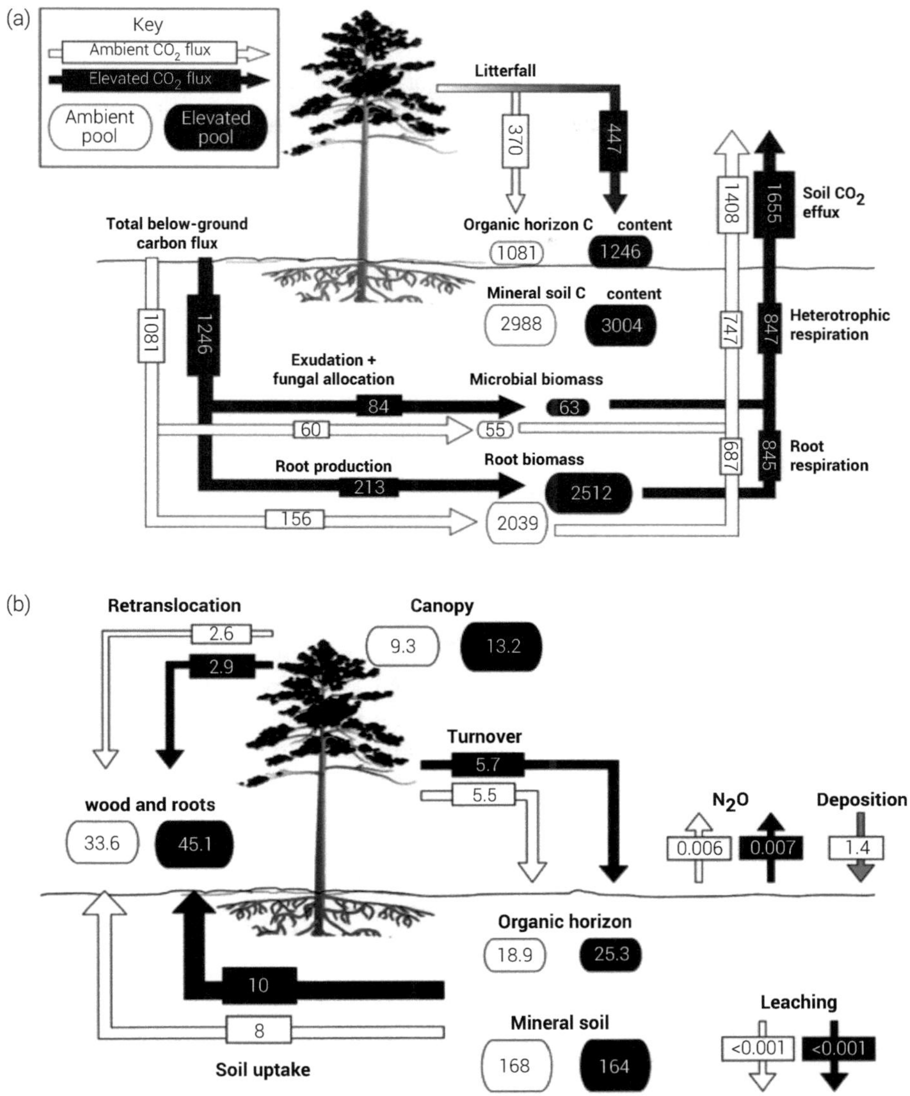

Figure 6.12 Elevated atmospheric CO_2 stimulates below-ground C flux and plant N uptake, which sustains the enhancement of forest productivity and tree C storage, but not an increase in soil C. Data show the impact of elevated atmospheric CO_2 on the ecosystem C (a) and N (b) budget of a low fertility pine forest, Duke Forest free-air CO_2 enrichment site, in North Carolina, USA. Ambient CO_2 is shown in white and elevated CO_2 in black, and ovals reflect C and N pools (g m^{-2}), whereas squares within arrows reflect C and N fluxes (g m^{-2} year^{-1}). For C (a), all terms were significantly different between ambient and elevated CO_2 aside from mineral soil C content, which did not change in response to elevated CO_2. For N (b), all terms were significantly different aside from turnover, N content of the organic N horizon, mineral soil N, N_2O, and leaching. From Drake et al. (2011) by permission of John Wiley and Sons.
Ecology Letters, Volume: 14, Issue: 4, Pages: 349–357. First published: 09 February 2011, DOI: (10.1111/j.1461-0248.2011.01593.x).

lack of soil C accumulation in these forests under elevated CO_2 (Phillips et al. 2012).

It should be noted, however, that under different conditions, especially when microbes are strongly N limited, enhanced plant root exudation can stimulate microbial biomass and immobilisation of N, thereby limiting N availability to plants, plant growth, and ultimately C transfer to soil (Díaz et al. 1993, De Graff et al. 2006) (Figure 6.13). Increased plant–microbial competition for N under elevated CO_2 can also exacerbate N constraints on soil microbes, suppressing microbial activity and organic matter decomposition, which in turn increases C accumulation in soil (Hu et al. 2001). Similarly, in P-limited grasslands and forests, enhanced microbial competition for P has been shown to constrain P availability for plant P uptake under elevated CO_2, thereby limiting CO_2 fertilisation of plant productivity (Keane et al. 2023, Jiang et al. 2024). Long-term elevated CO_2 has been shown to decrease the abundance of N-fixing and ammonium-oxidising genes in grassland, which reduces inorganic N availability, constraining microbial decomposition and potential microbial growth rate (Yang et al. 2019a).

While our understanding of the mechanisms by which elevated CO_2 impacts below-ground C flux and C storage has advanced greatly in recent years, major uncertainties still exit. One uncertainty concerns the mechanisms that explain variation in responses of soil C to elevated CO_2 across experiments; in other words, why does elevated CO_2 increase soil C storage in some situations, but decrease it in others? This issue was addressed in a recent synthesis of data from 108 elevated CO_2 experiments, which revealed that the effect of elevated CO_2 on SOC stocks was best explained by a negative relationship with plant biomass: in situations where plant biomass is strongly stimulated by elevated CO_2, SOC declines; conversely, when plant biomass is only weakly stimulated, SOC increases (Figure 6.14) (Terrer et al. 2021). This finding was taken to suggest that there is a trade-off between plant and soil C storage under elevated CO_2 that is explained by rhizosphere processes related to plant nutrient acquisition: plants increase their biomass under elevated CO_2 by mining the soil for nutrients via the priming effect, which accelerates SOM losses. This trade-off was also found to be mediated by symbiotic associations between plants and mycorrhizal fungi: at one end of the spectrum, ECM plants are more effective at increasing N uptake from soil via priming under elevated CO_2, leading to a greater plant biomass sink at the expense of SOM. At the other end of the spectrum, low nutrient availability constrains the plant biomass sink of AM plants under elevated CO_2, although increased below-ground C investment can contribute to C accrual (Figure 6.14) (Terrer et al. 2021).

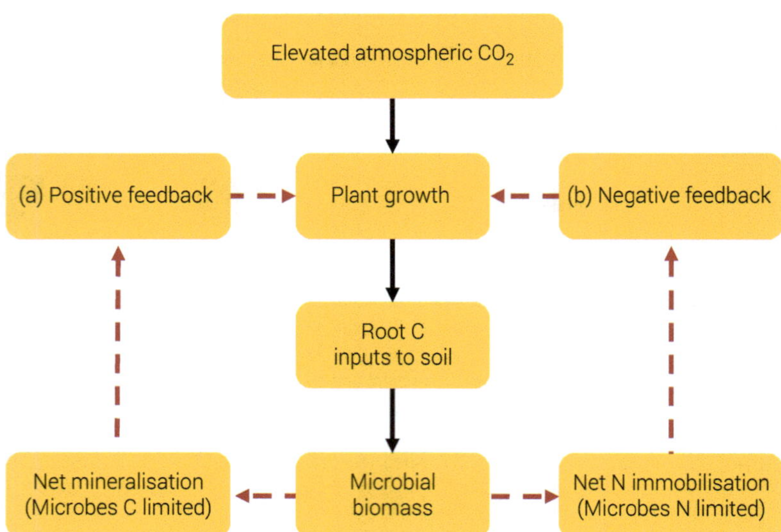

Figure 6.13 Schematic of potential impact of elevated atmospheric CO_2 on root C inputs to soil and microbial biomass, leading to either (a) positive or (b) negative feedback on plant nutrient availability and growth. Positive feedback occurs when the increased supply of root-derived C to soil leads to increased microbial biomass turnover and net N mineralisation, and uptake of N by plants. Negative feedback occurs when elevated CO_2 increases the flux of C to soil via roots, but an enlarged microbial biomass leads to net immobilisation of N, thereby limiting the availability of N to plants.

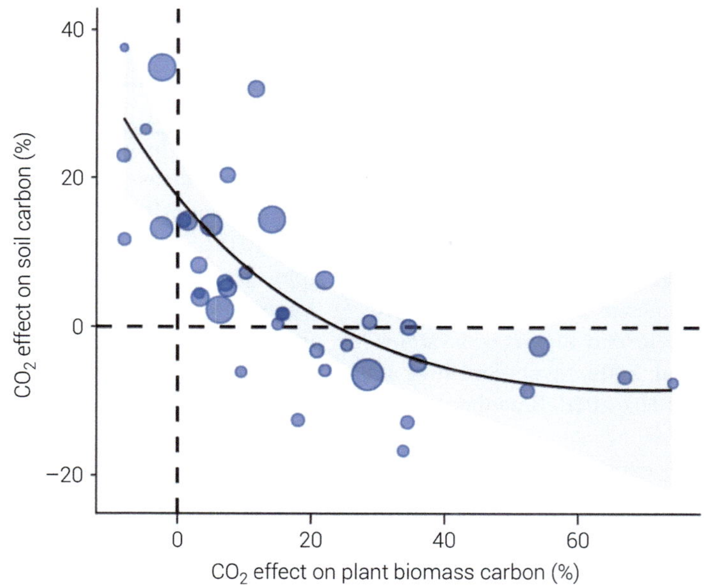

Figure 6.14 Synthesis of elevated CO_2 experiments show an inverse relationship between the effects of elevated CO_2 on plant biomass and SOC stocks. From Terrer et al. (2021) by permission of Springer Nature.

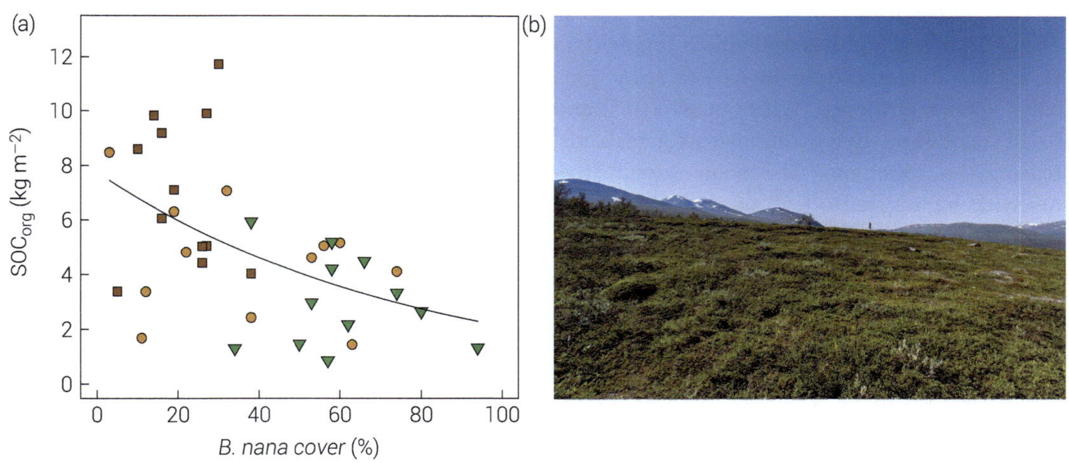

Figure 6.15 (a) Negative relationship between cover of the deciduous shrub *Betula nana* and SOC of the surface soil organic horizon of tundra heath (squares), shrub heath (circles), and shrub (triangles) sites near Abisko, Swedish Lapland. (b) The image shows subarctic tundra with large patches of *B. nana* (left) adjacent to more ericaceous-dominated vegetation (right). From Parker et al. (2015) with image by Tom Parker.
Global Change Biology, Volume: 21, Issue: 5, Pages: 2070–2081. First published: 04 November 2014, DOI: (10.1111/gcb.12793).

Plant community change

Climate change can impact below-ground communities and their functioning via shifts in vegetation composition and diversity that occur over a range of spatial and temporal scales (Bardgett et al. 2013). At regional scales, for example, climate change is responsible for shifts in vegetation composition of tropical rainforest (Engelbrecht et al. 2007, Comita and Engelbrecht 2014) and African savannas (Sankaran et al. 2005, Sankaran 2019) and in plant functional composition of the tundra biome (Bjorkman et al. 2018). Climate change is leading to major shifts in plant species ranges, with many species expanding their ranges to higher elevations and the poles (Walther et al. 2005, Hickling et al. 2006). The global advance in treelines, for example, has been attributed to climate warming (Harsch et al. 2009, Mayor et al. 2017), as has the upward movement of alpine plant species (Steinbauer et al. 2018) and widespread shrub encroachment that is occurring across arctic and alpine ecosystems worldwide (Wookey et al. 2009, Myers-Smith et al. 2011, Myers-Smith and Hik 2018, Broadbent et al. 2022).

Field manipulation experiments show that climate change causes vegetation change at a local scale, with reports of experimental warming, drought, and CO_2 enrichment causing shifts in plant community composition and diversity across a variety of ecosystems: for example, simulated warming increased the abundance of grasses relative to forbs across a range of Tibetan Plateau grasslands (Liu et al. 2018a) and enhanced the dominance of C4 over C3 plants in tallgrass prairie (Luo et al. 2009); simulated drought causes reductions in plant species diversity (White et al. 2014) and shifts in vegetation composition across grasslands, increasing the dominance of resource acquisitive plants with high specific leaf area (SLA) and leaf N content (Griffin-Nolan et al. 2019); and in situ CO_2 enrichment has been shown to both positively (Leadley et al. 1999, Niklaus et al. 2001) and negatively (Zavaleta et al. 2003) affect grassland plant diversity, and to increase the abundance of legumes (Newton et al. 1995, Ross et al. 2004), especially under low nutrient conditions (Hanley et al. 2004). Finally, warming was recently shown to intensify soil pathogen-induced negative plant–soil feedback of a temperate tree species, *Prunus padus*, with potential consequences for forest community dynamics in a warmer world (Liu and He 2021).

Such climate induced-changes in vegetation, and associated shifts in root systems, can impact soil biota and the biogeochemical processes they drive via a variety of mechanisms. For example, the expansion of deciduous shrubs in the Arctic has been linked to increased soil C turnover and loss from soil due to mining for nutrients via root-associated ECM fungi, which enhance the breakdown of SOM (Figure 6.15) (Parker et al. 2015, Street et al. 2020). Also, deciduous shrubs produce easily decomposable leaves, so their expansion can contribute to increased C loss by promoting rates of litter decomposition (Cornelissen et al. 2007). Conversely, the expansion of ericaceous shrubs into alpine grasslands, which is occurring rapidly across alpine regions of the world in response to climate (and land use) change (Rumpf et al. 2022), has been tentatively linked to the accumulation of soil C (Broadbent et al. 2022). The mechanisms underpinning this remain unclear, but it is potentially related to: (i) a shift in the fungal community towards dominance of ERM fungi, which have a suppressive effect on saprotrophic fungi and organic matter decomposition (Fanin et al. 2022); and (ii) production of ERM fungal necromass, which contributes to the formation of persistent SOM, especially when adsorbed to mineral surfaces to form MAOM (Clemmensen et al. 2015) or stabilised via root-derived condensed tannins to form stable complexes (Adamczyk et al. 2019) (Figure 6.16). Roots of ericaceous shrubs and ERM fungi also have the capacity to enhance C loss via priming (Adamczyk et al. 2019), but the importance of this pathway relative to the pathways of C gain following shrub expansion remains unclear.

Climate change-related shifts in grassland composition and diversity can modify C acquisition by photosynthesis, below-ground C allocation to roots and root exudates, and the quality of litter inputs to soil from shoots and roots, given that different plant species have distinct litter chemistries (Bardgett et al. 2013). Moreover, associated shifts in root traits, such as rooting depth and mycorrhizal status, can impact soil biota, the processes of C and nutrient cycling they drive, and physical properties of soil (Bardgett et al. 2014). For example, studies have shown that reduced precipitation (Debinski et al. 2010, Hoeppner and Dukes 2012) and warming (Weltzin et al. 2003, Hoeppner and Dukes 2012) select for deeper rooting, woody plant species, which increase below-ground C inputs and mycorrhizal colonisation, with likely consequences for soil biota and the biogeochemical processes they perform. It should be noted that

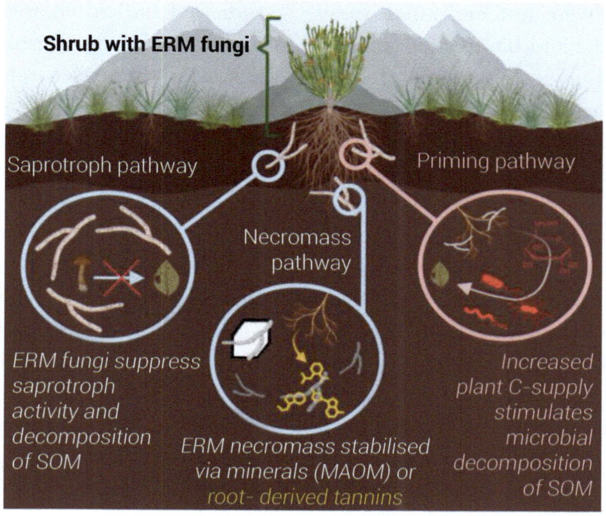

Figure 6.16 Schematic of the below-ground pathways by which ericaceous shrub expansion into alpine grasslands influences SOM stabilisation (blue) and decomposition (red). Image by Arthur Broadbent.

certain plant communities are resistant to simulated climate change (Grime et al. 2008, Hudson and Henry 2010), and there are also cases where below-ground communities are affected by climate manipulations, but above-ground communities are not (Cantarel et al. 2012, Yergeau et al. 2012), and vice versa (Cruz-Martinez et al. 2009). This suggests that above- and below-ground responses to climate change are not always coupled, perhaps because of differences in sensitivity and/or speed of response, with soil communities typically responding more rapidly to climate than do plant communities (Bardgett et al. 2013).

In some situations, climate change-driven range expansion of plants can decouple or disrupt existing ecological interactions between plants and associated soil organisms, with consequences for ecosystem structure and function. A major driver of this biotic decoupling is the unequal dispersal of plants and soil organisms, in that plants generally disperse faster than do soil organisms, including pathogens, mutualists, and decomposers (Van der Putten et al. 2010). Consequently, plant species that expand their ranges under climate change can be released from their native pathogens, and hence suffer less from soil pathogen-driven negative feedback, which benefits their growth and establishment in their new range (Van Grunsven et al. 2007, Engelkes et al. 2008, Morriën and Van der Putten 2013). However, range-expanding plants may be disadvantaged by the absence of specialist mutualists and/or decomposers in their new range, or by forming mycorrhizal relationships that benefit them less than those formed in their native range (Klironomos 2003). Range-expanding species may also disrupt co-evolved mutualistic networks of native species and other components of the soil food web, which could negatively impact the range expander as well as resident plant and soil communities (Morriën et al. 2012).

Before leaving the topic of indirect effects of climate change on below-ground communities, it is worth noting that large herbivores can modulate climate change effects on plant communities and below-ground processes. While still poorly understood, a growing number of studies show that vegetation responses to climate change can be constrained by herbivores, potentially moderating below-ground effects on soil C cycling. For example, studies in northern-temperate and arctic ecosystems show that woody plant expansion under warming is constrained by large herbivores (Post et al. 1999, Post and Pedersen 2008, Olofsson et al. 2009, Olofsson and Post 2018), and while below-ground responses were not measured, they are likely also constrained. Field-based experiments in subarctic tundra also demonstrate that herbivores can modulate both plant and soil microbial responses to warming. In one study, for example, simulated warming enhanced the growth of the dominant dwarf shrub *Vaccinium myrtillus* and reduced soil N availability and microbial N immobilisation only in the absence of simulated herbivory (Rinnan et al. 2009); in another study, soil fungal community responses to warming were found to be dependent on grazing history, being more pronounced under light than heavy grazing by reindeer (Ahonen et al. 2021). Reindeer grazing history was also found to change the mechanism by which warming stimulates soil C loss in subarctic tundra: under long-term light grazing, soil C loss, in response to simulated warming, was linked to higher abundance of dwarf shrubs and a greater activity of enzymes involved in organic matter breakdown, whereas under long-term heavy grazing, warming-induced C loss was linked to drier soil conditions and higher enzymatic activities (Ylänne et al. 2020). While many questions remain concerning the mechanisms involved and how they vary across ecosystems, these studies suggest that above-ground consumers are critical moderators of ecosystem responses to climate change and hence that they warrant further study.

Multiple global change factors

Terrestrial ecosystems not only are subject to climate change, but also are simultaneously impacted by other global change factors, such as intensification of land use, invasive species, soil pollution, and N deposition. Added to this, different aspects of climate change (i.e., warming, elevated CO_2, and climate extremes) co-occur, and have potential for additive, synergistic, or antagonistic impacts on below-ground communities and the biogeochemical cycles they drive. The potential for multiple global change factors to influence below-ground communities is a poorly studied topic (Rillig et al. 2019), but available evidence suggests that interactive effects on below-ground communities and the processes they drive are common (Zhou et al. 2016, Yue et al. 2017, Li et al. 2022). For example, a meta-analysis of data from field manipulation studies of individual and combined global change factors, including warming, elevated CO_2, N and P addition, increased rainfall, and drought, found that the combined effects of paired factors on soil microbial biomass were always additive, rather than synergistic or antagonistic (Yue et al. 2017). (Additive effects occur when the combined effect of different global change factors equates to the sum of the individual effects, whereas synergistic and antagonistic effects are when the combined effect is greater or weaker than the sum of the individual effects, respectively.) Similarly, a meta-analysis of data from 150 studies exploring interactive effects of global change factors on soil respiration found that effects of combined factors were always greater on average than those of single factors, and they were mostly additive in nature (Zhou et al. 2016).

Although large data syntheses reveal that interactive effects of global change factors on below-ground communities are common and mostly additive, responses vary tremendously depending on factors such as taxonomic group, the type of global change factor involved, and abiotic factors such as climate and soil conditions. For example, a meta-analysis of data from studies exploring individual and interactive effects of global change factors on soil fauna revealed that the magnitude of responses varied with latitude and elevation, likely due to differences in climate and soil conditions (Peng et al. 2022). Also, regarding broad taxonomic groups, responses of soil fauna to combined global change factors were found to be generally negligible (Peng et al. 2022). This contrasts with data syntheses based on soil microbial diversity, which show that responses are predominately negative and additive, although synergistic effects are sometimes found (Yang et al. 2021, Li et al. 2022). Moreover, laboratory studies show that increasing the number of simultaneous global change factors imposed on soil caused reductions in fungal diversity and decomposition (Rillig et al. 2019), suggesting that responses of soil microbes and microbial-mediated functions to multiple global factors can be directional and profound.

Interactions between global change factors have potential implications for processes of C cycling, and hence C storage in soil. The combination of elevated temperature and atmospheric CO_2, for example, has been shown to enhance microbial

Box 6.2 Anthropogenic N inputs and soil biota

Over the last century, anthropogenic N inputs to ecosystems have increased dramatically due to the copious use of fertilisers on agricultural land and atmospheric deposition of N emitted to the atmosphere from fossil fuel combustion (Galloway et al. 2008). While there is evidence that N availability is declining in many non-agricultural ecosystems (Mason et al. 2022), excessive anthropogenic N inputs continue to negatively impact ecosystems and human health worldwide (De Vries 2021). As with climate change, N enrichment can modify below-ground communities and their functioning via both *direct* and *indirect* pathways.

Direct below-ground effects: N enrichment is a major driver of soil microbial community structure and function, causing reductions in microbial biomass and respiration, the abundance of mycorrhizal and decomposer fungi, increased abundance of bacteria relative to fungi, and dominance of fast-growing copiotrophic bacteria (Ramirez et al. 2012, Leff et al. 2015). Soil animals are very sensitive to N enrichment, with meta-analysis revealing overall negative effects of N addition on the abundance of all nematode trophic groups, aside from bacterial feeders, but limited effects on Collembola and mites, across a range of ecosystems (Hu et al. 2022). However, negative effects of N on soil animals were found to depend on N addition rate, occurring at rates above 100 kg N ha^{-1} $year^{-1}$, and were greater in grassland and cropland than in forests (Hu et al. 2022). Soil enrichment with N has differential impacts on extracellular enzymes involved in decomposition processes, typically stimulating cellulases that degrade labile, high cellulose litter, but suppressing of the synthesis of ligninolytic enzymes by white rot fungi that decompose recalcitrant, high lignin litter (Frey et al. 2004, Allison et al. 2008). Therefore, ecosystems with more recalcitrant litter and many white rot fungi tend to respond negatively to N deposition, whereas ecosystems with rapidly cycling litter have different microbial communities that respond positively to N deposition (Waldrop et al. 2004).

Indirect below-ground effects: Because most plant communities are N limited, the addition of N substantially alters their productivity and composition, typically favouring fast-growing plant species over slow-growers, with associated reductions in plant diversity and enhancing net primary productivity (Midolo et al. 2019). While shifts in plant communities resulting from N addition vary depending on environmental context, they typically increase the amount and quality of plant litter entering soil, by both promoting fast-growing plant species and enhancing leaf N concentrations. In turn, this typically promotes the abundance of soil microbes and animals and causes a shift to bacterial-based rather than fungal-based soil food webs, potentially creating a positive feedback on N cycling and plant production, and this further promotes the growth of competitive, fast-growing plant species (Bardgett and Wardle 2010). The potential for such positive feedback on N cycling, however, varies across ecosystems depending on factors such as the capacity of soil microbial communities to immobilise additional N and the subsequent transfer of this N to stable, non-microbial organic matter pools, which can be considerable in some ecosystems (e.g., Zogg et al. 2000, Bardgett et al. 2003, Phoenix et al. 2005).

decomposition of peat more than when these two factors operated alone (Fenner et al. 2007), leading to even stronger positive feedback on peatland C loss (Freeman et al. 2004). Increased anthropogenic N input from the use of fertilisers and atmospheric deposition, which substantially impacts below-ground communities (Box 6.2), has been shown to negate the positive effect of increased precipitation on the formation of soil aggregates and SOC protection in semi-arid grassland due to suppression of fungal biomass (Bai et al. 2020) and to alleviate N limitation of plant growth under elevated atmospheric CO_2 (Finzi et al. 2002, Luo et al. 2004, De Graaff et al. 2006). Also, while long-term soil warming and N fertilisation was found to have opposing effects on soil C when applied alone, with warming causing soil C loss and N fertilisation leading to soil C gain, this effect was cancelled out when treatments were combined, thereby helping to maintain soil C stocks near control levels (Knorr et al. 2024). This was taken to suggest that when soils are exposed to both warming and N fertilisation, soil C inputs from plants are enhanced, which counterbalances C losses from soil (Knorr et al. 2024). Enhanced soil nutrient availability resulting from intensive grassland management has been shown to reduce the transfer of photosynthetic C below-ground following drought, with implications for the recovery of the soil food web and rates of soil C cycling (Chomel et al. 2022). Furthermore, intensive land management combined with drought has been shown to promote the growth of invasive plants, given their capacity to capitalise on the sizable flush of nutrients on rewetting of nutrient-rich soil (Schrama and Bardgett 2016). Through this mechanism, high N availability and drought likely promote the spread of invasives, with potential impacts on C cycling and storage in soil (Martin et al. 2017). In a similar way, warming in combination with N deposition has been shown to make alpine ecosystems more prone to invasive plants (Bowman et al. 2006, Porter et al. 2013), with potential consequences for below-ground communities and biogeochemical cycles.

Much is still to be learned about how combined global change factors influence below-ground communities, and the consequences for the structure and functioning of terrestrial ecosystems, and C cycle feedbacks. However, it is abundantly clear that combinations of global change factors can amplify, but also suppress or neutralise climate change-driven impacts on soil microbes and their functioning, and C cycle feedbacks. As such, a multifactor experimental approach is needed in future research to better understand how below-ground communities respond to combinations of global change factors and the consequences for above-ground biodiversity, ecosystem functioning, and C cycle feedbacks.

Conclusions

Climate change is a major driver of the composition and functioning of below-ground communities, and their interactions with plants, with potentially far-reaching consequences for biogeochemical cycles and C cycle feedback. Climate change can impact below-ground communities via a wide range of mechanisms, both direct and indirect, and responses among different groups of soil organisms and taxa are highly divergent, which complicates the picture. Below-ground responses to climate change are modified by other global change factors, such as N enrichment and land use, with additive effects being common, with potential for far-reaching consequences for ecosystem functioning. Evidence points to a high potential for climate change to disrupt the many ecological interactions that have evolved over time between plants and soil organisms, including mutualists, pathogens, consumers, and decomposers, with largely unknown, but potentially significant, consequences for ecosystem functioning and the C cycle.

While our understanding of climate change impacts on below-ground communities has advanced greatly over recent years, major research challenges remain. Among these include the need for improved understanding of the relative and interactive effects of warming and CO_2 enrichment on below-ground communities and their links to plants, and of the potential for other global change factors to modify these effects. There is also an urgent need for improved understanding of the influence of extreme climatic events such as drought, floods, fires, and altered freeze–thaw cycles on below-ground communities, which are becoming increasingly common and severe, and of the mechanisms that underpin their capacity to resist and recover from climate extremes. Finally, there is a need for improved understanding of the environmental factors that modify below-ground responses to climate change and a need to integrate this improved understanding into predictive C cycling models and land management for climate mitigation.

CHAPTER 7

Soil biota and sustainability goals

Key messages

- Soil ecology is at the heart of many sustainability challenges, including sustainable agriculture, ecosystem restoration, and climate change adaptation and mitigation.
- The capacity of management interventions to promote soil biodiversity and its contribution to sustainability goals is context dependent.
- Smart combinations of management interventions tailored to local soil and climatic conditions are needed to harness soil biodiversity benefits.
- Plant diversification and its capacity to enhance soil biodiversity and the ecological processes it underpins is of critical importance for meeting sustainability goals.

The previous chapters explored the vast diversity of life in soil, the ecological importance of soil organisms and their interactions with each other, plants, and their consumers, and how these ecological interactions are affected by global change. However, over the last decade there has been a growing awareness of the importance of soil biota for supporting the ecosystem services on which humans depend, such as food production, nutrient retention, and climate regulation. Indeed, soils and their biodiversity contribute to several of the United Nation's Sustainable Development Goals (SDGs) that seek to improve human life and environmental sustainability (Figure 7.1) (Keesstra et al. 2016): soil biodiversity supports food production, which is central to SDG 2's goal of food security and sustainability, and is key to SDG 13, which seeks to combat climate change, and SDG 15 on the restoration of degraded land and biodiversity gain. In fact, because soil biodiversity is intimately linked to the multiple ecosystem services on which society depends, it has been argued that most sustainability goals can be achieved only if soil biodiversity is placed at the centre of land management decisions (Bach et al. 2020).

Recognition of the importance of soil biodiversity for meeting sustainability goals is also reflected in a growing number of global initiatives that seek to put soil biodiversity at the heart of the sustainability agenda. These include the Global Soil Biodiversity Initiative (GSBI), which has the aim of bringing together researchers and policy makers to integrate soil biodiversity knowledge with actions for a sustainable future, and the Soil Biodiversity Observation Network (Soil BON), which seeks to make global observations on soil organisms available to ensure living soil resources are protected and managed to support essential human needs. Soils and their biodiversity are also filtering into policy on soil health, defined as the continued capacity of soil to function as a vital living ecosystem that sustains plants, animals, and humans (Lehmann et al. 2020b). The European Union (EU) Soil Strategy 2030, for example, was set up to combat declining soil health in Europe and includes the development of biological indicators of soil health. In England, incentives are being offered to farmers to protect and restore soils as part of the governments' Environmental Land Management Schemes, which also recognise the importance of soil biodiversity as a key component of soil health. Recognising the importance of soil biodiversity for human well-being, however, is not new. For example, Lady Eve Balfour, the author of *The Living Soil*, penned the line '*the health of soil, animal and man is one*

The Ecology of Soil. Richard D. Bardgett, Oxford University Press. © Richard D. Bardgett (2025). DOI: 10.1093/9780191858710.003.0007

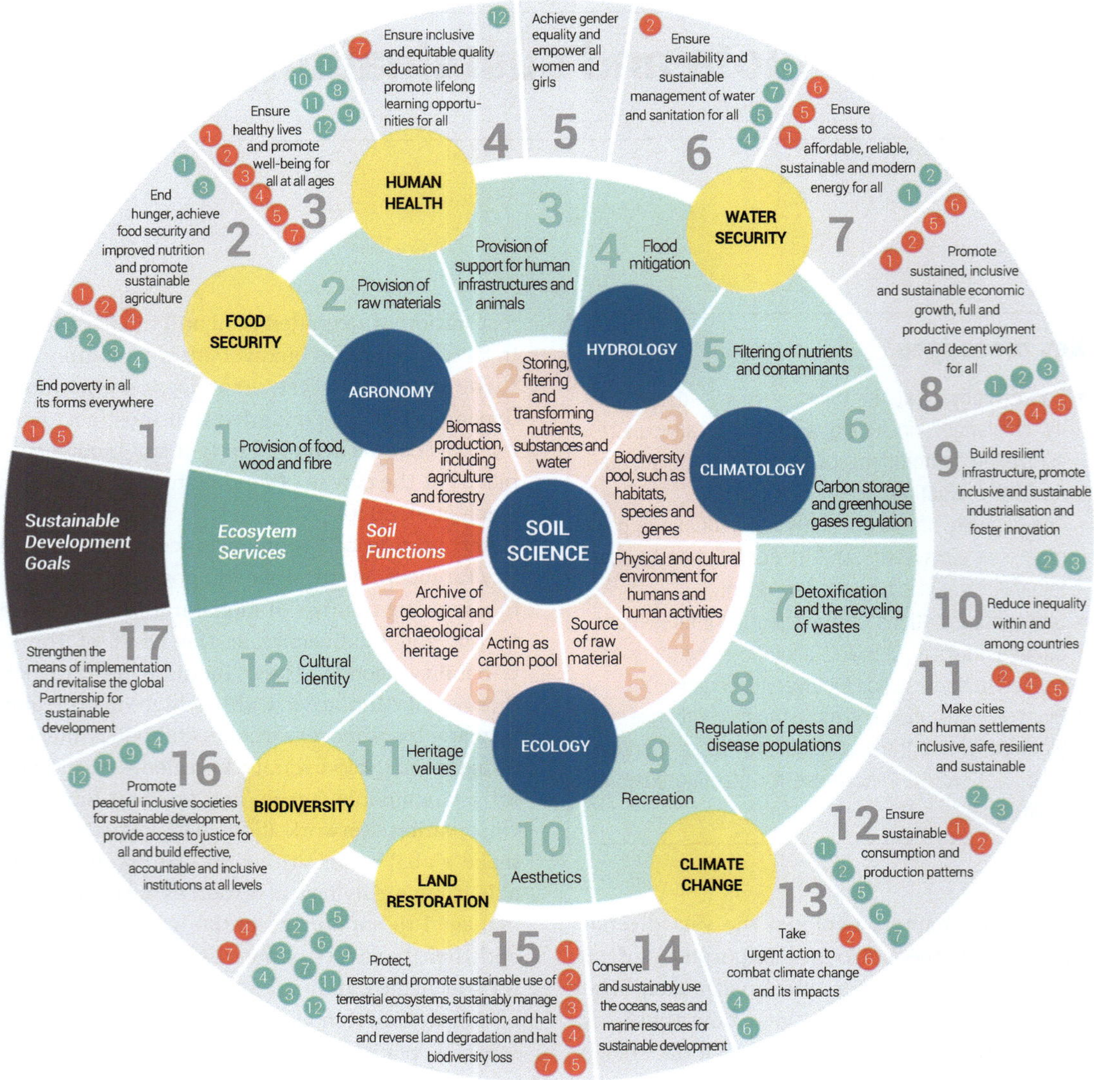

Figure 7.1 Contribution of soils to the United Nations SDGs. From Keesstra et al. (2016).

and indivisible' in 1943 (Balfour 1943), and, as mentioned in Chapter 2, Cleopatra (69–30 BC) declared the earthworm as sacred because of its contribution to soil fertility and Egyptian agriculture.

This last chapter addresses this shift in soil ecology research towards achieving sustainability goals. Specifically, the chapter considers how soil ecological knowledge can be harnessed to address sustainability challenges related to sustainable agriculture, ecosystem restoration, biodiversity gain, and climate change adaptation and mitigation. The main goal of the chapter is to highlight the importance of soil ecological research for the development of future sustainable land management strategies. It does not provide an exhaustive review of the topic; rather it seeks to illustrate some of the ways that soil ecological knowledge can contribute to the sustainability agenda.

Sustainable agriculture

The widespread adoption of intensive agricultural practices, such as continuous cultivation, overreliance on inorganic fertilisers and pesticides, the use of heavy machinery, and overgrazing by livestock, are major causes of soil degradation and environmental pollution, and lead to reductions in both soil biodiversity and the provision of ecosystem services. To address this, it is now widely recognised that sustainable (or regenerative) agricultural practices that focus on enhancement of soil health, which has been positively linked to cropland and grassland productivity (Romero et al. 2024), are urgently needed; central to this is the diversity of life in soil. As discussed throughout this book, complex soil biological communities, and their biotic interactions, regulate plant growth through a variety of mechanisms, including the transformation of organic matter into plant-available forms, the formation of symbiotic associations with roots, and the actions of plant growth-promoting microorganisms that can stimulate plant growth and tolerance to abiotic stress. Soil organisms also play a key role in plant defence, serving as biocontrol agents to soil-borne pests and pathogens, and are intimately linked to the formation of soil structure, which controls the flow of water and nutrients through soil and to plant roots. It therefore follows that the link between soil biota, the soil environment, and agricultural management is key to more sustainable or regenerative agriculture and the delivery of wider ecosystem services, such as nutrient and water retention, erosion control, and soil C sequestration. This section considers ways that land management can be adapted in agroecosystems to harness soil biota, and their associated beneficial functions, via the management of soil, crops, and soil amendments. These interventions are considered individually but will likely be most effective in achieving more sustainable agriculture when they are implemented alongside other measures, such as more precise fertiliser use to match crop needs, the use of soil-protective agricultural machinery, and targeted breeding to enhance nutrient use efficiency of high yielding crops (Gao and Cabrera Serrenho 2023, Gu et al. 2023). As such, smart combinations of interventions should be considered to fully harness the beneficial interplay between soil biota, crop yields, and other ecosystem services in agroecosystems (Hartmann and Six 2023). Many of the agricultural practices considered below also fall under the banner of regenerative agriculture, which lacks a clear definition but is focussed on the regeneration of soil health as an entry point to sustainable food production.

Soil tillage

Soil tillage, defined as the use of mechanical operations to create a favourable soil environment for seed germination and root development, has long been a central element of agricultural management. Yet, conventional tillage, which typically involves repeated annual ploughing and inversion of soil, causes major soil problems, adversely affecting soil structure, SOC content, and biodiversity, ultimately leading to increased soil erosion risk and loss of productive soil. This is because conventional tillage breaks down soil aggregates, disrupts soil pore networks, accelerates the decomposition of organic matter, and destroys fungal hyphal networks that play a key role in the formation of soil structure and stabilisation of SOC. The frequent passage of heavy machinery, which is common to conventional tillage, compacts the soil, which impairs root penetration and the movement of gases and water through soil. It also causes shifts in soil microbial communities, negatively affecting aerobic microbes and plant-associated mycorrhizal fungi, but favouring anaerobic prokaryotes and saprotrophic fungi (Longepierre et al. 2021) and plant pathogenic fungi (Srour et al. 2020). To overcome these problems, no-till or minimum tillage practices (often termed conservation tillage), where seeds are placed in furrows or slots with minimum soil disturbance and crop residues are left on the soil surface, are widely proposed as an alternative to conventional tillage, with benefits for both soil health and production costs.

No-till and reduced tillage can be associated with short-term reductions in crop yield, albeit limited and dependent on factors such as crop and soil type, and climate (Van der Putte et al. 2010). Nevertheless, benefits for soil health can be considerable, including improved soil physical structure (Cooper et al. 2021), soil water retention and infiltration, erosion control (Mondal and Chakraborty 2022), increased rooting density (Mondal et al. 2020), and increased SOC storage, albeit only in surface soil (Haddaway et al. 2017, Cai et al. 2022). Non-till and reduced tillage also have significant benefits for a wide range of functionally important taxonomic groups of soil organisms. An extensive meta-analysis using data from studies carried out across the world, for example, showed that stopping and reducing tillage significantly increases the biomass and abundance of earthworms compared to when soil is inverted by conventional tillage (Figure 7.2) (Briones and Schmidt 2017). These benefits of reduced tillage are most pronounced for surface-dwelling (i.e., epigeic) and deep-burrowing (i.e., anecic) earthworms, including the keystone

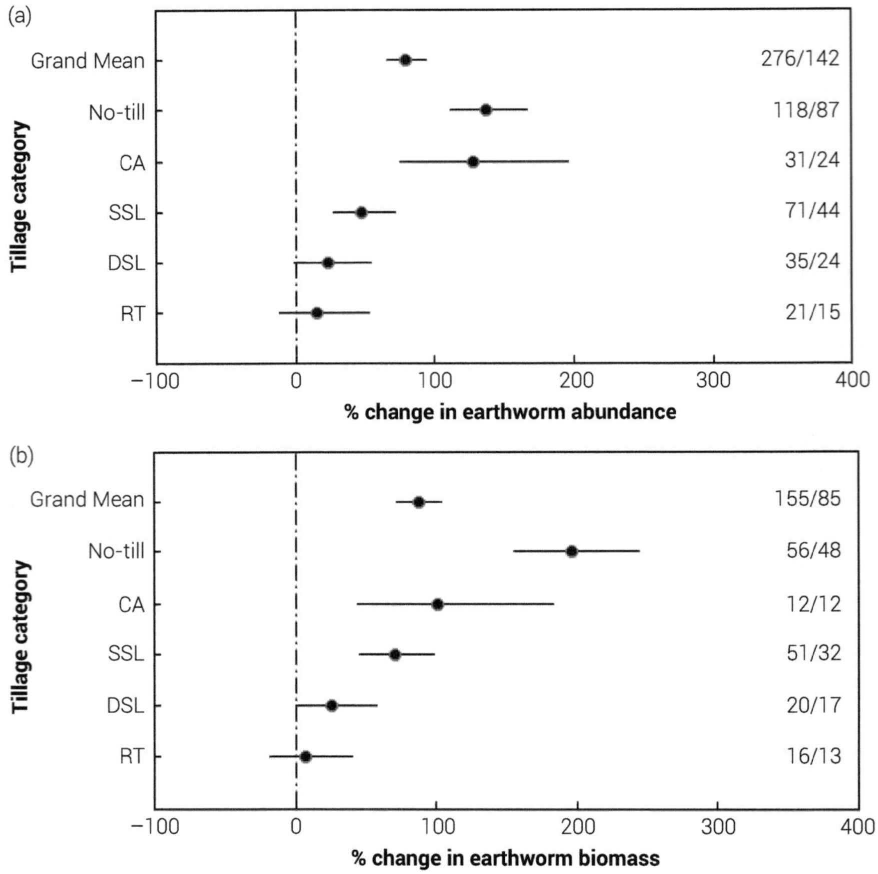

Figure 7.2 Impact of the different reduced/minimal tillage treatments on earthworm abundance (a) and biomass (b) as a percentage of the control (i.e., traditional tillage to 25 cm or more with a mouldboard plough). Reduced/minimal tillage categories include: RT, reduced tillage with shallow ploughing (usually < 15 cm); DSL, deep soil loosening to a depth of > 15 cm, but no soil inversion; SSL, shallow soil loosening where soil is loosened to a depth of no more than 15 cm, without inversion tillage; CA, conservation agriculture, which involves minimal soil disturbance and retention of crop stubble and residues; and no-till, with minimal or no soil disturbance other than a small incision to sow seeds. Data are mean effects and 95% confidence intervals, and sample sizes are on the right of each confidence interval: first the number of control–treatment pairs, followed (after the slash) by the number of studies from which the comparisons were derived. From Briones and Schmidt (2017) by permission of John Wiley and Sons.
Global Change Biology, Volume: 23, Issue: 10, Pages: 4396–4419. First published: 02 May 2017, DOI: (10.1111/gcb.13744).

species *Lumbricus terrestris*, and become stronger both with time (i.e., > 10 years) and when crop residues are left on the soil surface (Briones and Schmidt 2017). Meta-analyses also show that the abundance of other functionally important faunal groups is generally promoted by reduced tillage, including Collembola, mites, enchytraeids, and omnivorous nematodes (Betancur-Corredor et al. 2022), as is the biomass and enzymatic activity of microorganisms (Zuber and Villamil 2016). No-till and reduced tillage also benefit AM fungi, whose hyphal networks are disrupted by continuous tillage (Kabir 2005), oligotrophic prokaryotes that break-down complex organic matter (Babin et al. 2019), and microbial taxa that produce polysaccharides involved in the formation and stabilisation of soil aggregates, thereby improving soil structure (Cania et al. 2020).

It should be noted that conversion to no-till or reduced tillage has consequences for soil microorganisms involved in the production of greenhouse gases, especially denitrifiers that produce gaseous N_2O, a major contributor to N_2O emissions from agricultural land (Ward 2013). Studies report contrasting effects of no-till and reduced tillage on N_2O emissions, but they are generally larger than from conventional tillage systems, at least during initial years of conversion (Six et al. 2004, Van Kessel et al. 2013). This is because increases in soil moisture and the creation of anoxic hot spots from the decomposition of crop residues under no-till or reduced tillage together favour denitrifiers and N_2O production (Hartmann and Six 2023). Reduced tillage has also been shown to create more organic-rich niches in upper soil layers for fungal denitrifiers, increasing their diversity, abundance, and overall potential for fungal denitrification and gaseous N loss (Bösch et al. 2022). In the longer term, however, improvements in soil structure resulting from reduced tillage increase oxygen diffusion and soil drainage, thereby alleviating anoxic conditions and reducing N_2O emissions (Six et al. 2004, Van Kessel et al. 2013). Nevertheless, there remains much uncertainty over how and why no-till or reduced tillage impact denitrifier communities and N_2O emissions, and responses vary considerably depending on soil texture, crop type, climate, and fertiliser use (Van Kessel et al. 2013). Finding ways to constrain denitrifying communities and associated N_2O emissions in reduced tillage systems, for instance, via matching of fertiliser application to crop needs, adoption of precision farming for spatial planning of fertiliser use, and deep placement of fertiliser N, therefore represents an important challenge for future sustainable agriculture and climate mitigation, especially given the high global warming potential of N_2O and the fact that agriculture is one of its biggest sources (Paustian et al. 2016).

Crop management

Intensively managed crop monocultures have become a dominant feature of agricultural landscapes worldwide and are frequently associated with soil degradation, increased pest and disease risk, and environmental pollution. As an alternative to crop monocultures, it has been proposed that diversifying cropping systems through traditional practices of mixed planting (i.e., intercropping) and the rotation of crops could be an effective route to delivering sustainable agriculture, with additional benefits for soil health (Brooker et al. 2015, Isbell et al. 2017, Tamburini et al. 2020). As discussed in Chapter 4, higher levels of plant diversity in natural and semi-natural settings can increase plant productivity and nutrient recycling, with potential, albeit variable, effects on the abundance and functioning of soil biota at a local scale (De Deyn et al. 2011, Eisenhauer et al. 2013, Lange et al. 2015, Chen et al. 2019a). Similarly, in agricultural settings, there is growing evidence that, relative to monocultures, crop diversification practices can benefit a wide range of ecosystem services, including soil nutrient cycling and soil C storage, without compromising yield (Tamburini et al. 2020).

One of the most common forms of crop diversification is the growing of two or more crops at the same time and in the same place, termed intercropping (Figure 7.3), which is often associated with increases in crop yield, or overyielding (Li et al. 2021b). This overyielding is largely attributed to below-ground facilitative interactions between neighbouring crops that enhance soil resource use (Figure 7.4) (Li et al. 2014, Brooker et al. 2015) and associated increases in soil fertility that become stronger with time (Cong et al. 2015a).

Figure 7.3 Intercropping is often associated with increased crop yields. These images show hand-harvesting of intercropped potatoes (foreground) and garlic in China (a), pea and barley intercropping in Scotland (b), and wheat and maize intercropping in China (c). Images by Rob Brooker.

For example, crops that mobilise soil P by releasing phosphatases from their roots can enhance P nutrition for themselves, but also for neighbouring crops that cannot mobilise P (Li et al. 2014). Also, growing N-fixing legumes, such as faba bean (*Vicia faba*), alfalfa (*Medicago sativa*), and soyabean (*Glycine max*), alongside cereals such as maize can boost total crop yields by increasing N supply via N_2 fixation (Cong et al. 2015b, Duchene et al. 2017). Moreover, the release of exudates by maize roots has been shown to stimulate the activity of N_2 fixing bacteria in the rhizosphere of neighbouring faba beans and trigger the expression of genes related to nodulation and N_2 fixation in roots, thereby increasing N supply and total crop yield (Li et al. 2016, Hu et al. 2021a). Relative to monocultures, intercropping can also increase the abundance of earthworms (Schmidt et al. 2003), which are known to contribute significantly to global food production (Fonte et al. 2023), and the dominance of fungi relative to bacteria (Morugán-Coronado et al. 2022), with benefits for soil nutrient retention (De Vries and Bardgett 2012).

Another common practice of crop diversification is crop rotation, or the growing of a sequence of different crops in the same place over successive growing seasons. Crop rotation has been practiced for millennia as a means of keeping crop-specific pathogens in check and boosting soil fertility. Yet, it is gaining renewed interest, given its promise for enhancing crop yields and crop resilience to extreme weather conditions, whilst also reducing negative impacts of crop production on the environment (Bowles et al. 2020, Volsi et al. 2022, Smith et al. 2023). Additionally, evidence is mounting that increasing the diversity of crops in rotations (i.e., crop rotational diversity) can enhance grain yields, and that associated yield benefits increase over time and with low fertiliser N inputs, thereby reducing environmental pollution associated with high fertiliser use (Smith et al. 2023). A range of studies report beneficial effects of increasing crop rotational diversity on below-ground communities, with reports of increases in soil microbiome complexity and multifunctionality (Yang et al. 2023), disease suppressive bacteria (Peralta et al. 2018), bacterial metabolic activity (D'Acunto et al. 2018), decomposition of crop residues (McDaniel et al. 2014), and shifts in soil microbial communities related to increases in both soil aggregation and the C and N content of soil (Tiemann et al. 2015). A 6-year field study on the North China Plain, the food basket of

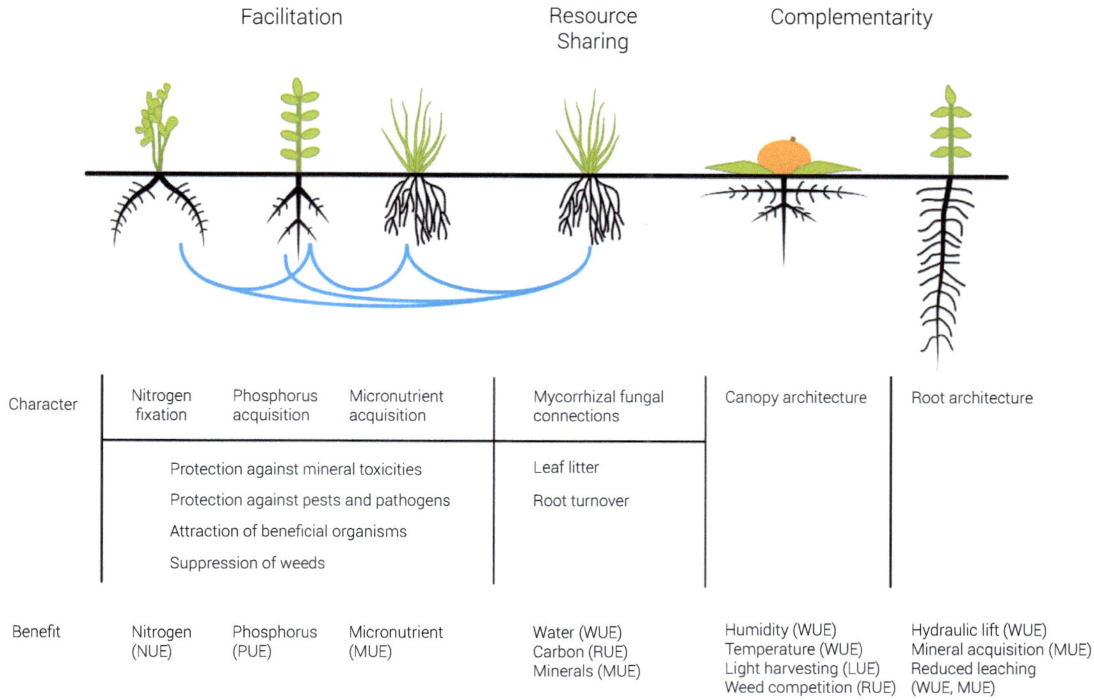

Figure 7.4 Mechanisms explaining greater yields of mixed cropping systems compared to monocultures, including facilitation, resource sharing, and niche complementarity. Facilitation occurs when mixing of crops increases the availability of water, phosphorus, micronutrients (e.g., Fe, Zn, Cu), or N (via N_2 fixation). Resource sharing operates via common mycorrhizal networks or nutrient recycling through leaf senescence and root turnover. Niche complementarity allows for maximal exploitation of soil resources when crop species with shallow and deep roots grow next to each other. These benefits lead to enhanced resource-use efficiencies for phosphorus (PUE), nitrogen (NUE), mineral nutrients (MUE), water (WUE), light (LUE), and assimilates (RUE). From Brooker et al. (2015) by permission of John Wiley and Sons. *New Phytologist*, Volume: 206, Issue: 1, Pages: 107–117. First published: 03 November 2014, DOI: (10.1111/nph.13132).

China, showed that diversified cropping rotations increased crop yields and soil health, measured using a soil health scoring method (i.e., Cornell Soil Health Assessment), and reduced soil N_2O emissions due to reduced fertiliser N use (Yang et al. 2024). As such, this study demonstrates how crop diversification can serve as an effective strategy for sustainable food production and improving soil health in one of the most intensively cultivated regions of the world.

Yet, when considered across a wide range of sites in Europe, the effects of increasing crop rotational diversity on soil microbial diversity and multifunctionality were found to be relatively weak (Garland et al. 2021). Increasing the number of crop species within a crop rotation was found to increase bacterial diversity across sites, but not that of fungi or archaea. Moreover, beneficial effects of crop species on bacterial diversity occurred between five and ten crop species, as they also did for crop yield, suggesting that the inclusion of more crop species in a crop rotation may no longer have benefits (Figure 7.5a). However, this study did show that the proportion of time a field was planted with a crop (i.e., crop cover time) over a 10-year crop rotation, including cash crops, cover crops, and grass–clover forage leys, had a much larger positive impact on bacterial diversity, soil multifunctionality, and crop yields, than does crop rotational diversity (Figure 7.5b). This suggests that there is potential to increase soil microbial diversity and multifunctionality by increasing the duration of crop cover in crop rotations, especially through the planting of cover crops and/or temporary grass–clover forage leys, which are known to benefit below-ground communities and their functioning, help reduce soil erosion and

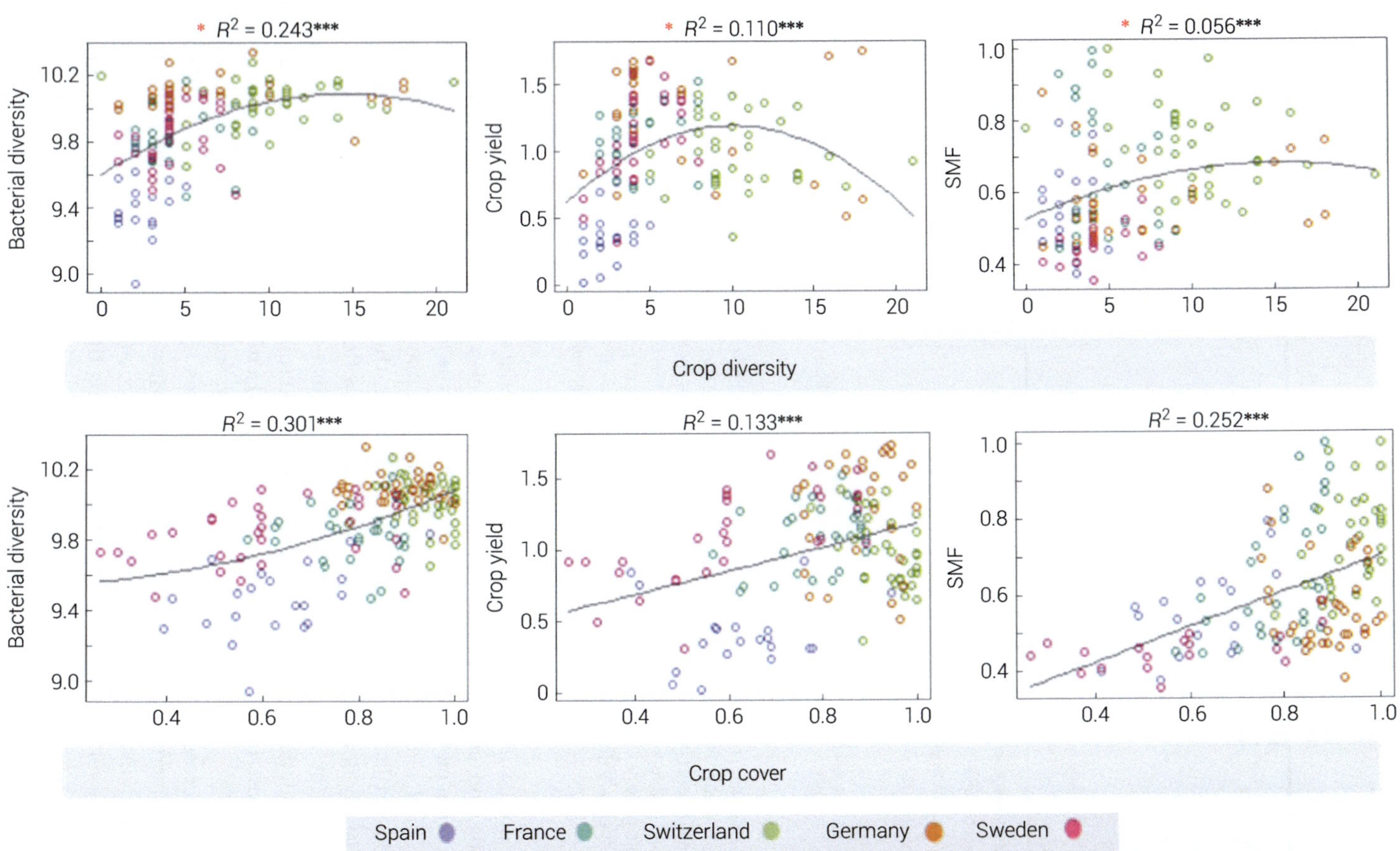

Figure 7.5 Impact of increasing the number of crop species within a crop rotation (i.e., crop diversity) (a) and crop cover time (i.e., proportion of time during a 10-year crop rotation planted with a cover crop) (b) on soil bacterial diversity, crop yield, and soil multifunctionality (SMF), measured in 155 cereal fields across five European countries. Soil bacterial diversity was measured using the Shannon diversity index, whereas crop yield is expressed as a standardised yield for each site using the grain yield divided by the average yield of that crop species across all five countries. From Garland et al. (2021) by permission of Springer Nature.

nutrient leaching, and improve soil structure (Garland et al. 2021, Hartmann and Six 2023). These findings suggest that benefits of increasing crop rotational diversity for below-ground communities and their functioning are context dependent, varying with crop type, climate, and historical management. (Garland et al. 2021).

As a final note on crop management, there has been much recent focus on the potential to harness beneficial plant–microbial interactions, especially those involving the microbiome associated with plant roots via crop engineering. As discussed in Chapters 2 and 3, a vast array of microbes associate with plant roots in the rhizosphere—the rhizosphere microbiome—that together function in a way that is analogous to the human gut microbiome, providing similar functional roles to their hosts (e.g., growth promotion, nutrient uptake, stress tolerance, and suppression of disease) (Trivedi et al. 2022). However, plant domestication has led to the selection of plant traits for increased crop yield, at the expense of beneficial interactions with plant-associated microbiomes, such as the rhizosphere microbiome (Figure 7.6) (Nerva et al. 2022). As such, modern crop cultivars typically benefit less from interactions with root-associated microbiomes, including mycorrhizal symbiosis, than do traditional cultivars and their wild ancestors (Hetrick et al. 1993, Sangabriel-Conde et al. 2014), especially when under high fertilisation (Martín-Robles et al. 2018). In response, there is now growing interest in finding ways to mitigate this trend and tailor the rhizosphere microbiome of crops to carry out functions that benefit the host plant, and hence promote sustainable yields.

An area of promise is breeding programmes targeted at root traits, including root architectural traits and root exudation, that promote the recruitment of beneficial rhizosphere microbiomes. Also, the development of new plant engineering techniques offers potential to manipulate genes for specialised microbiome assembly and/or enhanced

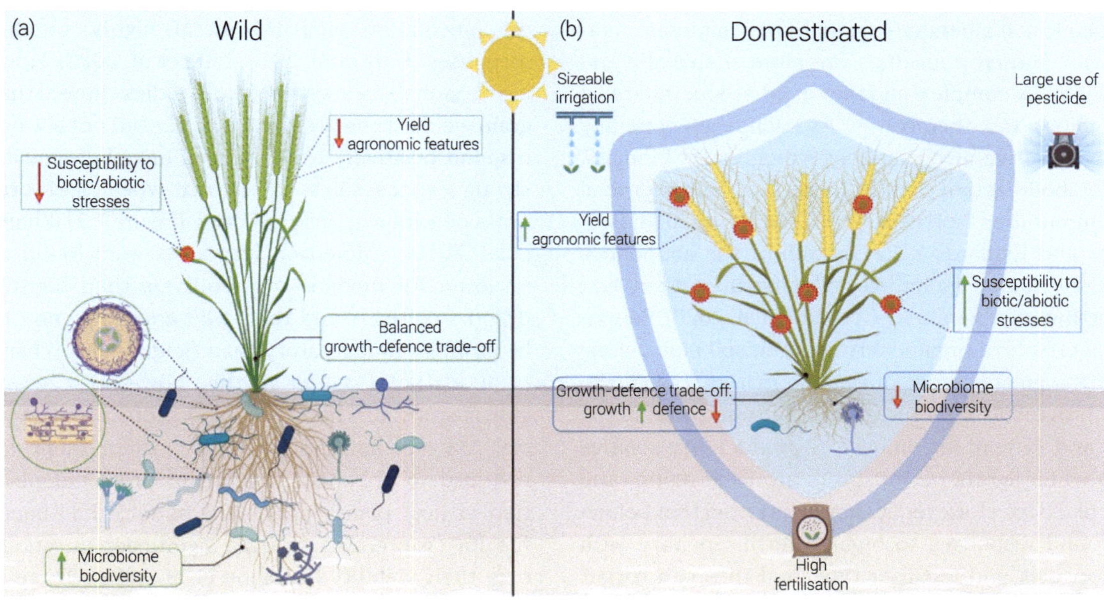

Figure 7.6 Comparison between wild and domesticated plant species. Wild relatives display better adaptation to environmental stresses (both biotic and abiotic), a balanced growth–defence trade-off, and a rich associated rhizosphere microbiome, but a low yield (a). Domesticated plant shows an improved productivity, but a reduced stress resilience and ability to recruit plant-associated microbes, and hence management interventions, such as irrigation, fertilisation, and pesticide application, are needed (b). From Nerva et al. (2022) by permission of Elsevier.
https://www.sciencedirect.com/science/article/pii/S136013852200156X.

microbial functioning, as does microbiome engineering, which involves the artificial selection of microbial communities with specific effects on host fitness (Trivedi et al. 2022). These approaches are in their infancy and, to be effective, a major research effort is needed to increase understanding of the mechanisms governing the selection of beneficial rhizosphere microbiomes, and to translate this knowledge into crop breeding for sustainable agriculture.

Soil amendments

Synthetic fertilisers have long been a mainstay of intensive agriculture, with major benefits for crop yields, but significant environmental costs such as reduced water and air quality (Gu et al. 2023). When used in high amounts, synthetic fertilisers can have detrimental effects on the diversity and functioning of the soil biota (De Vries et al. 2012a, Yang et al. 2020, 2022, Ma et al. 2021), effectively bypassing their role in the recycling and supply of nutrients to plants. In contrast to synthetic fertilisers, organic soil amendments, including manures, composts, and crop residues, are more chemically and physically complex, and they also harbour microbial communities themselves. As such, they generally have positive, albeit variable, effects on the biomass, metabolic activity, and diversity of soil microbial communities (Lori et al. 2017, Luo et al. 2018, Bebber and Richards 2022) as well as the abundance of soil fauna, including collembolans, mites, and earthworms (Betancur-Corredor et al. 2023). Beneficial effects of organic fertilisers on soil biota, along with a greater input of organic matter to soil, can improve soil structure (Hartmann and Six 2023), C and N content, and crop productivity relative to when synthetic fertilisers are used alone (Luo et al. 2018). However, it should be noted that belowground responses to organic fertilisers vary with crop, soil, and fertiliser type, and they can trigger increases in CO_2 and N_2O emissions from soil (Lazcano et al. 2021), albeit to a lesser extent than from soils receiving synthetic fertilisers (Skinner et al. 2019).

Another kind of soil amendment that is attracting much attention as a means of increasing crop yields while reducing reliance on agrochemicals is the use of microbial inoculants. These microbial inoculants are the focus of considerable investment from agrochemical companies and act through a range of mechanisms that can be broadly classified as those that: (i) increase plant nutrient availability, for example through the action of inoculants of P solubilising bacteria and fungi (termed plant growth promoting rhizobacteria or fungi), and symbiotic/free-living N-fixing bacteria; and (ii) improve tolerance to abiotic stresses, for example, through the production of phytohormones and other secondary metabolites that cause changes in root morphology, thereby improving plant–water use and tolerance to drought (Li et al. 2023), and protect plants against soil-borne pathogens (Li et al. 2024a). Regarding their effectiveness, a recent global-scale meta-analysis revealed that microbial inoculants generally have beneficial effects on crop yields and nutritional quality, although inoculant performance varies considerably with microbial inoculant species and crop type (Li et al. 2024a). Field experiments and meta-analyses on crop responses to AM fungal inoculation similarly reveal highly variable responses (Hart et al. 2018, Lutz et al. 2023). However, a global-scale synthesis of studies showed that grain yields of cereal crops generally, but not always, respond positively to AM fungal inoculation, with strong responses in sorghum, rice, wheat, and corn, but a neutral response in barley (Figure 7.7) (Zhang et al. 2019c). Moreover, responses were found to be lower for modern crop cultivars than for traditional cultivars and their wild ancestors, and to be stronger in laboratory than field studies (Zhang et al. 2019c). Counter to this, however, a recent global study of the effectiveness of 28 commercial AM inoculants found that most of them did not lead to enhanced mycorrhizal root colonisation of host plants and only a handful had benefits for crop growth, raising significant questions over their viability (Salomon et al. 2022). There is clearly much uncertainty over the effectiveness of AM fungal inoculants in agroecosystems, and much remains to be learned about the causes of variable responses, including why, in many cases, they fail (Hart et al. 2018, Zhang et al. 2019c). It has also been argued that the potential of AM fungal and other microbial inoculants needs to be placed in a

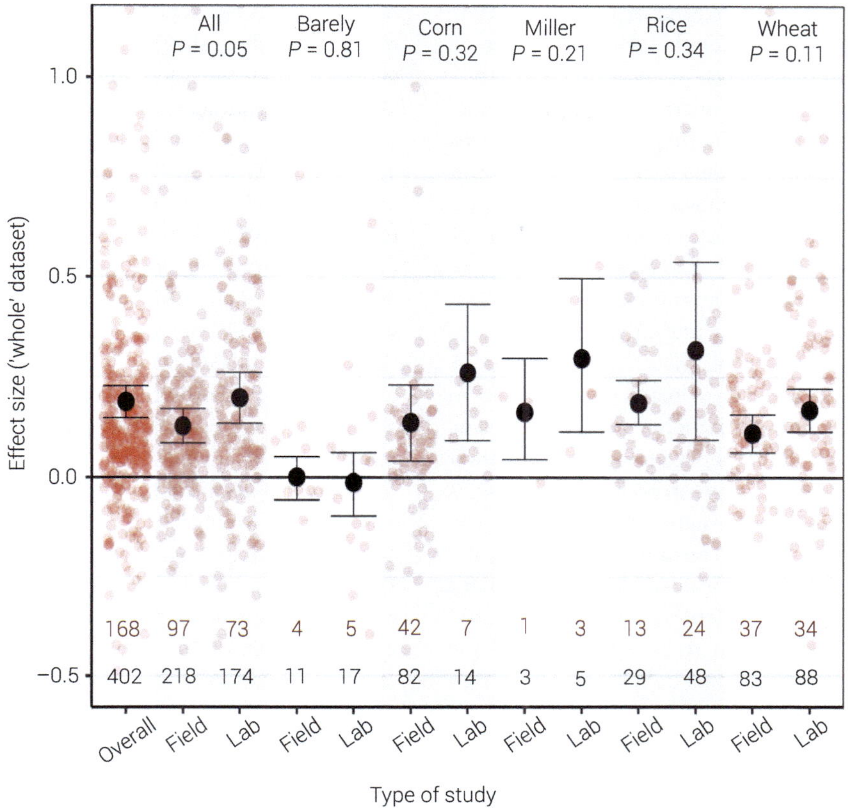

Figure 7.7 Meta-analysis of studies reporting AM fungal effects on grain yields of the whole data set (i.e. all) and of five investigated crops: corn, wheat, rice, barley, and millet. Effects are means and 95% confidence intervals, and study and trial numbers are in red and black fonts, respectively. From Zhang et al. (2019c) by permission of John Wiley and Sons.

broader agronomic context, considering the role of crop genotype, other soil biota, and environmental conditions, including soil type and climate (Ryan et al. 2019).

A particular concern with microbial inoculants relates to how they impact resident soil microbial communities and their functioning. A recent literature synthesis, for example, showed that out of 108 studies analysed, 86% found that microbial inoculants modified the resident soil microbial community in some way, and, in many cases, it did not return to its initial state in the longer term; in other words, inoculation impacts on resident microbial communities were persistent, even if inoculum survival was short-lived (Mawarda et al. 2020). Similarly, another quantitative literature synthesis reports that microbial inoculants significantly affect the diversity of resident soil microbial communities, albeit in variable ways and with stronger effects on bacterial than fungal communities (Cornell et al. 2021), triggering major shifts in structure and bacterial composition of soil microbial communities, while increasing microbial biomass (Li et al. 2024a). The mechanisms explaining impacts of microbial inoculants on resident microbial communities and their functioning are poorly understood, as are their legacy effects on soil functions, but they likely involve: (i) competition for resources, especially if the invader species has traits that make them competitively superior to resident species; (ii) direct antagonism, especially through antibiosis, i.e., the secretion of chemical compounds that kill or inhibit resident microbes; and (iii) indirect effects via changes in root growth and morphology, and patterns of root exudation, induced by microbial inoculants (Mawarda et al. 2020). However, very little is known about how, why, and under what circumstances microbial inoculants impact resident soil microbial communities and their functioning. It also remains unclear what the legacy effects of changes in resident microbial communities are for other soil trophic groups, above-ground biodiversity, and ecosystem services, such as SOC sequestration and nutrient recycling. Given these uncertainties, and difficulties in monitoring impacts of microbial inoculants in the field, the primary focus of microbiome engineering in agroecosystems should arguably be on promoting beneficial native microbial communities and their functioning, rather than on introducing new species with unpredictable performance and poorly understood non-target effects (Hartmann and Six 2023).

Biochar is another form of soil amendment that has gained much attention for improving soil health, whilst also helping to offset anthropogenic CO_2 emissions via increased SOC sequestration (Lehmann et al. 2021). Biochar is produced by high temperature pyrolysis of organic raw materials (e.g., wood, crop residues, organic wastes, and manures) in an oxygen-depleted environment. Being relatively recalcitrant and C rich, it has been suggested soil amendment with biochar can boost long-term SOC storage, whilst also bringing co-benefits for soil health, including improved soil aggregation, nutrient cycling, and water retention, which in turn boost crop yields. Recent meta-analyses of experimental studies support this, showing that, on average, biochar addition increases SOC (Bai et al. 2019), soil microbial biomass and the activity of extracellular enzymes involved in N and P cycling (Zhang et al. 2019a, Schmidt et al. 2021), soil N transformations and plant N uptake (Zhang et al. 2021), and crop yields, especially on acidic soils (i.e., pH < 5) (Jeffery et al. 2011). However, it should be noted that while these meta-analyses point to general benefits of biochar for soil properties and crop yields, responses among individual studies are highly variable, with many reporting neutral and even negative effects (Schmidt et al. 2021). The factors causing this variability are poorly understood, but likely relate to the type of biochar and application method, whether it is applied with fertiliser or not, soil and crop type, and historic land management (Schmidt et al. 2021). Moreover, while biochar generally increases soil microbial biomass, available evidence indicates that impacts on soil microbial diversity are highly variable (Li et al. 2020b), as are impacts on soil animals, which vary greatly with crop and soil type, and among taxa (McCormack et al. 2013). Biochar clearly has promise as a means of improving soil health, SOC storage, and crop yields, but the variable and context-dependent nature of its effects on soils suggest that more field tests are needed before firm recommendations on its use as an effective strategy for sustainable agriculture (and climate change mitigation) can be made.

A final mention should go to enhanced weathering, which involves the addition of crushed silicate rock (usually basalt) to agricultural soil (Figure 7.8). The main motivation for enhanced weathering in agricultural soils is its promise for C removal (Beerling et al. 2020). This is because accelerated chemical breakdown of crushed rock minerals by carbonic acid in soil solution leads to the release of base cations (e.g., Mg^{2+} and Ca^{2+}) and conversion of atmospheric CO_2 into bicarbonate ions (HCO_3^-) and carbonate anions (CO_3^{2-}). These stable dissolved inorganic alkaline forms are then flushed into rivers and the oceans, where they form a stable C reservoir (Beerling et al. 2020). The C removal potential of enhanced weathering was recently demonstrated in field trials in the US Corn Belt, where the application of crushed basalt over 4 years (50 t ha^{-1} $year^{-1}$) across a maize–soya crop rotation gave a conservative cumulative CO_2 removal potential of 10.5 ± 3.8 t CO_2 ha^{-1} (Beerling et al. 2024). Moreover, this was associated with increases in maize and soya yield and improved nutritional status, which was attributed to decreased soil acidification and increased soil fertility, in the form of increased availability of mineral nutrients for crops (Figure 7.9) (Beerling et al. 2024). It should be noted, however, that while enhanced rock weathering has potential to contribute to climate mitigation through C removal (Kantzas et al. 2022), its large-scale adoption does come with risks. These include environmental risks associated with the expansion of mining operations and the use of energy to crush and transport rocks, and the potential washing

Figure 7.8 Field trials in North Queensland, Australia, testing for the effects of crushed basalt application on sugarcane yields, soil fertility, and C sequestration. Image by Paul Nelson, James Cook University in Australia, part of the Leverhulme Centre for Climate Change Mitigation.

of silicate rock into rivers and coral reefs, with unknown impacts on aquatic biodiversity (Edwards et al. 2017). There is also a need for further testing of the longer-term potential co-benefits of enhanced weathering for climate mitigation, soil fertility, and crop yields across a range of soil and climatic conditions, including assessments of impacts on native soil organisms and their functioning, which remains largely unexplored.

Ecosystem restoration

The United Nations declaration of 2021–2030 as the UN Decade on Ecosystem Restoration has brought much global attention to the challenge of restoring degraded ecosystems (Figure 7.10). Ecosystem restoration is central to meeting national, regional, and global targets for biodiversity gain and climate mitigation (Strassburg et al. 2020). A key element of this is the restoration of degraded soils: recent estimates indicate that 33% of Earth's soils are moderately to highly degraded (FAO 2015), with significant amounts of soil being lost by erosion (Borrelli et al. 2017, Thaler et al. 2021). Indeed, it has been estimated that, globally, between 25 and 40 billion tonnes of soil are lost annually to soil erosion (Quinton et al. 2010) and that rates of human-induced soil erosion outpace rates of soil formation by more than an order of magnitude (Montgomery 2007). It was recently estimated that some 35% of soils across the US Corn Belt have lost their surface

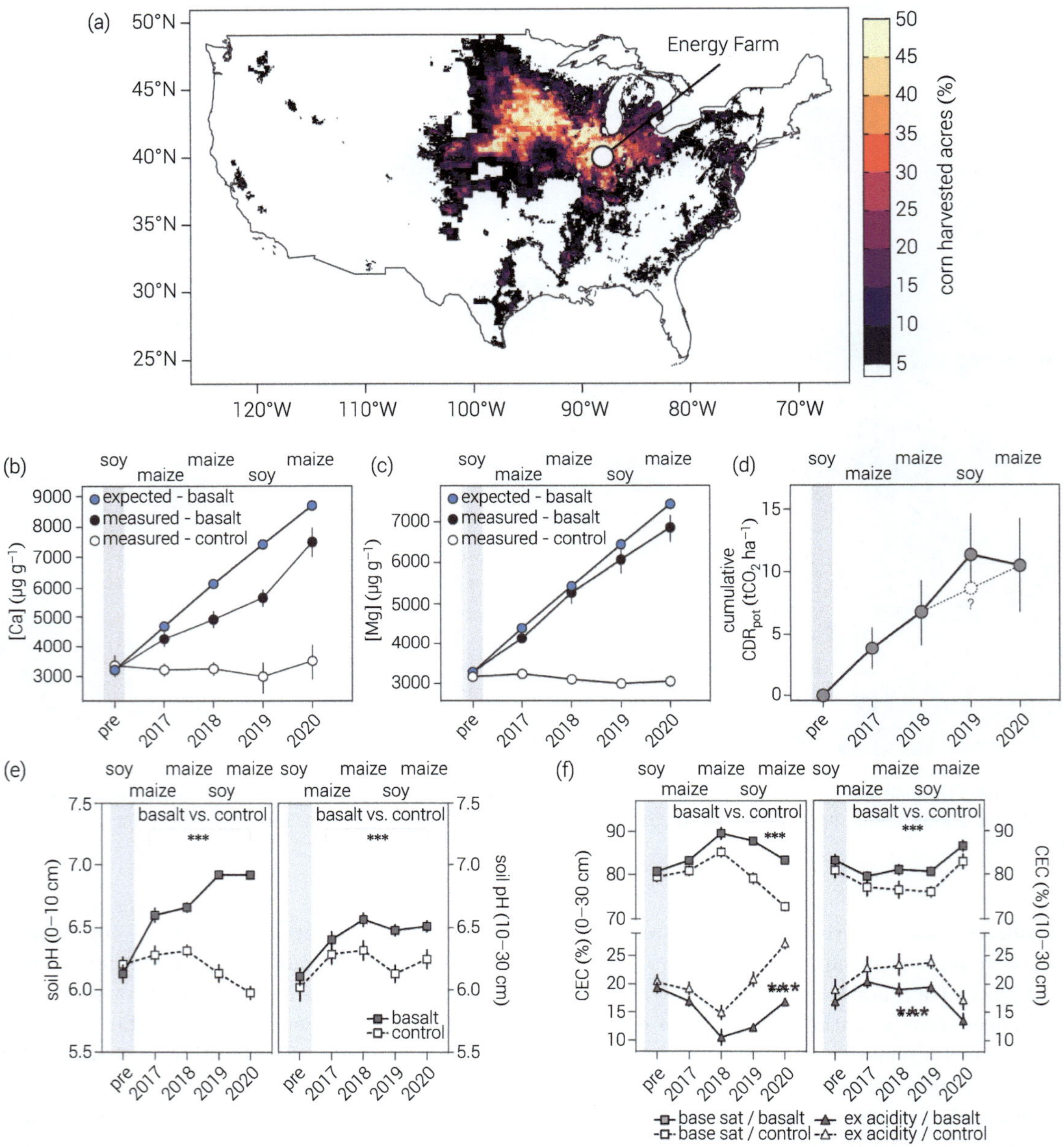

Figure 7.9 Carbon dioxide removal (CDR) potential and changes in soil biogeochemistry in response to enhanced weathering across a 4-year maize–soybean rotation in the US Corn Belt. (a) Harvested corn acres across the US Corn Belt, shown as regional percentage area (using data from Ray et al. 2012). The location of the field trial is shown with an open circle. Time-series of expected and observed soil Ca (b) and Mg (c) cation concentrations in treated and control plots. (d) Cumulative CDR potential (CDR_{pot}) over the 4-year (2016–2020) field trial, with the grey bar showing the pretreatment period (pre). '?' denotes a typical CDR_{pot} rate for maize in 2019. In response to enhanced weathering, soil pH (e) increased significantly at 0–10 cm and 10–30 cm depths, and soil cation exchange capacity (CEC) increased (f). Error bars are ± SEM, and statistical results are for repeated measures two-way ANOVAs with basalt vs. control as the main factor; asterisks indicate significant difference (***$P < 0.001$). From Beerling et al. (2024).

Figure 7.10 Soil degradation caused by a combination of overgrazing and climate change, especially recurring droughts and heavy rainfall events that trigger soil erosion. Images are from Golog, Qinghai-Tibetan Plateau (a), Lolldaiga Hills, Kenya (b), and Nyando, Kenya (c, d). Images by Richard Bardgett.

A-horizon to erosion, leading to declines in crop yield (Thaler et al. 2021). Erosion has been shown to be causing widespread thinning of agricultural soils globally, and under worst case scenarios, it is expected to lead to the complete loss of the most vulnerable soils (Evans et al. 2020). Moreover, studies show that soil degradation has significant negative impacts on soil microbial communities, reducing their diversity, network complexity, and ecosystem multifunctionality (Qiu et al. 2021, Luo et al. 2023). As such, some argue that soil degradation should be added to the planetary boundaries framework (Rockström et al. 2009), which identifies safe operating space for humanity based on nine Earth system processes; at present, soil and the enormous diversity of life that lives within it are not explicitly recognised within this framework (Kraamwinkel et al. 2021).

The causes of soil degradation are complex, but the principal drivers are intensive agricultural practices, including heavy cultivation, fertilisation, irrigation, overgrazing, deforestation, urbanisation, soil contamination, and climate change. What is clear is that soil biodiversity plays a central role in the restoration of degraded soils and the biodiversity and functioning of the ecosystems they support; soil is hence at the front line for meeting sustainability goals related to ecosystem restoration and biodiversity gain. This section considers recent advances in our understanding of how soil biological communities and their functioning, and interactions between plants and soil biota, can be harnessed for ecosystem restoration to meet biodiversity conservation and sustainability goals.

Priority effects and plant–soil feedback

Degraded soils are often depleted in organic matter and nutrients, have poor structure, are compacted, and, in some cases, are contaminated with pollutants. As such, the introduction of plants, by sowing or planting, that can tolerate these harsh soil conditions, and have subsequent positive effects on the reassembly and functioning of above- and below-ground communities, is key to restoration success (Weidlich et al. 2021). Of relevance here are priority effects, which in ecology refer to the order or timing of plant species arrival, or 'who arrives when' (Fukami 2015), and subsequent 'carry over' effects on the reassembly and functioning of communities after degradation (Weidlich et al. 2021). There are several examples of where priority effects have been successfully used to steer restoration in particular situations. These include the initial planting or seeding of fast-growing legumes in disused mine sites, which help to stabilise bare soil, reduce soil erosion, and increase soil N availability, thereby enhancing the subsequent growth and establishment of non-leguminous plants (Daws et al. 2019). Also, in a grassland experiment, the sowing of legumes before grasses consistently increased root productivity of the successive plant community, suggesting that below-ground effects can be key drivers of priority effects and their contribution to restoration (Weidlich et al. 2018). However, priority effects are not always positive: they have been shown to facilitate invasive plant species more strongly than natives, thereby impeding restoration efforts (Weidlich et al. 2021), and in a cross-site restoration experiment, priority effects of early-colonising plant species suppressed the growth and diversity of slow-growing target plant species typical of species-rich grasslands of high nature conservation value (Fry et al. 2017).

Priority effects are relevant to the use of nurse plants in restoration, which are widely used to facilitate the growth and establishment of target plant species planted or sown close to them, especially in harsh environments (Padilla and Pugnaire 2006, Weidlich et al. 2021). The mechanisms underpinning facilitative effects of nurse plants on close-by plants are varied, but mainly involve improvements in water and nutrient availability in the vicinity of the nurse plant and protection from herbivores (Padilla and Pugnaire 2006). For example, locally adapted nurse plants have been used to restore heavy metal polluted mine tailings by improving SOM and microbial-mediated processes of nutrient cycling, which, in turn, promotes the establishment of other plant species and speeds up restoration (Navarro-Cano et al. 2018). Similarly, the tussock-forming Alpha grass (*Stipa tenacissima*), which is common across southern Europe and northern Africa, has been shown to help restore degraded semi-arid steppe by creating a favourable soil environment, with higher SOM content and water availability for the growth of native shrubs (Maestre et al. 2001).

The mechanisms underpinning priority effects in restoration, especially those involving interactions between plants and soil biota, remain elusive, and vary depending on factors such as the identify of nurse and target plant species and the cause of ecosystem degradation. However, evidence is emerging that plant–soil feedbacks can play an important role in driving community reassembly during restoration both above- and below-ground. On former agricultural land taken out of production for nature conservation, for example, initially abundant, early successional plant species were found to display strong negative feedback from their soil microbial communities, which contributed to their replacement by late successional plant species (Kardol et al. 2006, Bauer et al. 2015). Further, soils taken from late successional grasslands have been shown to promote late successional plant species, suggesting that shifts in soil microbial communities over time favour these species, thereby speeding up habitat restoration (Kardol et al. 2006, Bauer et al. 2015). This is consistent with the finding that, over time, pathogenic fungi are replaced by beneficial AM fungi on former agricultural land, providing a potential mechanism for the switch from negative to positive feedback during succession and habitat restoration (Hannula et al. 2017).

Other evidence of priority effects being driven by plant–soil feedback comes from studies of degraded fens, where conditioning of degraded fen soil by plant species from intact fen meadows benefited the subsequent growth of desirable fen species more than when soils were conditioned with species from degraded fens (Brinkman et al. 2017). As such, it was proposed that growing typical fen meadow plant species in degraded fen soil led to priority effects that promoted the growth of other fen meadow species, thereby improving restoration outcomes (Brinkman et al. 2017). While these studies point to plant–soil feedbacks playing a key role in ecosystem restoration, more research is needed to identify plant species and traits that modify microbial communities of degraded soils in a way that facilitates the subsequent growth of target plant species and ecosystem functions and hence improves restoration success (Weidlich et al. 2021).

Biodiversity–productivity relationships

There is a vast body of literature demonstrating that plant diversity can enhance multiple ecosystem functions, both above- and below-ground, and the resistance of plant production to environmental perturbations, such as climate extremes (Tilman et al. 2014, Isbell et al. 2015, Jochum et al. 2020). This suggests that there is potential to apply knowledge on biodiversity–function relationships to the restoration of degraded ecosystems and regeneration of soil via creative uses of plant functional diversity. This idea is supported by a growing number of studies showing that higher plant diversity can help to regenerate the biodiversity of degraded soils as well as increase soil C storage, thereby contributing to climate mitigation. The planting of high diversity grassland mixtures on ex-agricultural degraded soil in the USA, for example, was found to increase soil nutrient and C stocks more that when the same plant species grew in monocultures (Furey and Tilman 2021). Moreover, this effect was found to be most pronounced when mixtures contained a full range of plant functional groups (i.e., legumes, grasses, and forbs) (Figure 7.11), suggesting that plant functional biodiversity is key to regenerating degraded soil (Furey and Tilman 2021). Experimental studies similarly show that high diversity grassland can increase plant C capture and allocation to roots and

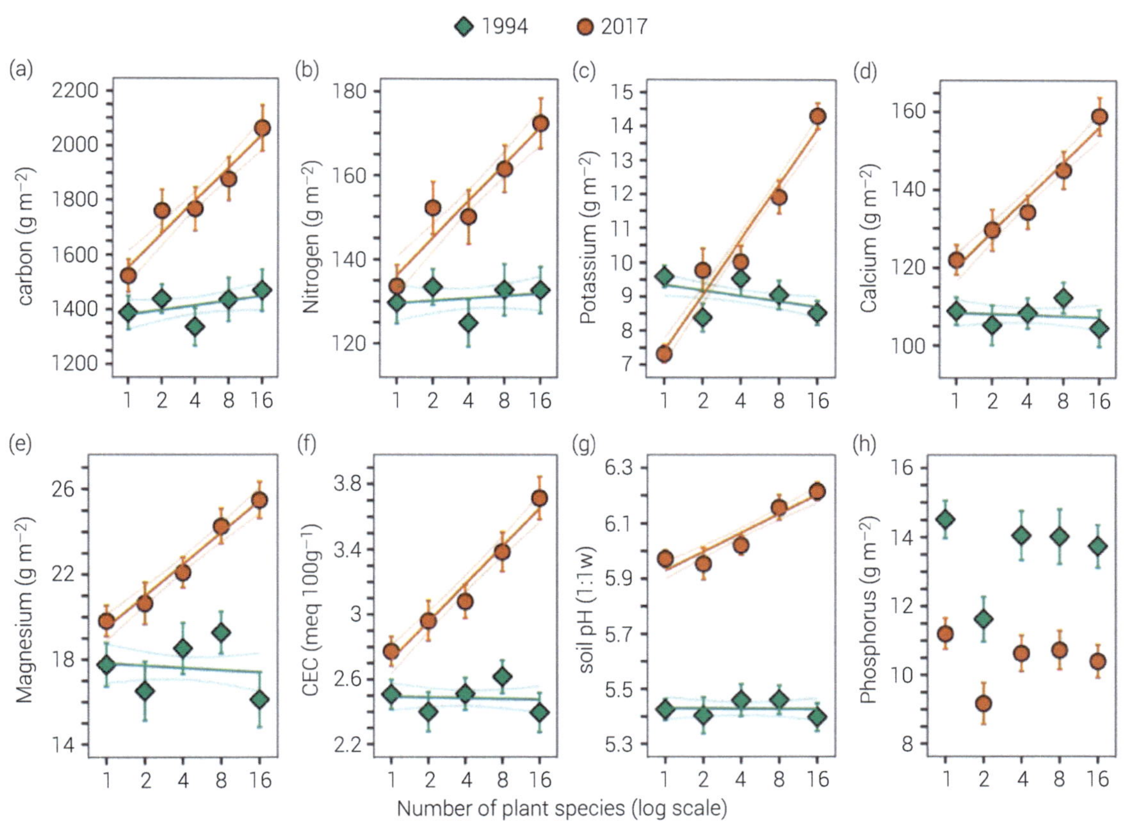

Figure 7.11 Plant species diversity regenerates the fertility of degraded sandy soil. Values are mean ±1 SE for soil chemical properties (g m^2) at 0–20 cm depth before planting in 1994 and 2017 for (a) total C, (b) total N, (c) exchangeable K, (d) exchangeable Ca, (e) exchangeable Mg, (f) CEC, (g) soil pH, and (h) extractable Bray P versus number of planted species (1, 2, 4, 8, or 16; log scale). Lines are linear regressions ± 1 SE (*n* = 154 plots). From Furey and Tilman (2021). https://www.pnas.org/doi/full/10.1073/pnas.2111321118.

mycorrhizal fungi (De Deyn et al. 2009, 2011, 2012), with potential benefits for SOC storage and nutrient retention (De Deyn et al. 2011, Lange et al. 2015, Yang et al. 2019b, Spohn et al. 2023) that strengthen over time (Lange et al. 2023), as well as reductions in N_2O emissions from soil (Dang et al. 2024). Planting high diversity mixtures in degraded soil has also been shown to increase soil aggregate stability, an important indicator of soil health (Gould et al. 2016), and, in many cases, to enhance the abundance and diversity of microbes and animals in soil (Liu et al. 2020, Zhang et al. 2022). Potential benefits of plant diversity for restoration are not restricted to grasslands: positive diversity–productivity relationships predominate in forests (Liang et al. 2016) and shrublands (Chen et al. 2018b), with benefits for SOC and nutrient storage (Chen et al. 2018b, Liu et al. 2018c, Mori et al. 2021).

Several mechanisms underpin the potential for plant diversity to regenerate degraded soil, which are touched on in Chapter 4. But plant functional diversity appears to play a predominant role. For instance, benefits of high diversity grassland for SOC and nutrient accumulation in degraded ex-agricultural soil were attributed to plant functional diversity, and specifically complementary effects of C4 grass and legume species within plant mixtures, i.e., C4 grasses efficiently take up N derived from N-fixing legumes, thereby increasing root biomass and inputs of C and N to soil (Fornara and Tilman 2008). Beneficial effects of plant diversity on physical properties of a degraded grassland soil were also attributed to plant functional traits, namely the inclusion within plant mixtures of grasses rich in fine roots, which play an integral role in developing soil aggregate stability by binding soil particles together (Gould et al. 2016). Grassland restoration experiments show that the inclusion of N-fixing legumes in species mixtures modifies microbial processes of C and N cycling, with co-benefits for soil C sequestration and hence climate mitigation. For example, in a long-term grassland restoration in northern England, biodiversity restoration practices, such as seeding with a locally sourced species mixture and cessation of fertiliser use, increased soil C and N storage, especially when these treatments were combined with sowing of the N-fixing legume *Trifolium pratense* (De Deyn et al. 2011). The mechanisms underpinning this increase in soil C and N accumulation with *T. pratense* addition are unclear, but it was proposed that they involve a combination of increased input of legume-derived C and N to soil, the suppression of microbial enzymes involved in the decomposition of recalcitrant organic matter, and improved soil structure, which protects organic C from microbial attack (De Deyn et al. 2011).

As well as accelerating ecosystem restoration, the sowing of species mixtures can enhance the capacity of soil microbial communities and their functioning to buffer climate extremes in restoring ecosystems, which as discussed in Chapter 6 are becoming more common and severe. An example of this comes from a restoration experiment on shallow, degraded ex-arable soils in southern England, which tested the idea that the sowing of plant species mixtures with deeper and more complex root systems would accelerate the restoration of microbial-mediated processes of nutrient and C cycling, and enhance their capacity to buffer drought, which is especially critical during early stages of restoration (Fry et al. 2018b). After 3 years of restoration, plant species mixtures with more complex root systems, based on variation in root architecture and rooting depth, were associated with greater soil multifunctionality, measured with a range of microbial-mediated soil functions related to C and nutrient cycling. Moreover, it was found that the ability of key microbial-mediated soil functions, including ecosystem respiration, N mineralisation, and nitrification, to resist a simulated summer drought was positively related to functional divergence of rooting depth of initially sown plant communities (Figure 7.12). This suggests that sowing a range of rooting strategies could be key to building the functioning and resistance of degraded soils to climate extremes during early stages of ecosystem restoration (Fry et al. 2018b). This finding is broadly consistent with results from recent field studies showing that grassland diversity restoration increases the resistance of ecosystem C fluxes to drought (Cole et al. 2019) and the resistance of ecosystem multifunctionality to recurring drought (Dietrich et al. 2024), and that plant functional diversity fosters resistance of soil multifunctionality to increasing aridity in drylands

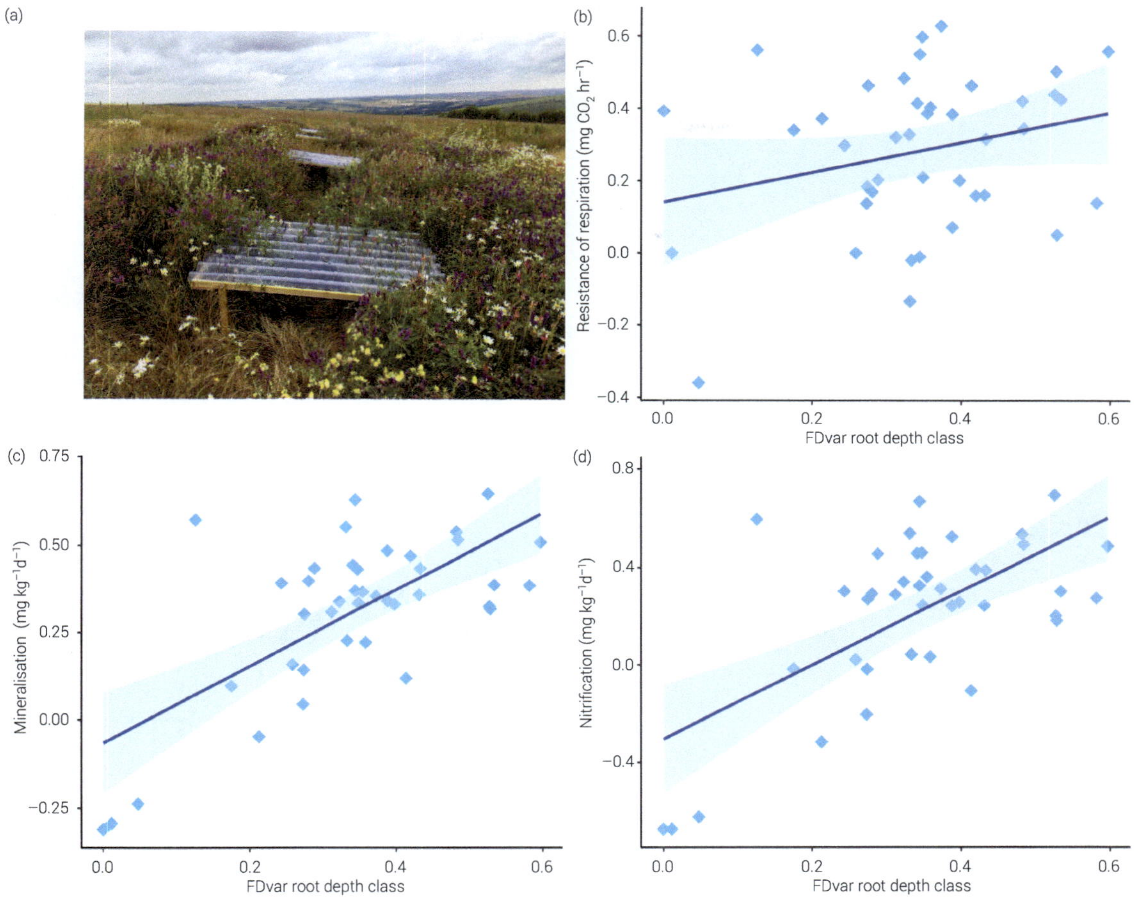

Figure 7.12 Increased variation in rooting depth enhances the resistance of microbial-mediated functions to drought in restoring grassland in southern England (a), including soil respiration (b), N mineralisation (c), and rates of nitrification (d). A resistance value of 1 represents full resistance (i.e., no difference from the control after drought), 0 denotes a 100% change, while negative numbers indicate that the droughted value is higher than the control value. Functional divergence of root depth class (FDvar) is bounded between 0 and 1. From Fry et al. (2018b). Image by Richard Bardgett.

(Valencia et al. 2015). As such, maintaining and enhancing plant functional diversity appears to be critical for restoring not only the functioning of heavily degraded soils and ecosystems, but also their capacity to buffer perturbations, such as climate extremes.

Soil transplants

An approach to ecosystem restoration that is gaining momentum is the use of soil transplants, which involves moving soil and associated microbial communities (usually as spreadable crumbs or soil slurry) from a nearby donor site to a site undergoing restoration management (Figure 7.13). An excellent example of this comes from a long-term (6-year) study in the Netherlands on soils that were degraded from multiple decades of arable farming. Here, the application of soil inoculum from nearby mature heathland and grassland donor sites was found to promote ecosystem restoration and steer the plant community towards these target communities (Wubs et al. 2016). Moreover, the impact of soil inoculation on plant and below-ground community composition was most pronounced when the topsoil layer was removed, suggesting that soil

Figure 7.13 Soil translocation for ecosystem restoration. Comparison of control (left) and soil amended plot (right) in restoring semi-dry grassland, Germany, 20 years post treatment (a); soil translocation near Brasilia, Brazil, for tropical dry forest regeneration (b); soil receiver site preparation to restore coastal sage scrub in Orange County, southern California, USA (c); and reference vegetation on a *Quercus rotundifolia* dehesa soil in Salamanca, Spain (d). Images (a)–(d) by Wolfgang von Brackel, Maxmiller Ferreira, Megan Lulow, and Ignacio Santa-Regina, respectively.

inoculation is most effective for restoring more degraded soils (Wubs et al. 2016). Inoculation with soil and native AM fungi from undisturbed old-growth grasslands in the USA has similarly been shown to accelerate plant succession and ecosystem restoration of degraded prairies (Koziol et al. 2022, Bever et al. 2003). Studies also show that soil inoculation with native mycorrhizal fungi facilitates the restoration of species-rich plant communities resembling reference sites from where the fungi are sourced (Koziol and Bever 2017, Neuenkamp et al. 2019), and the combined addition of native soil inocula and plant seeds was found to accelerate the recovery of grassland plant and below-ground communities that also became more connected over time (Wubs et al. 2019). Soil inoculation with native AM fungi and plant seeding has also been shown to enhance the restoration of species-rich vegetation on previously barren post-mining sites (Vahter et al. 2020), and biocrust inoculation is being used increasingly to help restore degraded drylands, both due to its critical role in stabilising soil and through increased soil input of C and N via photosynthesis and N fixation (Antoninka et al. 2020). The use of natural soil inocula sourced from dry, warm, and colder sites has also been shown to increase the survival of tree seedlings faced with drought, and heat and cold stress, respectively, pointing to microbial-mediated enhancement of tree resilience to climate change (Allsup et al. 2023).

The above examples all demonstrate that inoculation with native soil and its associated microbiome can be an effective tool for restoring degraded ecosystems and their biodiversity, both above- and below-ground. It should be noted, however, that effects of soil inoculation on ecosystem restoration and biodiversity are highly variable and depend on factors such as vegetation and soil type, restoration context, and the origin and amount of inoculum used (Maltz and Treseder 2015, Neuenkamp et al. 2019, Han et al. 2022). A recent meta-analysis of data from 46 restoration experiments across 17 countries, for example, showed that while soil translocation generally steers plant community development towards reference ecosystems more than when just plant propagules are added to degraded soils, responses among studies are extremely variable (Gerrits et al. 2023). This variability is explained by many factors such as soil type, the scale of the translocation, being more successful the larger the spatial scale, and the length of experiments, with restoration success increasing over time (Gerrits et al. 2023). Similarly, a recent synthesis of data from 27 different studies showed that plants inoculated with 'live' soils with wild microbiomes taken from reference ecosystems or native mycorrhizal spores from those ecosystems outperformed those that were not (Averill et al. 2022). As stressed by the authors, however, responses to 'live' soil inoculation are variable, and in some cases they fail, for instance, due to a particular restoration context or the absence of key members of the soil microbiome. Nevertheless, evidence suggests that native soil inocula sourced from reference ecosystems outperform microbial inocula from commercial sources (Maltz and Treseder 2015), which are typically composed of single microbial species or low diversity nonnative microbial consortia (Averill et al. 2022).

While soil transplant approaches appear to have promise as a tool for restoring degraded ecosystems, they are not without risk. A particular risk concerns damage to donor sites through excavating large amounts of soil for transplanting, especially if done at scale. Another risk concerns unintended consequences of soil inoculation for the diversity and functioning of resident microbial communities. As discussed earlier in this chapter, the use of microbial inoculants in agricultural soils can lead to persistent changes in the diversity and composition of resident microbial communities, even if inoculum survival is short-lived (Mawarda et al. 2020, Cornell et al. 2021). However, this is less of an issue when using soil transplants in ecosystem restoration, given that their associated microbiomes are native and well adapted to local conditions, and resident microbial communities of degraded soils are already depleted. The use of soil transplants in ecosystem restoration is a fast-developing area, and future research is needed to better understand when and where soil inoculation with native microbiomes is most effective, and how to maximise their effectiveness for facilitating restoration (Averill et al. 2022).

Climate change adaptation and mitigation

The role of soils and their biological communities in adaptation and climate mitigation is high on the scientific and political agenda. This is largely due to the potential for soils to sequester large quantities of atmospheric CO_2 into stable SOC pools, thereby increasing SOC stocks and helping to mitigate human-induced increases in atmospheric CO_2 (Lal 2004). Indeed, soils are the largest terrestrial organic C pool on Earth and even small increases in SOC storage could represent a substantial C sink potential (Friedlingstein et al. 2020), with potential co-benefits for soil fertility and food production (Ma et al. 2023). This recognition has led to various SOC sequestration schemes, including the '4 per 1000' initiative launched at COP 21 in Paris in 2015, with the bold aim of increasing C storage in agricultural soils by 0.4–1.2 gigatonnes (Gt) a year, which is equivalent to 5–15% of global fossil fuel emissions (Minasny et al. 2017, Rumpel et al. 2020). However, the climate mitigation potential of soil C sequestration is uncertain and much debated, largely because SOC is dynamic in space and time and strongly influenced by local factors such as soil type, climate, and land management (Don et al. 2024). Measures to increase SOC sequestration hence need to consider the dynamic nature of SOC and be adapted to local soil, climate, and management opportunities (Amelung et al. 2020). Nevertheless, SOC sequestration clearly offers a partial climate change mitigation strategy while low C technologies are being developed and adopted (Lal 2004, Paustian et al. 2016).

This section considers recent advances in our understanding of how improved management of agricultural soils can enhance soil C sequestration via modifying ecological processes involved in the formation and stabilisation of SOC, thereby contributing to climate change mitigation. This section also considers the potential for natural climate solutions such as tree planting to increase SOC, and how management of agricultural soils can contribute to climate change adaptation via enhancing the resistance and resilience of below-ground communities to climate extremes, which pose an increasing threat to soils and their functioning.

Sequestration and stabilisation of soil C

There is much uncertainty over the amount of organic C that soil can potentially store, especially given that rates of soil C sequestration saturate over time and vary greatly depending on soil type, climate, management interventions, and future climates. But recent estimates suggest that the protection and sequestration of SOC represents about 25% of the total mitigation potential of natural climate solutions (i.e., total potential of 23.8 Gt of CO_2 equivalents year^{-1}) and 47% of the mitigation potential of grasslands and croplands globally, largely through rebuilding SOC lost through centuries of agriculture (Bossio et al. 2020). Global data syntheses support this, showing that SOC losses associated with the conversion of forests to agricultural land average around 45% (Wei et al. 2014), whereas SOC losses associated with the conversion of grassland to cropland average 59% (Guo and Gifford 2002). Rebuilding SOC in degraded soils therefore represents an important climate mitigation strategy, especially given the extent of soil degradation globally and the potential for restoration practices to increase SOC (De Deyn et al. 2011, Chen et al. 2018b, Yang et al. 2019b, Spohn et al. 2023). An example of how large-scale changes in land management can rebuild soil C comes from France, where large areas of grassland were converted to arable land during the late 1900s, leading to significant declines in SOC. It has been predicted that restoring half of this arable land back to grassland (~ 90,000 ha) would eventually lead to an estimated increase of 16 megatonnes (Mt) of soil organic C, equivalent to 10% of the annual CO_2 emissions from fossil fuels in France (Soussana et al. 2004). As stressed by the authors, however, such a land use shift would imply major changes in livestock production systems and emissions of other greenhouse gases (i.e., CH_4 and N_2O) (Soussana et al. 2004).

Regarding agricultural soils, a host of 'climate smart' measures have been proposed to increase SOC sequestration (Paustian et al. 2016). These include the use of cover crops, deep-rooted crops, and crop diversification, the application of soil amendments such as biochar, and the adoption of no-till or reduced tillage, which, as discussed earlier, have potential co-benefits for soil fertility and food production when tailored to local conditions (Paustian et al. 2016, Amelung et al. 2020, Moinet et al. 2023). Given that around half of the global grassland area is degraded with depleted SOC stocks (Bardgett et al. 2021, Bai and Cotrufo 2022), it has been proposed that by restoring degraded grasslands an additional 2.3–7.3 Gt of CO_2 equivalents year^{-1} could be sequestered (Figure 7.14a), and on a global scale, improved grassland management, including conversion from cultivation to grassland, increasing plant diversity, sowing legumes and grasses, and fertilisation (Figure 7.14b) (Bai and Cotrufo 2022), could result in an increase in SOC stocks by some 0.47 Mg C ha^{-1} year^{-1} (Conant et al. 2017). However, as highlighted above, SOC is dynamic, with local factors potentially constraining SOC sequestration, and hence the climate change mitigation potential of soil. For instance, SOC saturates over time, thereby limiting the potential contribution of SOC sequestration to climate change mitigation (Moinet et al. 2023). Moreover, long-term SOC sequestration depends on the extent SOC is stabilised in mineral-associated (i.e., MAOM) forms, which are protected from microbial decomposition, and hence contribute more to the long-term SOC sequestration than labile, short-lived particulate forms (i.e., POM) (Lavallee et al. 2020). Recent evidence, however, indicates that there is considerable potential for SOC stabilisation in agricultural soils given that their mineralogical capacity for MAOM formation is undersaturated (Georgiou et al. 2022).

As touched on in Chapter 5, MAOM is formed by mineral adsorption and protection within small

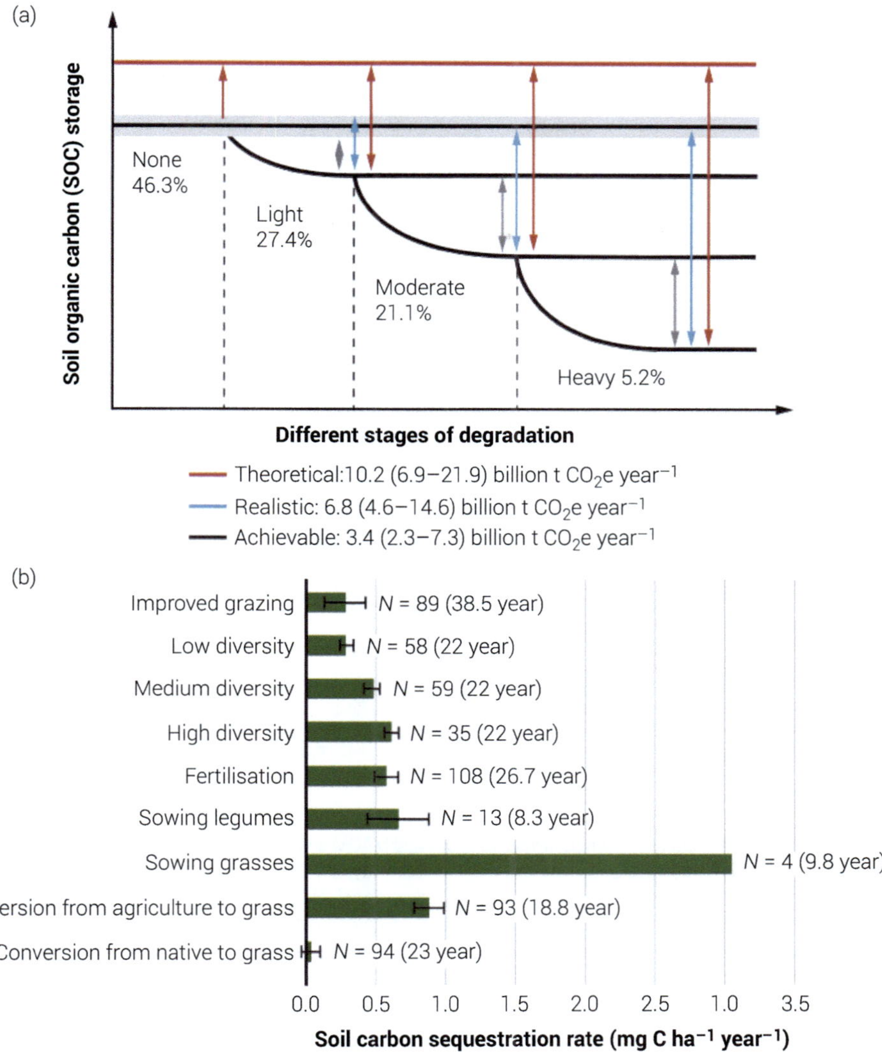

Figure 7.14 Soil C sequestration potentials for global grasslands according to the area of different stages of grassland degradation, ranging from none to heavy (a), and impacts of improved grassland management practices on SOC sequestration rate (mean ± SE) (b). (a) Data are means (with 95% confidence intervals in parentheses). 'Theoretical' SOC sequestration capacity refers to the estimate of restoring all soils to their natural capacity or even enhancing them through management interventions; 'realistic' refers to the optimistic value, accounting for social and economic constraints; and 'achievable' is the value of a pragmatic scenario. (b) The number of studies used to calculate the average is given for each type of management, along with the study duration (years). From Bai and Cotrufo (2022) by permission of the American Association for the Advancement of Science.

aggregates of relatively low molecular weight compounds produced by the depolymerisation of plant material, or from the microbial transformation of organic material and formation of microbial necromass (Buckeridge et al. 2022, Cotrufo and Lavallee 2022) (Figure 7.15). Essentially, microbes grow, die, and turnover while decomposing plant inputs, and generate a massive stock of non-living biomass, or necromass, that is then stabilised through mineral adsorption or protection within microaggregates, the so-called entombing effect (Liang et al. 2017). This process is of central

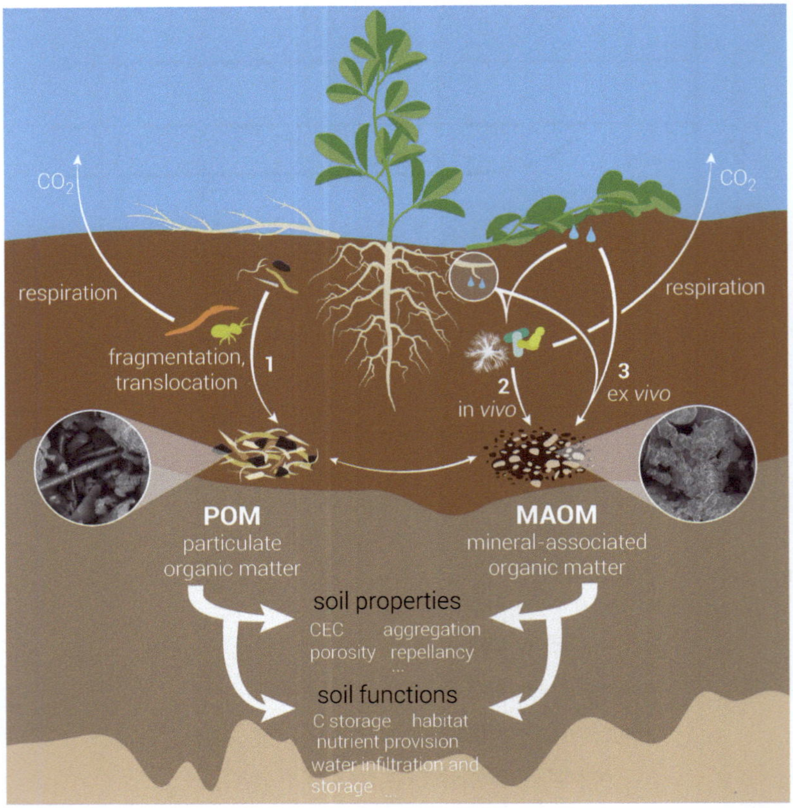

Figure 7.15 Schematic of the main formation pathways for POM and MAOM and links or soil properties and functions: fragmentation and translocation of plant-derived residues by soil biota is the main pathway for POM formation (1), whereas MAOM forms by direct mineral adsorption and protection within small aggregates of plant-derived soluble and low molecular weight compounds (*ex vivo*) (2), and microbial transformation of plant- and microbial-derived soluble organic compounds and the formation of microbial necromass, which is subsequently stabilised through mineral adsorption or protection within microaggregates, the so-called entombing effect (in vivo) (3). Compared to POM, MAOM typically lasts longer, contains less chemically complex compounds, and has a lower C:N ratio. It should be noted that POM and MAOM are not completely distinct, as elements and compounds can exchange between them. Image by J. M. Lavallee.

importance for longer-term SOC sequestration, given that microbial necromass constitutes a significant portion of total SOC, especially in cropland and grassland soils where it makes up around 50% of total topsoil SOC (Wang et al. 2021, Bai and Cotrufo 2022). However, the accumulation of microbial necromass in soil and its contribution to SOC varies considerably, being maximal in cold, moist (Bai and Cotrufo 2022), and acidic soils (Wang et al. 2021) on a global scale. This is likely because acidic, cold, and moist soils have high soil SOC stocks and C:N ratios, favouring the growth of fungi over bacteria (Fierer et al. 2009, Yu et al. 2022) and the accumulation of fungal necromass (Wang et al. 2021). Also, because fungal necromass decomposes more slowly than bacterial necromass (due to the presence of recalcitrant cell wall components such chitin and melanin), it persists longer; as such, around two-thirds of microbial necromass in global soils is of fungal origin (Wang et al. 2021), which underscores the key role of soil fungi in SOC sequestration.

Another microbial parameter that is of particular importance for MAOM formation is CUE, or microbial C use efficiency, which describes the partitioning of C used in microbial metabolism to growth versus respiration. This is because a high CUE reflects greater partitioning of C to microbial biomass production and microbial by-products, which in turn favours microbial necromass production and its association with minerals, leading to MAOM formation (Tao et al. 2023). Consistent with this, it has been proposed that high quality labile plant inputs (e.g., root exudates) that are used efficiently by microbes for biomass production will enhance the production of microbial necromass and formation of MAOM, thereby enhancing SOC stabilisation (Cotrufo et al. 2013, Castellano et al. 2015). However, labile plant inputs like root exudates can also stimulate the decomposition of existing stable SOM, leading to soil C loss, i.e., priming (see Chapter 3). As such, to understand the dynamics of SOC stabilisation requires consideration of the balance between microbial priming of MAOM and microbial entombing, i.e., incorporation of microbial necromass into MAOM (Liang et al. 2017, Dijkstra et al. 2021, Chari and Taylor 2022).

Mycorrhizal fungi play a key role in SOC sequestration via pathways that can lead to both SOC loss and gain (Frey 2019, Hawkins et al. 2023). Indeed, it has recently been estimated that some 13.12 Gt of CO_2 equivalents fixed by terrestrial plants is allocated below-ground, at least temporally, to mycorrhizal mycelium each year, equating to ~ 36% of annual CO_2 emissions from fossil fuels (Hawkins et al. 2023). As mentioned in Chapter 6, studies reveal that ERM fungi play a key role in SOC accumulation and stabilisation in boreal forest by suppressing saprotrophic fungi and organic matter decomposition (Fanin et al. 2022), and by producing fungal necromass (Clemmensen et al. 2015) that is adsorbed to mineral surfaces or stabilised via root-derived condensed tannins that form stable complexes (Adamczyk et al. 2019). Roots of ericaceous shrubs also have the capacity to enhance C loss by stimulating microbial growth and activity, and hence organic matter decomposition via the priming effect (Adamczyk et al. 2019). As such, the impact of ERM and other mycorrhizal fungi on the stabilisation of SOC depends on the balance of these opposing pathways.

Relatively few studies have explored how improved management could increase microbial necromass accumulation and its incorporation into MAOM. But it has been proposed that the amount of C stored in MAOM can be boosted by management that favours the transformation of plant-derived inputs, such as root exudates and litter, into simpler forms and microbial necromass, especially in soils with an abundance of reactive minerals (Angst et al. 2023). For example, increasing grassland plant diversity has been shown to boost SOC storage by increasing labile root inputs to soil (i.e., root exudates and litter), which favour faster rates of microbial growth, turnover, and necromass production, and ultimately the incorporation of larger amounts of microbial-derived C into the stable SOC pool (Prommer et al. 2020). Similarly, increasing understorey plant diversity grown with barley has been shown to promote positive associations within the barley rhizosphere microbial community, which in turn enhanced microbial CUE and C retention in soil (Domeignoz-Horta et al. 2024). Also, a global synthesis of data from field studies showed that management practices associated with regenerative farming, such as no-till and crop diversification (i.e., eliminating summer fallow in monocrop systems, increasing the number of annual crops grown per year, planting cover crops, and including perennial crops in rotation), generally increase both topsoil POM and MAOM (but not in subsoil), thereby promoting soil health and long-term C storage (Figure 7.16) (Prairie et al. 2023). Moreover, even greater benefits for topsoil MAOM accumulation were detected when crop diversification practices were combined with livestock management, suggesting the potential to increase SOC pools and their stabilisation via synergistic interactions between regenerative management practices (Prairie et al. 2023). As mentioned above, research on how best to manage soils to promote SOC stabilisation, and subsequent longer-term C storage, is in its infancy, and much remains to be learned. Moreover, it has been argued that effective management needs to consider both POM and MAOM dynamics; not only does POM make up a large proportion of total SOC and serve as a precursor of MAOM formation, but also it plays a key role in regulating below-ground communities and their roles in maintaining soil fertility (Angst et al. 2023).

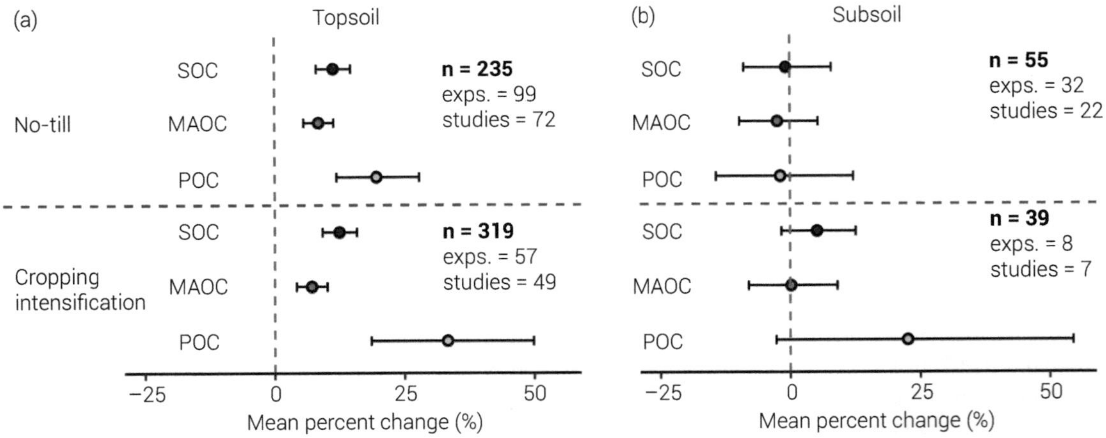

Figure 7.16 Soil C accumulation associated with regenerative farming practices. Responses are shown as percentage change in soil C concentrations (g C kg^{-1} soil) for total SOC, mineral-associated organic carbon (MAOC), and particulate organic carbon (POC) in topsoil (a) and subsoil (b). No-till systems are compared to conventional tillage systems, and cropping intensification includes eliminating summer fallow in monocrop systems, increasing the number of annual crops grown per year, planting cover crops, and incorporating perennial crops in rotation. The number of observations (*n*), experiments (exps.), and studies analysed for each variable is on the far right of the figure. From Prairie et al. (2023).

As a final note on the topic of soil C sequestration, increasing SOC and key groups of soil organisms involved in MAOM formation (such as saprotrophic and mycorrhizal fungi) can provide co-benefits for climate adaptation by enhancing soil resilience to climate extremes. A key element of climate change adaptation is building resilience, or the ability of soil functions to resist and recover from environmental perturbations, including extreme climate events. This issue is touched on in the above section on ecosystem restoration, where the critical role of plant diversity in restoring degraded soils and their capacity to buffer climate extremes is considered (Valencia et al. 2015, Cole et al. 2019, Dietrich et al. 2024). Yet, there is evidence that management targeted at shifting the distribution of biomass and energy flow to the fungal rather than the bacterial energy channel, which based on theory should stabilise the soil food web (see Box 6.1), can also enhance soil resilience to climate extremes.

Supporting this, it is well known that intensive grassland management increases the dominance of the bacterial over the fungal channel (Bardgett and McAlister 1999, De Vries et al. 2006), which in turn leads to faster rates of N mineralisation and greater loss of N in leachates from soil following dry–wet cycles (Gordon et al. 2008, De Vries et al. 2012a). Also, it has been shown that soil food webs of extensively managed grassland with greater biomass in the fungal energy channel retain nutrients more effectively and have lower greenhouse gas emissions following drought than do those of intensively managed cereal fields dominated by the bacterial energy channel (De Vries et al. 2012b). Moreover, a cross-site field study of grasslands in northern England showed that intensive grassland management disrupts the below-ground transfer of plant C to key components of the soil food web that confer resistance to drought, especially mycorrhizal fungi (Chomel et al. 2022). In contrast, extensive grassland management effectively buffered this negative effect of drought on below-ground C transfer, most likely because it promoted the transfer of recent photosynthates to key components of the fungal energy channel that confer resistance, i.e., mycorrhizal fungi and fungivorous mites (Chomel et al. 2022). Taken together, these studies suggest that management practices targeted at increasing the fungal relative to the bacterial energy channel in soil, such as reduced tillage, cessation of fertiliser use, and plant diversification, could help buffer soil systems against future climate extremes, whilst bringing co-benefits for SOC sequestration.

Afforestation

Afforestation, or the planting of trees in areas without recent tree cover, has become a key climate mitigation strategy in many parts of the world, given its promise to accelerate C sequestration in above-ground tree biomass and C transfers to soil via dead wood and tree litter (leaf and root) and root exudates. While tree planting can be controversial, for instance when targeted at natural grasslands, with negative impacts on native biodiversity and ecosystem functioning (Veldman et al. 2015), its potential to accelerate C sequestration in live above-ground tree biomass, when done appropriately, is considerable. It has been estimated, for example, that planted forests and harvested wood products uptake around 14% of economy-wide CO_2 emissions annually in the USA, and that live-tree above-ground C sequestration could be increased by ~ 20% per year by fully stocking currently understocked plantation forests (Domke et al. 2020). Moreover, natural forest regeneration of deforested land in the tropics, aided by planting of native trees, has potential to capture an estimated additional 42 Gt C in living above-ground tree biomass by 2100 (Lewis et al. 2019). However, many factors moderate the C storage potential of planted forests, including tree species and prior land use (Bukoski et al. 2022), future climate risks, such as fire, drought, and insect damage (Wu et al. 2023), reductions in surface albedo that could offset reforestation C sequestration potential (Rohatyn et al. 2022, Weber et al. 2024), and uncertainties about its impact on SOC storage.

Few studies have explored how tree planting impacts SOC over large spatial scales, but the emerging picture is that responses are context dependent. An analysis of 619 paired afforested and adjacent non-forested control sites across vast areas of northern China, for example, showed that responses depend on initial SOC content: afforestation increased SOC in C-poor soils, mostly in surface soil, but decreased SOC in C-rich soils, especially at depth (Figure 7.17) (Hong et al. 2020). As a result, the overall increase in SOC stocks in northern China, resulting from afforestation over the last

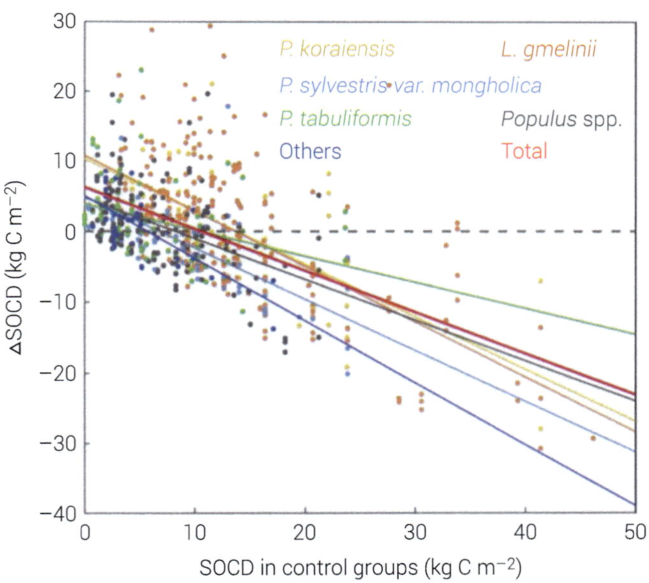

Figure 7.17 Afforestation increases soil organic carbon density (SOCD) in C-poor soils, mostly in surface soil, but decreases SOCD in C-rich soils, especially at depth. Data show relationships between ΔSOCD (i.e., the difference between SOCD in afforested plots and control plots, measured as kg C m^{-2}) and SOCD in control plots across the six groups of planted trees (i.e., *Pinus koraiensis*, *Larix gmelinii*, *P. sylvestris* var. *mongholica*, *P. tabuliformis*, and *Populus* spp.) and all groups pooled (i.e., Total). Data were collected from 619 control-and-afforested plot pairs in northern China. From Hong et al. (2020) by permission of Springer Nature.

three decades, was estimated to be relatively modest, at 234.9 ± 9.6 teregram (Tg) C (Hong et al. 2020). Similarly, planting of trees in C-rich organic soils across sites in Scotland was found to result in no net ecosystem C gain, and in some cases it even caused net ecosystem C loss due to enhanced SOC losses over decades, again raising the question about the merit of tree planting in C-rich soils (Friggens et al. 2020).

The reason why tree planting enhances C loss in C-rich soils is likely a combination of initial soil disturbance caused by planting, which enhances the decomposition of original SOM, and priming of SOM by tree roots and their associated mycorrhizal fungi, especially in nutrient-poor soils (Friggens et al. 2020, Hong et al. 2020). As discussed in Chapter 6, there is often a trade-off between plant and SOC storage, whereby trees increase their biomass by mining SOM for nutrients via the priming effect mediated by mycorrhizal fungi (Terrer et al. 2021). This leads to a greater plant biomass C sink at the expense of SOC accumulation, especially in high C soils and under elevated CO_2 (Terrer et al. 2021). In contrast, the build-up of C in C-poor soils is likely because microbial activity, and hence organic matter decomposition, is limited by low C availability, and C input from plant litter and tree roots is greater than the rate of decomposition, thus leading to SOC accumulation (Hong et al. 2020).

The above studies were mostly based on planting of tree monocultures. However, a promising strategy for increasing C capture and SOC storage is planting tree mixtures of native trees rather than monocultures. As touched on in the previous section on ecosystem restoration, positive diversity–plant productivity relationships predominate in forests (Liang et al. 2016), and higher tree diversity has been shown to be associated with greater below-ground C storage in root and SOC, effects that become stronger over time (Liu et al. 2018c). Moreover, planting native tree species mixtures in degraded tropical soils has been found to restore the diversity and functioning of multiple soil organism groups to levels found in local natural forests, whereas planting of monocultures of non-native Eucalyptus did not (Wu et al. 2021). Also, in some situations, simply allowing forest to regenerate naturally following abandonment of past land use can be more beneficial for SOC accumulation than tree planting. A recent large-scale comparison of active restoration by tree planting/seeding versus natural forest regeneration on soil C sequestration across landscapes in China sheds some light on this: active restoration sequestered more C in soil over decadal timescales than natural regeneration in C-poor soils, but the oppositive was the case in C-rich soils, where natural forest regeneration performed better (Tian et al. 2023). Moreover, while active restoration by tree planting/seeding increased surface soil C, natural regeneration was better for increasing subsoil C, which is generally a more stable C pool because of lower disturbance and temperatures (Tian et al. 2023). Many uncertainties exist over the factors that moderate effects of both tree planting and natural forest restoration on SOC, such as baseline soil properties, climate, past land use, and tree functional composition (Wallwork et al. 2022). But these results suggest that, in some situations, simply letting forest regenerate naturally on ex-agricultural land could bring greater benefits for SOC sequestration in the longer term and at lower cost (Tian et al. 2023), with potential co-benefits for biodiversity and ecosystem services (Benayas et al. 2009).

Agroforestry, or the incorporation of trees into agricultural systems, has also been proposed as a way of blending food production and climate mitigation, given its potential to increase C storage both above- and below-ground, whilst continuing to produce food (Figure 7.18). Relatively few studies have explored how agroforestry impacts SOC storage, and meta-analysis on the topic generally lacks sufficient data to consider key factors such as moderate responses, planting density, tree species choice, and management regime (Terasaki Hart et al. 2023). As such, syntheses on SOC responses to agroforestry paint a highly variable picture and leave much uncertainty. Nevertheless, a recent meta-analysis of available data from agroforestry studies showed that while responses were highly variable, the integration of trees in croplands increased SOC by on average 33%, and growing trees in grasslands, or silvopasture, increased SOC on average by 26% (Beillouin et al. 2023). However, in some cases, agroforestry can negatively affect SOC storage, for instance, when root exudates trigger priming of

Figure 7.18 Agroforestry research plots at James Hutton Institute's Glensaugh research farm, Scotland. The plots in the foreground (striped area) were planted in 2024 using a 'strip' design, where livestock graze in pasture strips between strips of trees, rather than the block design behind of older agroforestry plots planted in 1988. Image by Scot Ramsay.

SOC and hence SOC loss (Rolo et al. 2023). Also, the planting of trees in croplands and pastures can have negative side-effects, for instance through increased competition between trees and crops for soil water and nutrients (Beillouin et al. 2023). Moreover, benefits of agroforestry for SOC sequestration tend to occur in surface soil and labile SOC fractions (i.e., POM), and in the immediate vicinity of trees, thereby creating hot spots of SOC sequestration and soil biological activity in agricultural landscapes (Rolo et al. 2023).

Many questions remain about the effectiveness of tree planting as a strategy for increasing SOC sequestration, given that responses are highly context dependent. However, the available evidence suggests that planting of trees in C-poor soils and using mixtures of native tree species rather than monocultures offer the greatest potential for increasing SOC, with additional benefits for the diversity and functioning of soil biota. Evidence also suggests that benefits of tree planting for SOC tend to be more pronounced in croplands with C-poor soils than in grasslands (Laganiere et al. 2010), and that, in some cases, simply allowing forests to regenerate naturally could be the most effective way to build stable SOC stocks in the longer term, with added benefits for biodiversity and multiple ecosystem services.

Conclusions

It is hopefully evident from this chapter that the ecology of soil is at the heart of many sustainability challenges, including sustainable food production, ecosystem restoration, and climate change adaptation and mitigation. How soil biodiversity can be harnessed to address these sustainability goals is a fast-moving area, and our knowledge of how best to manage land to promote soil biodiversity, and the many ecosystem services it underpins, is growing fast. Moreover, there are many emerging opportunities for soil ecology to help meet sustainability challenges, for instance the use of 'wild' microbial inoculants to accelerate ecosystem restoration, the development of crop breeding programmes targeted at root traits that promote the recruitment of beneficial rhizosphere microbiomes, and the potential for plant diversification to enhance soil fertility, SOC sequestration, and the resistance of soil systems to climate extremes. However, focus on the potential for soil biota to contribute to sustainability goals is relatively new, which means that there are many unanswered questions about how best to manage the land to promote soil biodiversity and the ecosystem services that it underpins. Moreover, the effectiveness of different interventions for enhancing the soil biological processes that underpin food production, ecosystem restoration, and C sequestration are context dependent. As such, to be effective, interventions, whether targeted at sustainable food production, ecosystem restoration, or soil C sequestration and climate mitigation, need to be tailored to local site conditions, considering soil conditions, plant species, climate, and previous land use.

What is abundantly evident is that there are potential co-benefits from managing land in a way that is beneficial for multiple groups of soil organisms. An example of this is the use of plant diversity to accelerate the restoration of degraded soils, which benefits multiple soil organism groups, but also can lead to improvements in soil aggregation and C storage and help soil functions buffer against climate extremes. Similarly, crop diversification is an effective strategy for enhancing ecosystem multifunctionality, whilst not compromising crop yields. In general, management interventions that benefit the rich diversity of soil life will also likely have benefits for the ecosystem services that the land provides and the capacity of land to buffer future climate risks, such as climate extremes. This might be a generalisation, but as Lady Eve Balfour, penned, '*the health of soil, animal and man is one and indivisible*'.

CHAPTER 8

Some conclusions and future perspectives

It is often stated that we know little about life in soil. But the rich diversity of soil ecology knowledge contained in this book suggests that we do know a lot. In fact, the field of soil ecology has expanded so much over recent years that it is now hard to keep up with the stream of new papers published week on week reporting new and fascinating discoveries. This rapid and recent development of the field is reflected in the scope of this book, which hopefully does justice to our growing understanding of the ecology of soil, but also the many new and exciting areas of study that have come to prominence in recent years. It would be difficult to highlight them all, but a few advances are worthy of mention here.

An area that has witnessed recent advances, as covered in Chapter 2, has been the quantification of soil biodiversity, made possible, in part, by the increased accessibility of molecular techniques for interrogating soil life. From this has also emerged a deeper understanding of the abiotic and biotic factors that shape patterns of soil biodiversity at different spatial and temporal scales. We also know much more about the functional importance of soil biodiversity, as considered in Chapter 3, and the key roles that different taxonomic groups of soil organisms and their trophic interactions play in regulating biogeochemical cycles. Much is still to be learned, but compared to 20 years ago, we now have a better grasp of how changes in the diversity and structure of soil food webs might influence ecosystem functions and their stability in the face of global change.

Another topic that has greatly expanded in recent years concerns how plants shape the structure and functioning of the soil microbial communities, especially those in the rhizosphere, and feedback consequences for plant growth and community dynamics. As highlighted in Chapter 4, interest in this topic has boomed over the last two decades; as a result, we better understand how plant traits, including those of roots, impact below-ground communities and their functioning, and the biotic mechanisms that underpin plant–soil feedback and its role in regulating plant species coexistence and vegetation dynamics. Well-established trait-based frameworks from plant ecology have been adapted to explain how and why plant traits and below-ground communities and processes are coordinated, including the coordination of root traits and root-associated microbes across species and biomes. Trait-based frameworks have also enabled better prediction of how vegetation shifts due to global change impact on ecosystem services, including the sequestration of C in soil.

It has long been known that to fully understand how plants shape below-ground communities and their functioning requires consideration of the organisms that feed on plants. A particular conundrum, however, is understanding how and why herbivore effects on soil biota and ecosystem functioning vary so much from place to place, with studies reporting positive, negative, or neutral effects. This variability has been put down to factors such as soil fertility, climate, and vegetation type, which all undoubtedly play a role. But recent work has shed light on the importance of biotic factors, especially the type, size, and diversity of herbivores, as important drivers of herbivore effects on soil biota, and predators that indirectly affect below-ground communities by creating landscapes of fear. It is clear that large herbivores, both wild and domestic, play key roles in the sequestration of C in soil,

The Ecology of Soil. Richard D. Bardgett, Oxford University Press. © Richard D. Bardgett (2025). DOI: 10.1093/9780191858710.003.0008

with implications for land management and climate mitigation. For instance, evidence points to potential positive effects of livestock grazing on soil C storage at light intensities, and for wild herbivores playing key roles in enhancing the persistence of soil C.

The topic of soil ecology that arguably has seen the most research activity in recent years relates to climate change impacts on soils, as highlighted in Chapter 6. We now have a better grasp of the mechanisms by which climate change impacts below-ground communities, both direct and indirect, and the potential, albeit highly variable, consequences for biogeochemical cycles and C-cycle feedbacks. We know that below-ground responses to climate change are highly context dependent and are modified by other global change factors, such as N enrichment and land use, which can amplify, suppress, or even neutralise climate change-driven impacts on soil microbial processes and C-cycle feedbacks. Evidence is mounting that climate change can disrupt ecological interactions between plants and soil organisms that have evolved over time, but the mechanisms involved and consequences for ecosystems remain elusive. This uncertainty not only makes it difficult to predict future responses of soil biota and their functions to climate change, but also hampers the integration of soil ecology information into C-cycling models and land management decisions for climate mitigation.

The last topic to highlight is the contribution of soils and their biodiversity to meeting sustainability challenges. This is a relatively new topic, but it is fast moving and we are starting to gain a better understanding of the potential for soil ecology knowledge to be harnessed to help address sustainability challenges. As highlighted in Chapter 7, there are several emerging opportunities in this area, such as the use of 'wild' microbial inoculants to accelerate the restoration of degraded ecosystems, the development of crop breeding programmes targeted at root traits that promote beneficial rhizosphere microbiomes, and integrating crop diversification into farming systems to enhance soil fertility, soil C sequestration and its stabilisation, and resilience of soils to climate extremes. But what is also becoming clear is that to be effective, these and other interventions targeted at sustainable soils need to be tailored to local conditions; how this is done is not always clear.

Despite the recent growth in soil ecology research, we certainly don't know it all. In fact, substantial gaps in our knowledge of the below-ground world remain and generalisations are still hard to find. We have growing amounts of data on the ecology of soil and can draw on an ever-increasing number of case studies to illustrate what soil organisms do in certain contexts, and how they might respond to stresses or disturbances. But despite all this new knowledge, it is still difficult to explain why soil biodiversity varies from one place to another, or to predict what the consequences of a change in soil biodiversity, for instance resulting from climate or land use change, are for ecosystem function. Also, despite tremendous advances in technology, we still can't accurately quantify how many species of soil-dwelling organisms are present in soil and what they do, and for some taxonomic groups, such as viruses and tardigrades, we know virtually nothing about their diversity and functional roles. In fact, our knowledge of the diversity and distribution of soil organisms still lags far behind what we know about biodiversity above-ground, and what we do know is limited to relatively few ecosystems and taxonomic groups, with a strong bias to the northern hemisphere.

So, what are the big topics, or gaps in understanding, that will occupy soil ecologists over the coming decade? It is difficult to predict the future in any field, but ten partly overlapping challenges that emerge from this book are listed below, albeit in no particular order.

1. Broad-scale soil biodiversity patterns
   Global and regional maps are being produced for an ever-growing range of soil organisms. Yet, we still lack a good understanding of broad-scale soil biodiversity patterns at a regional or global scale, and the ecological and evolutionary factors driving these patterns. Significant geographic and taxonomic gaps in our understanding of broad-scale soil biodiversity patterns exist and, as a result, the data we do have are unequally distributed. Research in this area will therefore continue at pace over the

next decade, with an expansion of regional and global soil surveys to cover a broader range of soil organisms and increasing use of artificial intelligence (AI) tools, such as machine learning, for prediction of broad-scale soil biodiversity patterns. However, experimental studies distributed across multiple sites, while challenging, are needed to identify the abiotic and biotic drivers of regional and global soil biodiversity patterns, and the response of soil taxa to global change.

2. Temporal soil ecology
Much soil ecology research is still focussed on samples taken at single or relatively few timepoints. As such, we know little about how below-ground communities, and the functions they perform, change over different timescales, whether that be daily, seasonal or decadal, and how the rate and direction of change is modified by global change. We need an improved mechanistic understanding of how below-ground communities resist and recover from shocks to soil systems, such as climate extremes, and of legacy effects of these perturbations for plant communities and ecosystem functions. Over the next decade, increases in the accessibility of high throughput methods for profiling below-ground communities, combined with developments in sensors technology for measuring soil functions over time, will open the door to greater temporal resolution of experimental data and help track the dynamics of soil biota and their functioning in response to global change. Such work will improve our understanding of how below-ground communities respond to global change and what the consequences are for biogeochemical cycles and C-cycle feedbacks.

3. Soil biodiversity and ecosystem function
There have been many recent advances in our understanding of biodiversity–function relationships in soil, and there is no reason to expect the pace of this research topic to slow. For example, much remains to be learned about the mechanisms by which different facets of soil biodiversity influence ecosystem functions and their stability, considering the many biotic interactions that operate in soil. We know little about what different soil taxa do in soil and how fast they do it, or how biodiversity–function relationships in soil are moderated by soil fertility and the structural complexity of soil. Trait-based approaches offer a way forward and will no doubt be used more to unravel the role of functional diversity and/or the traits of dominant soil taxa as drivers of ecosystem functions and their response to global change. The next decade will see an increased use of metagenomics and metatranscriptomics to measure functional consequences of changes in soil microbial communities and identify microbial traits of importance for ecosystem functions and their stability in the face of perturbations linked to land use and climate change. Given that most biodiversity–function studies focus on soil microbes, there will be a renewed interest in the functional roles of soil animal communities across trophic levels in regulating ecosystem function, and the inclusion of soil food web information in biogeochemical models.

4. It's all about roots
Interest in roots and the role of root traits as drivers of ecosystem functions has increased considerably in recent years and will likely continue into the next decade. However, important questions remain about the coordination of root traits and below-ground communities across plant species and biomes, and how root traits shape rhizosphere communities and ecosystem processes, including the formation, persistence, and loss of SOC, and how this varies across ecosystems and in response to global change. Topics that will occupy soil ecologists over the next decade include how variation in roots and their symbionts regulate SOC sequestration and persistence by modifying decomposition processes and the formation of MAOM, and to what extent root exudation regulates SOC loss via the priming effect. The availability of root trait data is increasing rapidly, which opens the door to a more predictive understanding of how root trait variation

and shifts in root traits resulting from global change impact ecosystem properties and processes at local and regional scales. Roots are important for agricultural systems, and crop breeding for root traits to enhance nutrient and water acquisition and soil health will be a research focus in the future.

5. The underappreciated role of above-ground consumers
A lot of soil ecology research ignores plants that grow in and influence the soil, but also the above-ground animals that feed on those plants or on each other. However, interest in the importance of above-ground consumers as drivers of below-ground communities and the functions they perform has increased in recent years and will likely continue. One reason for this has been a recognition that large herbivores play important roles in soil C cycling, and that changes in grazing management can have implications for SOC sequestration and climate mitigation. Yet, uncertainties remain over how above-ground grazers impact below-ground communities and ecosystem function, how grazing management can be changed to promote ecosystem service multifunctionality, and how historical contingencies moderate herbivore impacts on soil biota. Studies addressing these uncertainties and the spatial effects of above-ground consumers on below-ground communities and ecosystem function will therefore become more prominent over the next decade.

6. Climate change and other stressors
A dominant theme of soil ecology over recent years has been exploring how global change, especially climate change, but also intensive agriculture, N enrichment, land contamination, and invasions impact below-ground communities, with ecosystem-level, and in some cases Earth system-level, consequences. But, as noted above, there is still much uncertainty surrounding the mechanisms involved, the context dependency of responses, and the potential for global change factors to interact and amplify below-ground responses. These topics, especially understanding and predicting climate change impacts, will almost certainly continue as dominant themes in soil ecology over the next decade. Additional challenges are the need for a deeper understanding of the influence of climate extremes on below-ground communities, including their capacity to resist and recover from them, and eco-evolutionary responses of below-ground communities to accelerating climate change. Finally, over the coming decade, soil ecologists will have much to contribute towards the integration of soil biota into predictive C cycling models and land management decisions for climate mitigation.

7. Soil C
The processes that regulate soil C storage, especially decomposition, have long been a core theme of soil ecology, with classic texts such as Swift, Heal, and Anderson's *Decomposition in Terrestrial Ecosystems* (1979) being entirely focussed on the topic. Yet, interest in SOC storage has reached another level, driven mainly by its potential to mitigate climate change, but also due to the emergence of new paradigms on the persistence of SOC as an ecosystem property. This interest shows no signs of slowing, as growing numbers of soil ecologists turn their attention to understanding the mechanisms that underpin SOC sequestration and stabilisation, and the development of models that include soil biotic processes in SOC projections. Leading on from this, there will be a continued focus on how best to integrate this new knowledge into land management strategies at a scale that matters for climate mitigation.

8. Below-ground theory
A long-standing question that is still very much open in soil ecology, and community ecology in general, is: are there any general rules or is it just a maze of unique cases? Compared to above-ground, below-ground ecology is still relatively theory free, with most effort being targeted at observational and experimental studies. But things are changing and existing ecological theories from plant and animal ecology are being used, for example, to explain

patterns of below-ground community organisation across space and time, and how the diversity and structure of multitrophic networks influence stability. However, transferring above-ground ecological theory to the soil is not straightforward, especially given the vast array of species and biotic interactions that operate within it and its enormous structural and chemical complexity. But as more theoretical ecologists turn their attention to soil, and soil ecologists become interested in ecological theory, new frameworks will be developed to help understand how soils and their biological communities work, how they will change in the future, and what the consequences of community change will be for the ecosystem function.

9. Technology
   The next decade will see major changes in the way soil ecologists collect data on below-ground communities and their functioning. New omics methods are being developed at a fast pace to interrogate below-ground communities and their function. But also sensor technologies, including eco-acoustics, are being developed to enable real-time measurement of below-ground communities and their activities at meaningful scales. Advances in remote sensing from airborne and satellite platforms will open the door to monitor soil properties at large spatial and temporal scales, and to use vegetation as a proxy for deriving spatial patterns in below-ground communities and how they change over time. Another area that is gaining much recent attention among soil ecologists is the use of AI tools for deciphering patterns in increasingly vast and diverse data sources. As already mentioned, the next decade will see an increased use of AI tools to map and explain patterns of soil biodiversity at regional and global scales, and to develop predictive models and enhance the analysis of real-time soil data. The rise of AI use in soil ecology offers many opportunities to address complex challenges, but, as stressed by others, it will be important to consider ethical issues, such as data integrity and transparency, that may arise from its use.

10. Activating soil ecology knowledge
    The next decade will likely see increased efforts to raise awareness of the importance of soil life for a healthy planet and to integrate soil ecology knowledge into sustainable land management decisions for agriculture, forestry, biodiversity restoration, and climate mitigation. This will not be easy, especially given the need for management interventions that target soil ecology to be tailored to local conditions. Also, for many interventions, such as the use of inoculants and other soil amendments, the jury is still out. But progress is already being made, and interest in these topics will only increase in future years, given the pressing nature of the sustainability challenges we face, and the fact that soil ecology is at their heart. Although a long-standing topic, there is a pressing need from a wide range of stakeholders, including farmers, land managers, and policy makers, to measure and monitor soil health, which will continue to drive interest in this topic. Put simply, there is still little consensus on which combinations of soil properties, whether biological, chemical, or physical, constitute a healthy soil and what the best measures are to measure it. Given these challenges, the next decade will see a new age of research to integrate soil ecology knowledge into, and help tailor, land management strategies to meet sustainability goals and to develop systems to classify and measure soil health.

As a final note, it is important to stress that the above challenges cannot be met by soil ecologists alone. We live in an age of collaboration, and fruitful collaborations between soil scientists, ecologists, technologists, mathematicians, plant scientists, microbiologists, social scientists, policy makers, and those who rely on and manage the soil will be key to delivering future ground-breaking research discoveries in soil ecology and solutions to sustainability challenges.

References

Abbas, F., Merlet, J., Morellet, N. et al. (2012) Roe deer may markedly alter forest nitrogen and phosphorus budgets across Europe. *Oikos*, 121, 1271–1278.

Abdalla, M., Hastings, A., Chadwick, D. R. et al. (2018) Critical review of the impacts of grazing intensity on soil organic carbon storage and other soil quality indicators in extensively managed grasslands. *Agriculture, Ecosystems and Environment*, 253, 62–81.

Aber, T., Bignell, D. E. and Higashi, M. (2000) *Termites: Evolution, Sociality, Symbioses, Ecology*. Springer Science + Business Media, Berlin.

Adamczyk, B., Sietiö, O. M., Straková, P. et al. (2019) Plant roots increase both decomposition and stable organic matter formation in boreal forest soil. *Nature Communications*, 10, 3982.

Adams, B. J., Bardgett, R. D., Ayres, E. et al. (2006) Diversity and distribution of Victoria Land biota. *Soil Biology and Biochemistry*, 38, 3003–3018.

Adams, B. J., Wall, D. H., Virginia, R. A. et al. (2014) Ecological biogeography of the terrestrial nematodes of Victoria Land, Antarctica. *ZooKeys*, 419, 29–71.

Adl, M. S. and Gupta, V. S. (2006) Protists in soil ecology and forest nutrient cycling. *Canadian Journal of Forest Research*, 36, 1805–1817.

Afzal, M. and Adams, W. A. (1992) Heterogeneity of soil mineral nitrogen in pasture grazed by cattle. *Soil Science Society of America Journal*, 56, 1160–1166.

Agrawal, A. A., Laforsch, C. and Tollrian, R. (1999) Transgenerational induction of defences in animals and plants. *Nature*, 401, 60–63.

Ahonen, S. H. K., Ylänne, H., Väisänen, M. et al. (2021) Reindeer grazing history determines the responses of subarctic soil fungal communities to warming and fertilization. *New Phytologist*, 232, 788–801.

Al-Mufti, M. M., Sydes, C. L., Furness, S. B. et al. (1977) A quantitative analysis of shoot phenology and dominance in herbaceous vegetation. *Journal of Ecology*, 65, 759–791.

Aleklett, K., Ohlsson, P., Bengtsson, M. and Hammer, E. C. (2021) Fungal foraging behaviour and hyphal space exploration in micro-structured Soil Chips. *The ISME Journal*, 15, 1782–1793.

Allison, S.D., Czimczik, C.I. and Treseder, K.K. (2008) Microbial activity and soil respiration under nitrogen addition in Alaskan boreal forest. *Global Change Biology*, 14, 1156–1168.

Allsup, C. M., George, I. and Lankau, R. A. (2023) Shifting microbial communities can enhance tree tolerance to changing climates. *Science*, 380, 835–840.

Alori, E. T., Glick, B. R. and Babalola, O. O. (2017) Microbial phosphorus solubilization and its potential for use in sustainable agriculture. *Frontiers in Microbiology*, 8, 971.

Amelung, W., Bossio, D., De Vries, W. et al. (2020) Towards a global-scale soil climate mitigation strategy. *Nature Communications*, 11, 5427.

Amundson, R., Guo, Y. and Gong, P. (2003) Soil diversity and land use in the United States. *Ecosystems*, 6, 470–482.

Andersen, K. M., Mayor, J. R. and Turner, B. L. (2017) Plasticity in nitrogen uptake among plant species with contrasting nutrient acquisition strategies in a tropical forest. *Ecology*, 98, 1388–1398.

Anderson, J. M. (1978) Inter- and intra-habitat relationships between woodland *Cryptostigmata* species diversity and the diversity of soil and litter microhabitats. *Oecologia*, 32, 341–348.

Andresen, E. and Levey, D. J. (2004) Effects of dung and seed size on secondary dispersal, seed predation, and seedling establishment of rain forest trees. *Oecologia*, 139, 45–54.

Andriuzzi, W. S. and Wall, D. H. (2018a) Grazing and resource availability control soil nematode body size and abundance–mass relationship in semi-arid grassland. *Journal of Animal Ecology*, 87, 1407–1417.

Andriuzzi, W. S. and Wall, D. H. (2018b) Soil biological responses to, and feedbacks on, trophic rewilding. *Philosophical Transactions of the Royal Society B: Biological Sciences*, 373, 20170448.

Angst, G., Mueller, C. W., Prater, I. et al. (2019) Earthworms act as biochemical reactors to convert labile plant compounds into stabilized soil microbial necromass. *Communications Biology*, 2, 441.

Angst, G., Mueller, K. E., Castellano, M. J. et al. (2023) Unlocking complex soil systems as carbon sinks: multi-pool management as the key. *Nature Communications*, 14, 2967.

Anthony, M. A., Bender, S. F. and Van der Heijden, M. G. A. (2023) Enumerating soil biodiversity. *Proceedings of the National Academy of Sciences*, 120, e2304663120.

Antoninka, A., Faist, A., Rodriguez-Caballero, E. et al. (2020) Biological soil crusts in ecological restoration: emerging research and perspectives. *Restoration Ecology*, 28, 53–58.

Arellano-Caicedo, C., Ohlsson, P., Bengtsson, M. et al. (2021) Habitat geometry in artificial microstructure affects bacterial and fungal growth, interactions, and substrate degradation. *Communications Biology*, 4, 1226.

Argüello, A., O'Brien, M. J., Van der Heijden, M. G. A. et al. (2016) Options of partners improve carbon for phosphorus trade in the arbuscular mycorrhizal mutualism. *Ecology Letters*, 19, 648–656.

Asner, G. P. and Levick, S. R. (2012) Landscape-scale effects of herbivores on treefall in African savannas. *Ecology Letters*, 15, 1211–1217.

Atkins, J. L., Long, R. A., Pansu, J. et al. (2019) Cascading impacts of large-carnivore extirpation in an African ecosystem. *Science*, 364, 173–177.

Atwood, T. B., Valentine, S. A., Hammill, E. et al. (2020) Herbivores at the highest risk of extinction among mammals, birds, and reptiles. *Science Advances*, 6, eabb8458.

Augustine, D. J. and Frank, D. A. (2001) Effects of migratory grazers on spatial heterogeneity of soil nitrogen properties in a grassland ecosystem. *Ecology*, 82, 3149–3162.

Aupic-Samain, A., Baldy, V., Delcourt, N. et al. (2021) Water availability rather than temperature control soil fauna community structure and prey–predator interactions. *Functional Ecology*, 35, 1550–1559.

Austin, A. T. (2011) Has water limited our imagination for arid land biogeochemistry? *Trends in Ecology and Evolution*, 26, 229–235.

Austin, A. T., Vivanco, L., González-Arzac, A. and Pérez, L. I. (2014) There's no place like home? An exploration of the mechanisms behind plant litter–decomposer affinity in terrestrial ecosystems. *New Phytologist*, 204, 307–314.

Averill, C., Anthony, M. A., Baldrian, P. et al. (2022) Defending Earth's terrestrial microbiome. *Nature Microbiology*, 7, 1717–1725.

Averill, C., Cates, L. L., Dietze, M. C. and Bhatnagar, J. M. (2019) Spatial vs. temporal controls over soil fungal community similarity at continental and global scales. *The ISME Journal*, 13, 2082–2093.

Ayres, E., Dromph, K. M. and Bardgett, R. D. (2006) Do plant species encourage soil biota that specialise in the rapid decomposition of their litter? *Soil Biology and Biochemistry*, 38, 183–186.

Ayres, E., Dromph, K. M., Cook, R. et al. (2007) The influence of below-ground herbivory and defoliation of a legume on nitrogen transfer to neighbouring plants. *Functional Ecology*, 21, 256–263.

Ayres, E., Heath, J., Possell, M. et al. (2004) Tree physiological responses to above-ground herbivory directly modify below-ground processes of soil carbon and nitrogen cycling. *Ecology Letters*, 7, 469–479.

Ayres, E., Steltzer, H., Berg, S. and Wall, D. H. (2009) Soil biota accelerate decomposition in high-elevation forests by specializing in the breakdown of litter produced by the plant species above them. *Journal of Ecology*, 97, 901–912.

Baas Becking, L. G. M. (1934) *Geobiologie of Inleiding tot de Milieukunde.* W. P. Van Stockum & Zoon, The Hague.

Babikova, Z., Gilbert, L., Bruce, T. J. A. et al. (2013) Underground signals carried through common mycelial networks warn neighbouring plants of aphid attack. *Ecology Letters*, 16, 835–843.

Babin, D., Deubel, A., Jacquiod, S. et al. (2019) Impact of long-term agricultural management practices on soil prokaryotic communities. *Soil Biology and Biochemistry*, 129, 17–28.

Bach, E. M., Ramirez, K. S., Fraser, T. D. and Wall, D. H. (2020) Soil biodiversity integrates solutions for a sustainable future. *Sustainability*, 12, 2662.

Bach, E. M., Williams, R. J., Hargreaves, S. K. et al. (2018) Greatest soil microbial diversity found in microhabitats. *Soil Biology and Biochemistry*, 118, 217–226.

Bacher, M. G., Fenton, O., Bondi, G. et al. (2018) The impact of cattle dung pats on earthworm distribution in grazed pastures. *BMC Ecology*, 18, 59.

Badri, D. V., Chaparro, J. M., Zhang, R. et al. (2013) Application of natural blends of phytochemicals derived from the root exudates of *Arabidopsis* to the soil reveal that phenolic-related compounds predominantly modulate the soil microbiome. *Journal of Biological Chemistry*, 288, 4502–4512.

Baetz, U. and Martinoia, E. (2014) Root exudates: the hidden part of plant defense. *Trends in Plant Science*, 19, 90–98.

Bagchi, S. and Ritchie, M. E. (2010) Introduced grazers can restrict potential soil carbon sequestration through impacts on plant community composition. *Ecology Letters*, 13, 959–968

Bahram, M., Hildebrand, F., Forslund, S. K. et al. (2018) Structure and function of the global topsoil microbiome. *Nature*, 560, 233–237.

Bai, T., Wang, P., Hall, S. J. et al. (2020) Interactive global change factors mitigate soil aggregation and carbon change in a semi-arid grassland. *Global Change Biology*, 26, 5320–5332.

Bai, X., Huang, Y., Ren, W. et al. (2019) Responses of soil carbon sequestration to climate-smart agriculture practices: a meta-analysis. *Global Change Biology*, 25, 2591–2606.

Bai, Y. and Cotrufo, M. F. (2022) Grassland soil carbon sequestration: current understanding, challenges, and solutions. *Science*, 377, 603–608.

Bai, Y., Wu, J., Clark, C. M. et al. (2012) Grazing alters ecosystem functioning and C:N:P stoichiometry of grasslands along a regional precipitation gradient. *Journal of Applied Ecology*, 49, 1204–1215.

Bailey, D. W. and Provenza F. D. (2008) Mechanisms determining large-herbivore distribution. In: *Resource Ecology: Spatial and Temporal Dynamics of Foraging* (Prins, H. H. T. and Van Langevelde, F., eds), pp. 7–28. Springer Netherlands, Dordrecht.

Bakker, E. S., Gill, J. L., Johnson, C. N. et al. (2016) Combining paleo-data and modern exclosure experiments to assess the impact of megafauna extinctions on woody vegetation. *Proceedings of the National Academy of Sciences*, 113, 847–855.

Bakker, E. S., Olff, H., Boekhoff, M. et al. (2004) Impact of herbivores on nitrogen cycling: contrasting effects of small and large species. *Oecologia*, 138, 91–101.

Baldwin, I. T., Olson, R. K. and Reiners, W. A. (1983) Protein binding phenolics and the inhibition of nitrification in subalpine balsam fir soils. *Soil Biology and Biochemistry*, 15, 419–423.

Balfour, E. B. (1943) *The Living Soil*. Faber and Faber, London.

Banerjee, S., Walder, F., Büchi, L. et al. (2019) Agricultural intensification reduces microbial network complexity and the abundance of keystone taxa in roots. *The ISME Journal*, 13, 1722–1736.

Banerjee, S., Zhao, C., Garland, G. et al. (2024) Biotic homogenization, lower soil fungal diversity and fewer rare taxa in arable soils across Europe. *Nature Communications*, 15, 327.

Baraniya, D., Nannipieri, P., Kublik, S. et al. (2018) The impact of the diurnal cycle on the microbial transcriptome in the rhizosphere of barley. *Microbial Ecology*, 75, 830–833.

Barberán, A., McGuire, K. L., Wolf, J. A. et al. (2015) Relating belowground microbial composition to the taxonomic, phylogenetic, and functional trait distributions of trees in a tropical forest. *Ecology Letters*, 18, 1397–1405.

Barberán, A., Ramirez, K. S., Leff, J. W. et al. (2014) Why are some microbes more ubiquitous than others? Predicting the habitat breadth of soil bacteria. *Ecology Letters*, 17, 794–802.

Bardgett, R. D. (2005) *The Biology of Soil: A Community and Ecosystem Approach*. Oxford University Press, Oxford.

Bardgett, R. D. (2017) Plant trait-based approaches for interrogating belowground function. *Biology and Environment: Proceedings of the Royal Irish Academy*, 117B, 1–13.

Bardgett, R. D. (2023) The hidden majority in soil. *Proceedings of the National Academy of Sciences*, 120, e2312358120.

Bardgett, R. D. and Caruso, T. (2020) Soil microbial community responses to climate extremes: resistance, resilience and transitions to alternative states. *Philosophical Transactions of the Royal Society B*, 375, 20190112.

Bardgett, R. D. and Chan, K. F. (1999) Experimental evidence that soil fauna enhance nutrient mineralization and plant nutrient uptake in montane grassland ecosystems. *Soil Biology and Biochemistry*, 31, 1007–1014.

Bardgett, R. D. and Griffiths, B. S. (1997) Ecology and biology of soil protozoa, nematodes and microarthropods. In: *Modern Soil Microbiology* (Van Elsas, J. D., Trevors, J. T. and Wellington, E., eds.), pp. 129–163. Marcell Dekker, New York.

Bardgett, R. D. and McAlister, E. (1999) The measurement of soil fungal:bacterial biomass ratios as an indicator of ecosystem self-regulation in temperate meadow grasslands. *Biology and Fertility of Soils*, 29, 282–290.

Bardgett, R. D. and Van der Putten, W. H. (2014) Belowground biodiversity and ecosystem functioning. *Nature*, 515, 505–511.

Bardgett, R. D. and Wardle, D. A. (2003) Herbivore mediated linkages between above-ground and below-ground communities. *Ecology*, 84, 2258–2268.

Bardgett, R. D. and Wardle, D. A. (2010) *Aboveground–Belowground Linkages: Biotic Interactions, Ecosystem Processes, and Global Change*. Oxford University Press, Oxford.

Bardgett, R. D., Anderson, J. M., Behan-Pelletier, B. et al. (2001a) The influence of soil biodiversity on hydrological pathways and the transfer of materials between terrestrial and aquatic ecosystems. *Ecosystems*, 4, 421–429.

Bardgett, R. D., Bowman, W. D., Kaufmann, R. and Schmidt, S. K. (2005a) A temporal approach to linking aboveground and belowground ecology. *Trends in Ecology and Evolution*, 20, 634–641.

Bardgett, R. D., Bullock, J. M., Lavorel, S. et al. (2021) Combatting global grassland degradation. *Nature Reviews Earth and Environment*, 2, 720–735.

Bardgett, R. D., Denton, C. S. and Cook, R. (1999a) Belowground herbivory promotes soil nutrient transfer and root growth in grassland. *Ecology Letters*, 2, 357–360.

Bardgett, R. D., Frankland, J. C. and Whittaker, J. B. (1993a) The effects of agricultural practices on the soil biota of some upland grasslands. *Agriculture, Ecosystems and Environment*, 45, 25–45.

Bardgett, R. D., Freeman, C. and Ostle, N. J. (2008) Microbial contributions to climate change through carbon cycle feedbacks. *The ISME Journal*, 2, 805–814.

Bardgett, R. D., Jones, A. C., Jones, D. L. et al. (2001b) Soil microbial community patterns related to the history and intensity of grazing in sub-montane ecosystems. *Soil Biology and Biochemistry*, 33, 1653–1664.

Bardgett, R. D., Manning, P., Morriën, E. and De Vries, F. T. (2013) Hierarchical responses of plant–soil interactions to climate change: consequences for the global carbon cycle. *Journal of Ecology*, 101, 334–343.

Bardgett, R. D., Mawdsley, J. L., Edwards, S. et al. (1999b) Plant species and nitrogen effects on soil biological properties of temperate upland grasslands. *Functional Ecology*, 13, 650–660.

Bardgett, R. D., Mommer, L. and De Vries, F. T. (2014) Going underground: root traits as drivers of ecosystem processes. *Trends in Ecology and Evolution*, 29, 692–699.

Bardgett, R. D., Richter, A., Bol, R. et al. (2007) Heterotrophic microbial communities use ancient carbon following glacial retreat. *Biology Letters*, 3, 487–490.

Bardgett, R. D., Smith, R. S., Shiel, R. S. et al. (2006) Parasitic plants indirectly regulate below-ground properties in grassland ecosystems. *Nature*, 439, 969–972.

Bardgett, R. D., Speir, T. W., Ross, D. J. et al. (1994) Impact of pasture contamination by copper, chromium and arsenic timber preservative on soil microbial properties and nematodes. *Biology and Fertility of Soils*, 18, 71–79.

Bardgett, R. D., Streeter, T. C. and Bol, R. (2003) Soil microbes compete effectively with plants for organic nitrogen inputs to temperate grasslands. *Ecology*, 84, 1277–1287.

Bardgett, R. D., Streeter, T. C., Cole, L. and Hartley, I. R. (2002) Linkages between soil biota, nitrogen availability, and plant nitrogen uptake in a mountain ecosystem in the Scottish Highlands. *Applied Soil Ecology*, 19, 121–134.

Bardgett, R. D., Whittaker, J. B. and Frankland, J. C. (1993b) The diet and food preferences of *Onychiurus procampatus* (Collembola) from upland grassland soils. *Biology and Fertility of Soils*, 16, 296–298.

Bardgett, R. D., Yeates, G. W. and Anderson, J. M. (2005b) Patterns and determinants of soil biological diversity. In: *Biological Diversity and Function in Soils* (Bardgett, R. D., Usher, M. B. and Hopkins, D. W., eds), pp. 100–118. Cambridge University Press, Cambridge.

Barnard, R. L., Osborne, C. A. and Firestone, M. K. (2013) Responses of soil bacterial and fungal communities to extreme desiccation and rewetting. *The ISME Journal*, 7, 2229–2241.

Barnes, C. J., Van der Gast, C. J., McNamara, N. P. et al. (2018) Extreme rainfall affects assembly of the root-associated fungal community. *New Phytologist*, 220, 1172–1184.

Barneze, A. S., van Groenigen, J. W., Philippot, L. et al. (2023) Plant communities can attenuate flooding induced N_2O fluxes by altering nitrogen cycling microbial communities and plant nitrogen uptake. *Soil Biology and Biochemistry*, 185, 109142.

Barnosky, A. D., Matzke, N., Tomiya, S. et al. (2011) Has the Earth's sixth mass extinction already arrived? *Nature*, 471, 51–57.

Barto, E. K. and Rillig, M. C. (2010) Does herbivory really suppress mycorrhiza? A meta-analysis. *Journal of Ecology*, 98, 745–753.

Bastida, F., Eldridge, D. J., Abades, S. et al. (2020) Climatic vulnerabilities and ecological preferences of soil invertebrates across biomes. *Molecular Ecology*, 29, 752–761.

Bastida, F., Eldridge, D. J., García, C. et al. (2021) Soil microbial diversity–biomass relationships are driven by soil carbon content across global biomes. *The ISME Journal*, 15, 2081–2091.

Bates, S. T., Berg-Lyons, D., Caporaso, J. G. et al. (2011) Examining the global distribution of dominant archaeal populations in soil. *The ISME Journal*, 5, 908–917.

Bates, S. T., Clemente, J. C., Flores, G. E. et al. (2013) Global biogeography of highly diverse protistan communities in soil. *The ISME Journal*, 7, 652–659.

Batterman, S. A., Hedin, L. O., Van Breugel, M. et al. (2013) Key role of symbiotic dinitrogen fixation in tropical forest secondary succession. *Nature*, 502, 224–227.

Bauer, J. T., Mack, K. M. L. and Bever, J. D. (2015) Plant–soil feedbacks as drivers of succession: evidence from remnant and restored tallgrass prairies. *Ecosphere*, 6, 158.

Bebber, D. P. and Richards, V. R. (2022) A meta-analysis of the effect of organic and mineral fertilizers on soil microbial diversity. *Applied Soil Ecology*, 175, 104450.

Becklin, K. M., Pallo, M. L. and Galen, C. (2012) Willows indirectly reduce arbuscular mycorrhizal fungal colonization in understorey communities. *Journal of Ecology*, 100, 343–351.

Beer, C., Zimov, N., Olofsson, J. et al. (2020) Protection of permafrost soils from thawing by increasing herbivore density. *Scientific Reports*, 10, 4170.

Beerling, D. J., Epihov, D. Z., Kantola, I. B. et al. (2024) Enhanced weathering in the US Corn Belt delivers carbon removal with agronomic benefits. *Proceedings of the National Academy of Sciences*, 121, e2319436121.

Beerling, D. J., Kantzas, E. P., Lomas, M. R. et al. (2020) Potential for large-scale CO2 removal via enhanced rock weathering with croplands. *Nature*, 583, 242–248.

Beerling, D. J., Leake, J. R., Long, S. P. et al. (2018) Farming with crops and rocks to address global climate, food and soil security. *Nature Plants*, 4, 138–147.

Behan-Pelletier, V. M. (1978) *Diversity, Distribution and Feeding Habits of North American Arctic Soil Acari*. Unpublished Ph.D. thesis, McGill University, Montreal.

Beillouin, D., Corbeels, M., Demenois, J. et al. (2023) A global meta-analysis of soil organic carbon in the Anthropocene. *Nature Communications*, 14, 3700.

Bellingham, P. J., Walker, L. R. and Wardle, D. A. (2001) Differential facilitation by a nitrogen-fixing shrub during primary succession influences relative performance of canopy tree species. *Journal of Ecology*, 89, 861–875.

Belnap, J. (2003) The world at your feet: desert biological soil crusts. *Frontiers in Ecology and the Environment*, 1, 181–189.

Belnap, J. and Lange, O. L. (2003) *Biological Soil Crusts: Structure, Function, and Management*. Springer, Berlin.

Belovsky, G. E. and Slade, J. B. (2000) Insect herbivory accelerates nutrient cycling and increases plant production. *Proceedings of the National Academy of Sciences*, 97, 14412–14417.

Benayas, J. M. R., Newton, A. C., Diaz, A. and Bullock, J. M. (2009) Enhancement of biodiversity and ecosystem services by ecological restoration: a meta-analysis. *Science*, 325, 1121–1124.

Bengtsson, G. and Rundgren, S. (1983) Respiration and growth of a fungus *Mortierella isabellina*, in response to grazing by *Onychiurus armatus* (Collembola). *Soil Biology and Biochemistry*, 15, 469–473.

Bengtsson, G., Erlandsson, A. and Rundgren, S. (1988) Fungal odour attracts soil Collembola. *Soil Biology and Biochemistry*, 20, 25–30.

Beniston, M. (2012) Is snow in the Alps receding or disappearing? *Wiley Interdisciplinary Reviews: Climate Change*, 3, 349–358.

Bennett, J. A. and Klironomos, J. (2019) Mechanisms of plant–soil feedback: interactions among biotic and abiotic drivers. *New Phytologist*, 222, 91–96.

Bennett, J. A., Maherali, H., Reinhart, K. O. et al. (2017) Plant–soil feedbacks and mycorrhizal type influence temperate forest population dynamics. *Science*, 355, 181–184.

Bérard, A., Sassi, M. B., Kaisermann, A. and Renault, P. (2015) Soil microbial community responses to heat wave components: drought and high temperature. *Climate Research*, 66, 243–264.

Berendsen, R. L., Pieterse, C. M. and Bakker, P. A. H. M. (2012) The rhizosphere microbiome and plant health. *Trends in Plant Science*, 17, 478–486.

Berg, M. P. and Bengtsson, J. (2007) Temporal and spatial variability in soil food web structure. *Oikos*, 116, 1789–1804.

Bergmann, J., Weigelt, A., Van der Plas, F. et al. (2020) The fungal collaboration gradient dominates the root economics space in plants. *Science Advances*, 6, eaba3756.

Betancur-Corredor, B., Lang, B. and Russell, D. J. (2022) Reducing tillage intensity benefits the soil micro- and mesofauna in a global meta-analysis. *European Journal of Soil Science*, 73, e13321.

Betancur-Corredor, B., Lang, B. and Russell, D. J. (2023) Organic nitrogen fertilization benefits selected soil fauna in global agroecosystems. *Biology and Fertility of Soils*, 59, 1–16.

Bevacqua, E., Maraun, D., Vousdoukas, M. I. et al. (2019) Higher probability of compound flooding from precipitation and storm surge in Europe under anthropogenic climate change. *Science Advances*, 5, eaaw5531.

Bever, J. D. (1994) Feedback between plants and their soil communities in an old field community. *Ecology*, 75, 1965–1977.

Bever, J. D. (2015) Preferential allocation, physio-evolutionary feedbacks, and the stability and environmental patterns of mutualism between plants and their root symbionts. *New Phytologist*, 205, 1503–1514.

Bever, J. D., Schultz, P. A., Miller, R. M. et al. (2003) Prairie mycorrhizal fungi inoculant may increase native plant diversity on restored sites (Illinois). *Ecological Restoration*, 21, 311–312.

Bever, J. D., Westover, K. M. and Antonovics, J. (1997) Incorporation of the soil community into plant population dynamics: the utility of the feedback approach. *Journal of Ecology*, 85, 561–573.

Bezemer, T. M. and Van Dam, N. M. (2005) Linking aboveground and belowground interactions via induced plant defenses. *Trends in Ecology and Evolution*, 20, 617–624.

Bezemer, T. M., De Deyn, G. B., Bossinga, T. M. et al. (2005) Soil community composition drives aboveground plant–herbivore–parasitoid interactions. *Ecology Letters*, 8, 652–661.

Bickel, S. and Or, D. (2020) Soil bacterial diversity mediated by microscale aqueous-phase processes across biomes. *Nature Communications*, 11, 116.

Bignell, D. E., Roisin, Y. and Lo, N. (2011) *Biology of Termites: A Modern Synthesis*. Springer, Dordrecht.

Bjorkman, A. D., Myers-Smith, I. H., Elmendorf, S. C. et al. (2018) Plant functional trait change across a warming tundra biome. *Nature*, 562, 57–62.

Blackshaw, R. and Kerry, B. (2008) Root herbivory in agricultural ecosystems. In: *Root Feeders: An Ecosystem Perspective* (Johnson, S. N. and Murray, P. J., eds), pp. 35–53. CAB International, Wallingford.

Blair, A. C., Hanson, B. D., Brunk, G. R. et al. (2005) New techniques and findings in the study of a candidate allelochemical implicated in invasion success. *Ecology Letters*, 8, 1039–1047.

Blankinship, J. C., Niklaus, P. A. and Hungate, B. A. (2011) A meta-analysis of responses of soil biota to global change. *Oecologia*, 165, 553–565.

Blazewicz, S. J., Hungate, B. A., Koch, B. J. et al. (2020) Taxon-specific microbial growth and mortality patterns reveal distinct temporal population responses to rewetting in a California grassland soil. *The ISME Journal*, 14, 1520–1532.

Bloemers, G. F., Hodda, M., Lambshead, P. J. D. et al. (1997) The effects of forest disturbance on diversity of tropical soil nematodes. *Oecologia*, 111, 575–582.

Blomqvist, M. M., Olff, H., Blaauw, M. B. et al. (2000) Interactions between above-and belowground biota: importance for small-scale vegetation mosaics in a grassland ecosystem. *Oikos*, 90, 582–598.

Boag, B. and Yeates, G. W. (1998) Soil nematode biodiversity in terrestrial ecosystems. *Biodiversity and Conservation*, 7, 617–630.

Boddey, R. M., Macedo, R., Tarré, R. M. et al. (2004) Nitrogen cycling in *Brachiaria* pastures: the key to understanding the process of pasture decline. *Agriculture, Ecosystems and Environment*, 103, 389–403.

Boddy, L. (1999) Saprotrophic cord-forming fungi: meeting the challenge of heterogeneous environments. *Mycologia*, 91, 13–32.

Bohlen, P. J. (2002) Earthworms. In: *Encyclopaedia of Soil Science* (Lall, R., ed.), pp. 370–373. Marcell Dekker, New York.

Bohlen, P. J., Scheu, S., Hale, C. M. et al. (2004) Non-native invasive earthworms as agents of change in northern temperate forests. *Frontiers in Ecology and the Environment*, 2, 427–435.

Bokhorst, S., Huiskes, A., Convey, P. et al. (2008) Climate change effects on soil arthropod communities from the Falkland Islands and the Maritime Antarctic. *Soil Biology and Biochemistry*, 40, 1547–1556.

Bokhorst, S., Metcalfe, D. B. and Wardle, D. A. (2013) Reduction in snow depth negatively affects decomposers but impact on decomposition rates is substrate dependent. *Soil Biology and Biochemistry*, 62, 157–164.

Bokhorst, S., Pedersen, S. H., Brucker, L. et al. (2016) Changing Arctic snow cover: a review of recent developments and assessment of future needs for observations, modelling, and impacts. *Ambio*, 45, 516–537.

Boller, B. C. and Nösberger, J. (1987) Symbiotically fixed nitrogen from field-grown white and red clover mixed with ryegrass at low levels of ^{15}N fertilisation. *Plant and Soil*, 104, 219–226.

Bond-Lamberty, B., Bailey, V. L., Chen, M. et al. (2018) Globally rising soil heterotrophic respiration over recent decades. *Nature*, 560, 80–83.

Bonkowski, M. (2004) Protozoa and plant growth: the microbial loop in soil revisited. *New Phytologist*, 162, 617–631.

Bonkowski, M. and Brandt, F. (2002) Do soil protozoa enhance plant growth by hormonal effects? *Soil Biology and Biochemistry*, 34, 1709–1715.

Bonkowski, M., Geoghegan, I. E., Birch, A. N. E. and Griffiths, B. S. (2001) Effects of soil decomposer invertebrates (protozoa and earthworms) on an aboveground phytophagous insect (cereal aphid) mediated through changes in the host plant. *Oikos*, 95, 441–450.

Bonkowski, M., Villenave, C. and Griffiths, B. (2009) Rhizosphere fauna: the functional and structural diversity of intimate interactions of soil fauna with plant roots. *Plant and Soil*, 321, 213–233.

Boone, R. B., Burnsilver, S. B., Worden, J. S. et al. (2008) Large-scale movements of large herbivores. In: *Resource Ecology: Spatial and Temporal Dynamics of Foraging* (Prins, H. H. T. and Van Langevelde, F., eds), pp. 187–206. Springer Netherlands, Dordrecht.

Borrelli, P., Robinson, D. A., Fleischer, L. R. et al. (2017) An assessment of the global impact of 21st century land use change on soil erosion. *Nature Communications*, 8, 2013.

Bösch, Y., Jones, C. M., Finlay, R. et al. (2022) Minimizing tillage modifies fungal denitrifier communities, increases denitrification rates and enhances the genetic potential for fungal, relative to bacterial, denitrification. *Soil Biology and Biochemistry*, 170, 108718.

Bossio, D. A., Cook-Patton, S. C., Ellis, P. W. et al. (2020) The role of soil carbon in natural climate solutions. *Nature Sustainability*, 3, 391–398.

Bossuyt, H., Six, J. and Hendrix, P. F. (2005) Protection of soil carbon by microaggregates within earthworm casts. *Soil Biology and Biochemistry*, 37, 251–258.

Bouffaud, M. L., Renoud, S., Moënne-Loccoz, Y. and Muller, D. (2016) Is plant evolutionary history impacting recruitment of diazotrophs and *nifH* expression in the rhizosphere? *Scientific Reports*, 6, 21690.

Bowker, M. A., Maestre, F. T. and Escolar, C. (2010) Biological crusts as a model system for examining the biodiversity–ecosystem function relationship in soils. *Soil Biology and Biochemistry*, 42, 405–417.

Bowles, T. M., Mooshammer, M., Socolar, Y. et al. (2020) Long-term evidence shows that crop-rotation diversification increases agricultural resilience to adverse growing conditions in North America. *One Earth*, 2, 284–293.

Bowman, W. D., Gartner, J. R., Holland, K. and Wiedermann, M. (2006) Nitrogen critical loads for alpine vegetation and terrestrial ecosystem response: are we there yet? *Ecological Applications*, 16, 1183–1193.

Bowman, W. D., Steltzer, H., Rosenstiel, T. N. et al. (2004) Litter effects of two co-occurring alpine species on plant growth, microbial activity and immobilization of nitrogen. *Oikos*, 104, 336–344.

Bradford, M. A. (2016) Re-visioning soil food webs. *Soil Biology and Biochemistry*, 102, 1–3.

Bradford, M. A., Jones, T. H., Bardgett, R. D. et al. (2002) Impacts of soil faunal community composition on model grassland ecosystems. *Science*, 298, 615–618.

Bradford, M. A., McCulley, R. L., Crowther, T. W. et al. (2019) Cross-biome patterns in soil microbial respiration predictable from evolutionary theory on thermal adaptation. *Nature Ecology and Evolution*, 3, 223–231.

Bradford, M. A., Wieder, W. R., Bonan, G. B. et al. (2016) Managing uncertainty in soil carbon feedbacks to climate change. *Nature Climate Change*, 6, 751–758.

Bradford, M. A., Wood, S. A., Bardgett, R. D. et al. (2014) Discontinuity in the responses of ecosystem processes and multifunctionality to altered soil community composition. *Proceedings of the National Academy of Sciences*, 111, 14478–14483.

Bragazza, L., Parisod, J., Buttler, A. and Bardgett, R. D. (2013) Biochemical plant–soil microbe feedbacks in response to climate warming in peatlands. *Nature Climate Change*, 3, 273–277.

Brand, R. H. and Dunn, C. P. (1998) Diversity and abundance of springtails (Insecta: Collembola) in native and restored tallgrass prairies. *American Midland Naturalist*, 139, 235–242.

Brinkman, E. P., Raaijmakers, C. E., De Boer, W. and Van der Putten, W. H. (2017) Changing soil legacies to direct restoration of plant communities. *AoB Plants*, 9, plx038.

Briones, M. J. I. and Schmidt, O. (2017) Conventional tillage decreases the abundance and biomass of earthworms and alters their community structure in a global meta-analysis. *Global Change Biology*, 23, 4396–4419.

Briones, M. J. I., Ineson P. and Poskitt, J. (1998) Climate change and *Cognettia sphagnetorum*: effects on carbon dynamics in organic soils. *Functional Ecology*, 12, 528–535.

Broadbent, A. A. D., Bahn, M., Pritchard, W. J. et al. (2022) Shrub expansion modulates belowground impacts of changing snow conditions in alpine grasslands. *Ecology Letters*, 25, 52–64.

Broadbent, A. A. D., Newbold, L. K., Pritchard, W. J. et al. (2024) Climate change disrupts the seasonal coupling of plant and soil microbial nutrient cycling in an alpine ecosystem. *Global Change Biology*, 30, e17245.

Broadbent, A. A. D., Snell, H. S. K., Michas, A. et al. (2021) Climate change alters temporal dynamics of alpine soil microbial functioning and biogeochemical cycling via earlier snowmelt. *The ISME Journal*, 15, 2264–2275.

Brooker, R. and van der Wal, R. (2003) Can soil temperature direct the composition of high Arctic plant communities? *Journal of Vegetation Science*, 14, 535–542.

Brooker, R. W., Bennett, A. E., Cong, W. F. et al. (2015) Improving intercropping: a synthesis of research in agronomy, plant physiology and ecology. *New Phytologist*, 206, 107–117.

Brown, V. K. and Gange, A. C. (1992) Secondary plant succession: how is it modified by insect herbivory? *Vegetatio*, 101, 3–13.

Buckeridge, K. M. and Grogan, P. (2008) Deepened snow alters soil microbial nutrient limitations in arctic birch hummock tundra. *Applied Soil Ecology*, 39, 210–222.

Buckeridge, K. M., Creamer, C. and Whitaker, J. (2022) Deconstructing the microbial necromass continuum to inform soil carbon sequestration. *Functional Ecology*, 36, 1396–1410.

Bueno, C. G., Moora, M., Gerz, M. et al. (2017) Plant mycorrhizal status, but not type, shifts with latitude and elevation in Europe. *Global Ecology and Biogeography*, 26, 690–699.

Bukoski, J. J., Cook-Patton, S. C., Melikov, C. et al. (2022) Rates and drivers of aboveground carbon accumulation in global monoculture plantation forests. *Nature Communications*, 13, 4206.

Bump, J. K., Peterson, R. O. and Vucetich, J. A. (2009a) Wolves modulate soil nutrient heterogeneity and foliar nitrogen by configuring the distribution of ungulate carcasses. *Ecology*, 90, 3159–3167.

Bump, J. K., Webster, C. R., Vucetich, J. A. et al. (2009b) Ungulate carcasses perforate ecological filters and create biogeochemical hotspots in forest herbaceous layers allowing trees a competitive advantage. *Ecosystems*, 12, 996–1007.

Butenko, K. O., Gongalsky, K. B., Korobushkin, D. I. et al. (2017) Forest fires alter the trophic structure of soil nematode communities. *Soil Biology and Biochemistry*, 109, 107–117.

Cai, A., Han, T., Ren, T. et al. (2022) Declines in soil carbon storage under no tillage can be alleviated in the long run. *Geoderma*, 425, 116028.

Callaway, R. M. and Ridenour, W. M. (2004) Novel weapons: invasive success and the evolution of increased competitive ability. *Frontiers in Ecology and the Environment*, 2, 436–443.

Callaway, R. M., Cipollini, D., Barto, K. et al. (2008) Novel weapons: invasive plant suppresses fungal mutualists in America but not in its native Europe. *Ecology*, 89, 1043–1055.

Callaway, R. M., Ridenour, W. M., Laboski, T. et al. (2005) Natural selection for resistance to the allelopathic effects of invasive plants. *Journal of Ecology*, 93, 576–583.

Callaway, R. M., Thelen, G. C., Rodriguez, A. and Holben, W. E. (2004) Soil biota and exotic plant invasion. *Nature*, 427, 731–733.

Cameron, E. K., Martins, I. S., Lavelle, P. et al. (2018) Global gaps in soil biodiversity data. *Nature Ecology and Evolution*, 2, 1042–1043.

Campbell, J. F., Lewis, E. E., Stock, S. P. et al. (2003) Evolution of host search strategies in entomopathogenic nematodes. *Journal of Nematology*, 35,142–145.

Camy, C., Dreyer, E., Delatour, C. and Marçais, B. (2003) Responses of the root rot fungus *Collybia fusipes* to soil waterlogging and oxygen availability. *Mycological Research*, 107, 1103–1109.

Canarini, A., Fuchslueger, L., Schnecker, J. et al. (2024) Soil fungi remain active and invest in storage compounds during drought independent of future climate conditions. *Nature Communications*, 15, 10410.

Canarini, A., Schmidt, H., Fuchslueger, L. et al. (2021) Ecological memory of recurrent drought modifies soil processes via changes in soil microbial community. *Nature Communications*, 12, 5308.

Cania, B., Vestergaard, G., Suhadolc, M. et al. (2020) Site-specific conditions change the response of bacterial producers of soil structure-stabilizing agents such as exopolysaccharides and lipopolysaccharides to tillage intensity. *Frontiers in Microbiology*, 11, 568.

Cantarel, A. A. M., Bloor, J. M. G., Pommier, T. et al. (2012) Four years of experimental climate change modifies the microbial drivers of N_2O fluxes in an upland grassland ecosystem. *Global Change Biology*, 18, 2520–2531.

Cardinale, B. J., Duffy, J. E., Gonzalez, A. et al. (2012) Biodiversity loss and its impact on humanity. *Nature*, 486, 59–67.

Carey, J. C., Tang, J., Templer, P. H. et al. (2016) Temperature response of soil respiration largely unaltered with experimental warming. *Proceedings of the National Academy of Sciences*, 113, 13797–13802.

Carline, K. A., Jones, H. E. and Bardgett, R. D. (2005) Large herbivores affect the stoichiometry of nutrients in a regenerating woodland ecosystem. *Oikos*, 110, 453–460.

Carmona, C. P., Bueno, C. G., Toussaint, A. et al. (2021) Fine-root traits in the global spectrum of plant form and function. *Nature*, 597, 683–687.

Carney, K. M., Hungate, B. A., Drake, B. G. and Megonigal, J. P. (2007) Altered soil microbial community at elevated CO_2 leads to loss of soil carbon. *Proceedings of the National Academy of Sciences*, 104, 4990–4995.

Caruso, T., Hogg, I. D., Nielsen, U. N. et al. (2019a) Nematodes in a polar desert reveal the relative role of biotic interactions in the coexistence of soil animals. *Communications Biology*, 2, 63.

Caruso, T., Schaefer, I., Monson, F. and Keith, A. M. (2019b) Oribatid mites show how climate and latitudinal gradients in organic matter can drive large-scale biodiversity patterns of soil communities. *Journal of Biogeography*, 46, 611–620.

Casalla, R. and Korb, J. (2019) Termite diversity in Neotropical dry forests of Colombia and the potential role of rainfall in structuring termite diversity. *Biotropica*, 51, 165–177.

Castellano, M. J., Mueller, K. E., Olk, D. C. et al. (2015) Integrating plant litter quality, soil organic matter stabilization, and the carbon saturation concept. *Global Change Biology*, 21, 3200–3209.

Ceja-Navarro, J. A., Wang, Y., Ning, D. et al. (2021) Protist diversity and community complexity in the rhizosphere of switchgrass are dynamic as plants develop. *Microbiome*, 9, 96.

Cesarz, S., Reich, P. B., Scheu, S. et al. (2015) Nematode functional guilds, not trophic groups, reflect shifts in soil food webs and processes in response to interacting global change factors. *Pedobiologia*, 58, 23–32.

Ceulemans, T., Bodé, S., Bollyn, J. et al. (2017) Phosphorus resource partitioning shapes phosphorus acquisition and plant species abundance in grasslands. *Nature Plants*, 3, 16224.

Chalupský, J. (1995) Long-term study of Enchytraeidae (Oligochaeta) in man-impacted mountain forest soils in the Czech Republic. *Acta Zoologica Fennica*, 196, 318–320.

Chapin, F. S. (1980) Mineral nutrition of wild plants. *Annual Review of Ecology and Systematics*, 11, 233–260.

Chapin III, F.S., Matson, P.A. and Mooney, H.A. (2002) *Principles of Terrestrial Ecosystem Ecology*. New York, NY: Springer New York.

Chapin, F. S., Moilanen L. and Kielland, K. (1993) Preferential use of organic nitrogen for growth by a non-mycorrhizal arctic sedge. *Nature*, 361, 150–153.

Chapin, F. S., Walker, L. R., Fastie, C. L. and Sharman, L. C. (1994) Mechanisms of primary succession following deglaciation at Glacier Bay, Alaska. *Ecological Monographs*, 64, 149–175.

Chari, N. R. and Taylor, B. N. (2022) Soil organic matter formation and loss are mediated by root exudates in a temperate forest. *Nature Geoscience*, 15, 1011–1016.

Chari, N. R., Tumber-Dávila, S. J., Phillips, R. P. et al. (2024) Estimating the global root exudate carbon flux. *Biogeochemistry*, 167, 895–908.

Chen, C., Chen, H. Y. H., Chen, X. and Huang, Z. (2019a) Meta-analysis shows positive effects of plant diversity on microbial biomass and respiration. *Nature Communications*, 10, 1332.

Chen, J., Luo, Y., García-Palacios, P. et al. (2018a) Differential responses of carbon-degrading enzyme activities to warming: implications for soil respiration. *Global Change Biology*, 24, 4816–4826.

Chen, L., Swenson, N. G., Ji, N. et al. (2019b) Differential soil fungus accumulation and density dependence of trees in a subtropical forest. *Science*, 366, 124–128.

Chen, S., Wang, W., Xu, W. et al. (2018b) Plant diversity enhances productivity and soil carbon storage. *Proceedings of the National Academy of Sciences*, 115, 4027–4032.

Chen, X., Chen, H. Y. H. and Chang, S. X. (2022) Meta-analysis shows that plant mixtures increase soil phosphorus availability and plant productivity in diverse ecosystems. *Nature Ecology and Evolution*, 6, 1112–1121.

Chen, Y. A. and Forschler, B. T. (2016) Elemental concentrations in the frass of saproxylic insects suggest a role in micronutrient cycling. *Ecosphere*, 7, e01300.

Chen, Y., Qin, W., Zhang, Q. et al. (2024) Whole-soil warming leads to substantial soil carbon emission in an alpine grassland. *Nature Communications*, 15, 4489.

Cheng, L., Chen, W., Adams, T. S. et al. (2016) Mycorrhizal fungi and roots are complementary in foraging within nutrient patches. *Ecology*, 97, 2815–2823.

Cheng, W., Parton, W. J., Gonzalez-Meler, M. A. et al. (2014) Synthesis and modeling perspectives of rhizosphere priming. *New Phytologist*, 201, 31–44.

Chiang, F., Mazdiyasni, O. and AghaKouchak, A. (2021) Evidence of anthropogenic impacts on global drought frequency, duration, and intensity. *Nature Communications*, 12, 2754.

Cho, J. C. and Tiedje, J. M. (2000) Biogeography and degree of endemicity of fluorescent *Pseudomonas* strains in soil. *Applied Environmental Microbiology*, 66, 5448–5456.

Chomel, M., Guittonny-Larchevêque, M., Fernandez, C. et al. (2016) Plant secondary metabolites: a key driver of litter decomposition and soil nutrient cycling. *Journal of Ecology*, 104, 1527–1541.

Chomel, M., Lavallee, J. M., Alvarez-Segura, N. et al. (2022) Intensive grassland management disrupts below-ground multi-trophic resource transfer in response to drought. *Nature Communications*, 13, 6991.

Chomel, M., Lavallee, J. M., Alvarez-Segura, N. et al. (2019) Drought decreases incorporation of recent plant photosynthate into soil food webs regardless of their trophic complexity. *Global Change Biology*, 25, 3549–3561.

Chung, Y. A. (2023) The temporal and spatial dimensions of plant–soil feedbacks. *New Phytologist*, 237, 2012–2019.

Clarholm, M. (1981) Protozoan grazing of bacteria in soil—impact and importance. *Microbial Ecology*, 7, 343–350.

Clarholm, M. (1985) Interactions of bacteria, protozoa and plants leading to mineralization of soil nitrogen. *Soil Biology and Biochemistry*, 17, 181–187.

Classen, A. T., Sundqvist, M. K., Henning, J. A. et al. (2015) Direct and indirect effects of climate change on soil microbial and soil microbial–plant interactions: what lies ahead? *Ecosphere*, 6, 1–21.

Clemmensen, K. E., Finlay, R. D., Dahlberg, A. et al. (2015) Carbon sequestration is related to mycorrhizal fungal community shifts during long-term succession in boreal forests. *New Phytologist*, 205, 1525–1536.

Cleveland, C. C., Townsend, A. R., Schimel, D. S. et al. (1999) Global patterns of terrestrial biological nitrogen (N_2) fixation in natural ecosystems. *Global Biogeochemical Cycles*, 13, 623–645.

Cole, A. J., Griffiths, R. I., Ward, S. E. et al. (2019) Grassland biodiversity restoration increases resistance of carbon fluxes to drought. *Journal of Applied Ecology*, 56, 1806–1816.

Cole, L., Bardgett, R. D., Ineson, P. and Adamson, J. K. (2002) Relationships between enchytraeid worms (Oligochaeta), climate change, and the release of dissolved organic carbon from blanket peat in northern England. *Soil Biology and Biochemistry*, 34, 599–607.

Cole, L., Bradford, M. A., Shaw, P. J. and Bardgett, R. D. (2006) The abundance, richness and functional role of soil meso- and macrofauna in temperate grassland—a case study. *Applied Soil Ecology*, 33, 186–198.

Cole, L., Buckland, S. M. and Bardgett, R. D. (2008) Influence of disturbance and nitrogen addition on plant and soil animal diversity in grassland. *Soil Biology and Biochemistry*, 40, 505–514.

Coleman, D. C., Reid, C. P. P. and Cole, C. V. (1983) Biological strategies of nutrient cycling in soil systems. *Advances in Ecological Research*, 13, 1–55.

Comita, L. S. and Engelbrecht, B. M. J. (2014) Drought as a driver of tropical tree species regeneration dynamics and distribution patterns. In: *Forests and Global Change* (Coomes, D. A., Burslem, D. F. R. P. and Simonson, W. D., eds), pp. 216–308. Cambridge University Press, Cambridge.

Conant, R. T., Cerri, C. E., Osborne, B. B. and Paustian, K. (2017) Grassland management impacts on soil carbon stocks: a new synthesis. *Ecological Applications*, 27, 662–668.

Conant, R. T., Drijber, R. A., Haddix, M. L. et al. (2008) Sensitivity of organic matter decomposition to warming varies with its quality. *Global Change Biology*, 14, 868–877.

Conant, R. T., Ryan, M. G., Ågren, G. I. et al. (2011) Temperature and soil organic matter decomposition rates—synthesis of current knowledge and a way forward. *Global Change Biology*, 17, 3392–3404.

Conen, F., Leifeld, J., Seth, B. and Alewell, C. (2006) Warming mobilises young and old soil carbon equally. *Biogeosciences*, 3, 515–519.

Cong, W. F., Hoffland, E., Li, L. et al. (2015a) Intercropping enhances soil carbon and nitrogen. *Global Change Biology*, 21, 1715–1726.

Cong, W. F., Hoffland, E., Li, L. et al. (2015b) Intercropping affects the rate of decomposition of soil organic matter and root litter. *Plant and Soil*, 391, 399–411.

Connell, J. H. (1978) Diversity in tropical rain forests and coral reefs: high diversity of trees and corals is maintained only in a nonequilibrium state. *Science*, 199, 1302–1310.

Conrad, R. (2007) Microbial ecology of methanogens and methanotrophs. *Advances in Agronomy*, 96, 1–63.

Cooper, H. V., Sjögersten, S., Lark, R. M. et al. (2021) Long-term zero-tillage enhances the protection of soil carbon in tropical agriculture. *European Journal of Soil Science*, 72, 2477–2492.

Cordero, I., Leizeaga, A., Hicks, L. C. et al. (2023) High intensity perturbations induce an abrupt shift in soil microbial state. *The ISME Journal*, 17, 2190–2199.

Cornelissen, J. H. C. (1996) An experimental comparison of leaf decomposition rates in a wide range of temperate plant species and types. *Journal of Ecology*, 84, 109–114.

Cornelissen, J. H. C. and Thompson, K. (1997) Functional leaf attribute predict litter decomposition rate in herbaceous plants. *New Phytologist*, 135, 573–582.

Cornelissen, J. H. C., Van Bodegom, P. M., Aerts, R. et al. (2007) Global negative vegetation feedback to climate warming responses of leaf litter decomposition rates in cold biomes. *Ecology Letters*, 10, 619–627.

Cornell, C., Kokkoris, V., Richards, A. et al. (2021) Do bioinoculants affect resident microbial communities? A meta-analysis. *Frontiers in Agronomy*, 3, 753474.

Cortez, J., Hameed, R. and Bouché, M. D. (1989) C and N transfer in soil with or without earthworms fed with ^{14}C- and ^{15}N-labelled wheat straw. *Soil Biology and Biochemistry*, 21, 491–497.

Cortois, R., Schröder-Georgi, T., Weigelt, A. et al. (2016) Plant–soil feedbacks: role of plant functional group and plant traits. *Journal of Ecology*, 104, 1608–1617.

Cotrufo, M. F. and Lavallee, J. M. (2022) Soil organic matter formation, persistence, and functioning: a synthesis of current understanding to inform its conservation and regeneration. *Advances in Agronomy*, 172, 1–66.

Cotrufo, M. F., Ineson, P. and Scott, A. (1998) Elevated CO_2 reduces the nitrogen concentration of plant tissues. *Global Change Biology*, 4, 43–54.

Cotrufo, M. F., Wallenstein, M. D., Boot, C. M. et al. (2013) The Microbial Efficiency–Matrix Stabilization (MEMS) framework integrates plant litter decomposition with soil organic matter stabilization: do labile plant inputs form stable soil organic matter? *Global Change Biology*, 19, 988–995.

Coughenour, M. B. (1991) Biomass and nitrogen responses to grazing of upland steppe on Yellowstone's northern winter range. *Journal of Applied Ecology*, 28, 71–82.

Courtright, E. M., Wall, D. H. and Virginia, R. A. (2001) Determining habitat suitability for soil invertebrates in an extreme environment: the McMurdo Dry Valleys, Antarctica. *Antarctic Science*, 13, 9–17.

Cox, P. M., Betts, R. A., Jones, C. D. et al. (2000) Acceleration of global warming due to carbon-cycle feedbacks in a coupled climate model. *Nature*, 408, 184–187.

Crawley, M. J. (1983) *Herbivory: The Dynamics of Animal–Plant Interactions*. Blackwell Scientific Publications, Oxford.

Creel, S. and Christianson, D. (2008) Relationships between direct predation and risk effects. *Trends in Ecology and Evolution*, 23, 194–201.

Crews T. E. (1999) The presence of nitrogen fixing legumes in terrestrial communities: evolutionary vs. ecological considerations. *Biogeochemistry*, 46, 233–246.

Crews, T. E., Kitayama K., Fownes, J. H. et al. (1995) Changes in soil phosphorus fractions and ecosystem dynamics across a long chronosequence in Hawaii. *Ecology*, 76, 1407–1424.

Crews, T. E., Kurina, L. M. and Vitousek, P. M. (2001) Organic matter and nitrogen accumulation and nitrogen fixation during early ecosystem development in Hawaii. *Biogeochemistry*, 52, 259–279.

Crocker, R. L. and Major, J. (1955) Soil development in relation to vegetation and surface age at Glacier Bay, Alaska. *Journal of Ecology*, 43, 427–448.

Crowther, T. W. and Bradford, M. A. (2013) Thermal acclimation in widespread heterotrophic soil microbes. *Ecology Letters*, 16, 469–477.

Crowther, T. W., Boddy, L. and Jones, T. H. (2011) Outcomes of fungal interactions are determined by soil invertebrate grazers. *Ecology Letters*, 14, 1134–1142.

Crowther, T. W., Todd-Brown, K. E. O., Rowe, C. W. et al. (2016) Quantifying global soil carbon losses in response to warming. *Nature*, 540, 104–108.

Cruz-Martinez, K., Suttle, K. B., Brodie, E. L. et al. (2009) Despite strong seasonal responses, soil microbial consortia are more resilient to long-term changes in rainfall than overlying grassland. *The ISME Journal*, 3, 738–744.

Curtis, P. S. and Wang, X. (1998) A meta-analysis of elevated CO_2 effects on woody plant mass, form, and physiology. *Oecologia*, 113, 299–313.

Curtis, T. P. and Sloan, W. T. (2005) Exploring microbial diversity—a vast below. *Science*, 309, 1331–1333.

Curtis, T. P., Sloan, W. T. and Scannell, J. W. (2002) Estimating prokaryotic diversity and its limits. *Proceedings of the National Academy of Sciences*, 99, 10494–10499.

D'Acunto, L., Andrade, J. F., Poggio, S. L. and Semmartin, M. (2018) Diversifying crop rotation increased metabolic soil diversity and activity of the microbial community. *Agriculture, Ecosystems and Environment*, 257, 159–164.

Dahl, M.B., Söllinger, A., Sigurðsson, P. et al. (2023) Long-term warming-induced trophic downgrading in the soil microbial food web. *Soil Biology and Biochemistry*, 181, p.109044.

Dai, L., Guo, X., Ke, X. et al. (2019) Moderate grazing promotes the root biomass in *Kobresia* meadow on the northern Qinghai–Tibet Plateau. *Ecology and Evolution*, 9, 9395–9406.

Daims, H., Lebedeva, E., Pjevac, P. et al. (2015) Complete nitrification by *Nitrospira* bacteria. *Nature*, 528, 504–509.

Dang, P., Zhang, M., Chen, X. et al. (2024) Plant diversity decreases greenhouse gas emissions by increasing soil and plant carbon storage in terrestrial ecosystems. *Ecology Letters*, 27, e14469.

Daniell, T. J., Husband, R., Fitter, A. H. and Young, J. P. W. (2001) Molecular diversity of arbuscular mycorrhizal fungi colonising arable crops. *FEMS Microbiology Ecology*, 36, 203–209.

D'Antonio, C. M. and Vitousek, P. M. (1992) Biological invasions by exotic grasses, the grass/fire cycle, and global change. *Annual Review of Ecology and Systematics*, 23, 63–87.

Darrouzet-Nardi, A., Steltzer, H., Sullivan, P. F. et al. (2019) Limited effects of early snowmelt on plants, decomposers, and soil nutrients in Arctic tundra soils. *Ecology and Evolution*, 9, 1820–1844.

Darwin, C. R. (1881) *The Formation of Vegetable Mould Through the Action of Worms and Observations on Their Habits*. John Murray, London.

Daufresne, T. (2021) A consumer-driven recycling theory for the impact of large herbivores on terrestrial ecosystem stoichiometry. *Ecology Letters*, 24, 2598–2610.

Davidson, E. A. and Janssens, I. A (2006) Temperature sensitivity of soil carbon decomposition and feedbacks to climate change. *Nature*, 440, 165–173.

Davies, A. B. and Asner, G. P. (2019) Elephants limit aboveground carbon gains in African savannas. *Global Change Biology*, 25, 1368–1382.

Daws, M. I., Grigg, A. H., Tibbett, M. and Standish, R. J. (2019) Enduring effects of large legumes and phosphorus fertiliser on Jarrah forest restoration 15 years after bauxite mining. *Forest Ecology and Management*, 438, 204–214.

Dawson, W. and Schrama, M. (2016) Identifying the role of soil microbes in plant invasions. *Journal of Ecology*, 104, 1211–1218.

Day, N. J., Dunfield, K. E., Johnstone, J. F. et al. (2019) Wildfire severity reduces richness and alters composition of soil fungal communities in boreal forests of western Canada. *Global Change Biology*, 25, 2310–2324.

Debinski, D. M., Wickham, H., Kindscher, K. et al. (2010) Montane meadow change during drought varies with background hydrologic regime and plant functional group. *Ecology*, 91, 1672–1681.

Decaëns, T., Jiménez, J. J., Gioia, C. et al. (2006) The values of soil animals for conservation biology. *European Journal of Soil Biology*, 42, 23–38.

De Deyn, G. B., Cornelissen, J. H. C. and Bardgett, R. D. (2008) Plant functional traits and soil carbon sequestration in contrasting biomes. *Ecology Letters*, 11, 516–531.

De Deyn, G. B., Quirk, H. and Bardgett, R. D. (2011) Plant species richness, identity and productivity differentially influence key groups of microbes in grassland soils of contrasting fertility. *Biology Letters*, 7, 75–78.

De Deyn, G. B., Quirk, H., Oakley, S. et al. (2012) Increased plant carbon translocation linked to overyielding in grassland species mixtures. *PLoS ONE*, 7, e45926.

De Deyn, G. B., Quirk, H., Yi, Z. et al. (2009) Vegetation composition promotes carbon and nitrogen storage in model grassland communities of contrasting soil fertility. *Journal of Ecology*, 97, 864–875.

De Deyn, G. B., Raaijmakers, C. E., Van Ruijven, J. et al. (2004) Plant species identity and diversity effects on different trophic levels of nematodes in the soil food web. *Oikos*, 106, 576–586.

De Deyn, G. B., Raaijmakers, C. E., Zoomer, H. R. et al. (2003) Soil invertebrate fauna enhances grassland succession and diversity. *Nature*, 422, 711–713.

De Graff, M. A., Van Groenigen, K. J., Six, J. et al. (2006) Interactions between plant growth and soil nutrient cycling under elevated CO_2: a meta-analysis. *Global Change Biology*, 12, 2077–2091.

Delgado-Baquerizo, M., Bardgett, R. D., Vitousek, P. M. et al. (2019) Changes in belowground biodiversity during ecosystem development. *Proceedings of the National Academy of Sciences*, 116, 6891–6896.

Delgado-Baquerizo, M., Eldridge, D. J., Hamonts, K. et al. (2018a) Experimentally testing the species–habitat size relationship on soil bacteria: a proof of concept. *Soil Biology and Biochemistry*, 123, 200–206.

Delgado-Baquerizo, M., Eldridge, D. J., Liu, Y. R. et al. (2021) Global homogenization of the structure and function in the soil microbiome of urban greenspaces. *Science Advances*, 7, eabg5809.

Delgado-Baquerizo, M., Maestre, F. T., Reich, P. B. et al. (2016) Microbial diversity drives multifunctionality in terrestrial ecosystems. *Nature Communications*, 7, 10541.

Delgado-Baquerizo, M., Oliverio A. M. Brewer, T. E. et al. (2018b) A global atlas of the dominant bacteria found in soil. *Science*, 359, 320–325.

Delgado-Baquerizo, M., Reich, P. B., Bardgett, R. D. et al. (2020a) The influence of soil age on ecosystem structure and function across biomes. *Nature Communicatons*, 11, 4721.

Delgado-Baquerizo, M., Reich, P. B., Trivedi, C. et al. (2020b) Multiple elements of soil biodiversity drive ecosystem functions across biomes. *Nature Ecology and Evolution*, 4, 210–220.

DeLuca, T. H., Zackrisson, O., Gundale, M. J. and Nilsson, M. C. (2008) Ecosystem feedbacks and nitrogen fixation in boreal forests. *Science*, 320, 1181.

DeLuca, T. H., Zackrisson, O., Nilsson, M. C. and Sellstedt, A. (2002) Quantifying nitrogen-fixation in feather moss carpets of boreal forests. *Nature*, 419, 917–920.

De Mazancourt, C., Loreau, M. and Abbadie, L. (1999) Grazing optimization and nutrient cycling: potential impact of large herbivores in a savannah system. *Ecological Applications*, 9, 784–797.

Derksen, C., Brown, R., Mudryk, L. and Luojus, K. (2015) Arctic–terrestrial snow. State of the climate in 2014. *Bulletin of the American Meteorological Society*, 96, 133–135.

de Ruiter, P. C., Neutel, A. M. and Moore, J. C. (1995) Energetics, patterns of interaction strengths, and stability in real ecosystems. *Science*, 269, 1257–1260.

De Vries, F. T. and Bardgett, R. D. (2012) Plant–microbial linkages and ecosystem nitrogen retention: lessons for sustainable agriculture. *Frontiers in Ecology and the Environment*, 10, 425–432.

De Vries, F. T. and Caruso, T. (2016) Eating from the same plate? Revisiting the role of labile carbon inputs in the soil food web. *Soil Biology and Biochemistry*, 102, 4–9.

De Vries, F. T. and Shade, A. (2013) Controls on soil microbial community stability under climate change. *Frontiers in Microbiology*, 4, 265.

De Vries, F. T., Bloem, J., Quirk, H. et al. (2012a) Extensive management promotes plant and microbial nitrogen retention in temperate grassland. *PLoS ONE*, 7, e51201.

De Vries, F. T., Griffiths, R. I., Bailey, M. et al. (2018) Soil bacterial networks are less stable under drought than fungal networks. *Nature Communications*, 9, 3033.

De Vries, F. T., Hoffland, E., van Eekeren, N. et al. (2006) Fungal/bacterial ratios in grasslands with contrasting nitrogen management. *Soil Biology and Biochemistry*, 38, 2092–2103.

De Vries, F. T., Lau, J., Hawkes, C. and Semchenko, M. (2023) Plant–soil feedback under drought: does history shape the future? *Trends in Ecology and Evolution*, 38, 708–718.

De Vries, F. T., Liiri, M. E., Bjørnlund, L. et al. (2012b) Land use alters the resistance and resilience of soil food webs to drought. *Nature Climate Change*, 2, 276–280.

De Vries, F. T., Manning, P., Tallowin, J. R. B. et al. (2012c) Abiotic drivers and plant traits explain landscape-scale patterns in soil microbial communities. *Ecology Letters*, 15, 1230–1239.

De Vries, F. T., Thébault, E., Liiri, M. et al. (2013) Soil food web properties explain ecosystem services across European land use systems. *Proceedings of the National Academy of Sciences*, 110, 14296–14301.

De Vries, W. (2021) Impacts of nitrogen emissions on ecosystems and human health: a mini review. *Current Opinion in Environmental Science & Health*, 21, 100249.

Díaz, S., Cabido, M. and Casanoves, F. (1998) Plant functional traits and environmental filters at a regional scale. *Journal of Vegetation Science*, 9, 113–122.

Díaz, S., Grime, J. P., Harris, J. and McPherson, E. (1993) Evidence of a feedback mechanism limiting plant response to elevated carbon-dioxide. *Nature*, 364, 616–617.

Díaz, S., Kattge, J., Cornelissen, J. H. C. et al. (2016) The global spectrum of plant form and function. *Nature*, 529, 167–171.

Didden, W. A. M. (1993) Ecology of terrestrial Enchytraeidae. *Pedobiologia*, 37, 2–29.

Dietrich, P., Ebeling, A., Meyer, S. T. et al. (2024) Plant diversity and community age stabilize ecosystem multifunctionality. *Global Change Biology*, 30, e17225.

Dijkstra, F. A., Carrillo, Y., Pendall, E. and Morgan, J. A. (2013) Rhizosphere priming: a nutrient perspective. *Frontiers in Microbiology*, 4, 216.

Dijkstra, F.A., Zhu, B. and Cheng, W. (2021) Root effects on soil organic carbon: a double-edged sword. *New Phytologist*, 230, 60–65.

Ding, J., Delgado-Baquerizo, M., Wang, J. T. and Eldridge, D. J. (2021) Ecosystem functions are related to tree diversity in forests but soil biodiversity in open woodlands and shrublands. *Journal of Ecology*, 109, 4158–4170.

Dlamini, T. C. and Haynes, R. J. (2004) Influence of agricultural land use on the size and composition of earthworm communities in northern KwaZulu-Natal, South Africa. *Applied Soil Ecology*, 27, 77–88.

Doblas-Miranda, E., Sánchez-Piñero, F. and González-Megías, A. (2007) Soil macroinvertebrate fauna of a Mediterranean arid system: composition and temporal changes in the assemblage. *Soil Biology and Biochemistry*, 39, 1916–1925.

Dokuchaev, V. V. (1879) Abridged historical account and critical examination of the principal soil classifications existing. *Transactions of the Petersburg Society of Naturalists*, 1, 64–67.

Dokuchaev, V. V. (1883) *The Russian Chernozem*. Report to the Free Economic Society. Imperial University of St. Petersburg, St. Petersburg, Russia.

Domeignoz-Horta, L. A., Cappelli, S. L., Shrestha, R. et al. (2024) Plant diversity drives positive microbial associations in the rhizosphere enhancing carbon use efficiency in agricultural soils. *Nature Communications*, 15, 8065.

Domeignoz-Horta, L. A., Pold, G., Liu, X. J. A. et al. (2020) Microbial diversity drives carbon use efficiency in a model soil. *Nature Communications*, 11, 3684.

Domke, G. M., Oswalt, S. N., Walters, B. F. and Morin, R. S. (2020) Tree planting has the potential to increase carbon sequestration capacity of forests in the United States. *Proceedings of the National Academy of Sciences*, 117, 24649–24651.

Don, A., Hagen, C., Grüneberg, E. and Vos, C. (2019) Simulated wild boar bioturbation increases the stability of forest soil carbon. *Biogeosciences*, 16, 4145–4155.

Don, A., Seidel, F., Leifeld, J. et al. (2024) Carbon sequestration in soils and climate change mitigation—definitions and pitfalls. *Global Change Biology*, 30, e16983.

Dong, Y., Wang, Z., Sun, H. et al. (2018) The response patterns of arbuscular mycorrhizal and ectomycorrhizal symbionts under elevated CO_2: a meta-analysis. *Frontiers in Microbiology*, 9, 1248

Dooley, S. R. and Treseder, K. K. (2012) The effect of fire on microbial biomass: a meta-analysis of field studies. *Biogeochemistry*, 109, 49–61.

Dottori, F., Szewczyk, W., Ciscar, J. C. et al. (2018) Increased human and economic losses from river flooding with anthropogenic warming. *Nature Climate Change*, 8, 781–786.

Dove, N. C. and Hart, S. C. (2017) Fire reduces fungal species richness and in situ mycorrhizal colonization: a meta-analysis. *Fire Ecology*, 13, 37–65.

Drake, J. E., Gallet-Budynek, A., Hofmockel, K. S. et al. (2011) Increases in the flux of carbon belowground stimulate nitrogen uptake and sustain the long-term enhancement of forest productivity under elevated CO_2. *Ecology Letters*, 14, 349–357.

Drigo, B., Van Veen, J. A. and Kowalchuk, G. A. (2009) Specific rhizosphere bacterial and fungal groups respond differently to elevated atmospheric CO_2. *The ISME Journal*, 3, 1204–1217.

Dror, D. and Klein, T. (2022) The effect of elevated CO_2 on aboveground and belowground carbon allocation and eco-physiology of four species of angiosperm and gymnosperm forest trees. *Tree Physiology*, 42, 831–847.

Duchene, O., Vian, J. F. and Celette, F. (2017) Intercropping with legume for agroecological cropping systems: complementarity and facilitation processes and the importance of soil microorganisms. A review. *Agriculture, Ecosystems & Environment*, 240, 148–161.

Duke, S. O., Blair, A. C., Dayan, F. E. et al. (2009) Is (-)-catechin a novel weapon of spotted knapweed (*Centaurea stoebe*)? *Journal of Chemical Ecology*, 35, 141–153.

Dupouey, J. L., Dambrine, E., Laffite, J. D. and Moares, C. (2002) Irreversible impact of past land use on forest soils and biodiversity. *Ecology*, 83, 2978–2984.

Du Toit, J. T. and Cumming, D. H. M. (1999) Functional significance of ungulate diversity in African savannas and the ecological implications of the spread of pastoralism. *Biodiversity and Conservation*, 8, 1643–1661.

Edwards, C. A. (2004) *Earthworm Ecology*. 2nd Edition. CRC Press, Boca Raton, Florida.

Edwards, C. A. and Bohlen, P. J. (1996) *Biology and Ecology of Earthworms*. 3rd Edition. Springer Science & Business Media, London.

Edwards, C. A., Hendrix, P. F. and Arancon, N. Q. (2020) *Biology and Ecology of Earthworms*. 4th Edition, Springer Science and Business Media, London.

Edwards, D. P., Lim, F., James, R. H. et al. (2017) Climate change mitigation: potential benefits and pitfalls of enhanced rock weathering in tropical agriculture. *Biology Letters*, 13, 20160715.

Edwards, J., Johnson, C., Santos-Medellín, C. et al. (2015) Structure, variation, and assembly of the root-associated microbiomes of rice. *Proceedings of the National Academy of Sciences*, 112, E911–E920.

Eggleton, P. and Bignell, D. E. (1995) Monitoring the response of tropical insects to changes in the environment: troubles with termites. In: *Insects in a Changing Environment* (Harrington, R. and Stork, N. E., eds), pp. 434–497. Academic Press, London.

Eggleton, P., Bignell, D. E., Hauser, S. et al. (2002) Termite diversity across an anthropogenic disturbance gradient in the humid forest zone of West Africa. *Agriculture, Ecosystems and Environment*, 90, 189–202.

Eggleton, P., Bignell, D. E., Sands, W. A. et al. (1996) The diversity, abundance and biomass of termites under differing levels of disturbance in the Mbalmayo Forest Reserve, southern Cameroon. *Philosophical Transactions of the Royal Society of London B: Biological Sciences*, 351, 51–68.

Egidi, E., Delgado-Baquerizo, M., Plett, J. M. et al. (2019) A few Ascomycota taxa dominate soil fungal communities worldwide. *Nature Communications*, 10, 2369.

Eilers, K. G., Debenport, S., Anderson, S. and Fierer, N. (2012) Digging deeper to find unique microbial communities: the strong effect of depth on the structure of bacterial and archaeal communities in soil. *Soil Biology and Biochemistry*, 50, 58–65.

Eisenhauer, N., Cesarz, S., Koller, R. et al. (2012) Global change belowground: impacts of elevated CO_2, nitrogen, and summer drought on soil food webs and biodiversity. *Global Change Biology*, 18, 435–447.

Eisenhauer, N., Dobies, T., Cesarz, S. et al. (2013) Plant diversity effects on soil food webs are stronger than those of elevated CO_2 and N deposition in a long-term grassland experiment. *Proceedings of the National Academy of Sciences*, 110, 6889–6894.

Eisenhauer, N., Ferlian, O., Craven, D. et al. (2019) Ecosystem responses to exotic earthworm invasion in northern North American forests. *Research Ideas and Outcomes*, 5, e34564.

Eissenstat, D. M. (1992) Costs and benefits of constructing roots of small diameter. *Journal of Plant Nutrition*, 15, 763–782.

Eldridge, D. J. (1993) Effects of ants on sandy soils in semiarid eastern Australia—local distribution of nest entrances and their effect in infiltration of water. *Australian Journal of Soil Research*, 31, 509–518.

el Zahar Haichar, F., Santaella, C., Heulin, T. and Achouak, W. (2014) Root exudates mediated interactions belowground. *Soil Biology and Biochemistry*, 77, 69–80.

Engelbrecht, B. M. J., Comita, L. S., Condit, R. et al. (2007) Drought sensitivity shapes species distribution patterns in tropical forests. *Nature*, 447, 80–82.

Engelkes, T., Morriën, E., Verhoeven, K. J. F. et al. (2008) Successful range-expanding plants experience less above-ground and below-ground enemy impact. *Nature*, 456, 946–948.

Epihov, D. Z., Saltonstall, K., Batterman, S. A. et al. (2021) Legume–microbiome interactions unlock mineral nutrients in regrowing tropical forests. *Proceedings of the National Academy of Sciences*, 118, e2022241118.

Escalas, A., Hale, L., Voordeckers, J. W. et al. (2019) Microbial functional diversity: from concepts to applications. *Ecology and Evolution*, 9, 12000–12016.

Essl, F., Lenzner, B., Bacher, S. et al. (2020) Drivers of future alien species impacts: an expert-based assessment. *Global Change Biology*, 26, 4880–4893.

Estes, J. A., Terborgh, J., Brashares, J. S. et al. (2011) Trophic downgrading of planet Earth. *Science*, 333, 301–306.

Ettema, C. H. and Wardle, D. A. (2002) Spatial soil ecology. *Trends in Ecology and Evolution*, 17, 177–183.

Ettema, C. H., Rathbun, S. L. and Coleman, D. C. (2000) On spatiotemporal patchiness and the coexistence of five species of Chronogaster (Nematoda: Chronogasteridae) in a riparian wetland. *Oecologia*, 125, 444–452.

Evans, D. L., Quinton, J. N., Davies, J. A. et al. (2020) Soil lifespans and how they can be extended by land use and management change. *Environmental Research Letters*, 15, 0940b2.

Evans, S. E. and Wallenstein, M. D. (2012) Soil microbial community response to drying and rewetting stress: does historical precipitation regime matter? *Biogeochemistry*, 109, 101–116.

Evans, S. E. and Wallenstein, M. D. (2014) Climate change alters ecological strategies of soil bacteria. *Ecology Letters*, 17, 155–164.

Fang, C., Smith, P., Moncrieff, J. B. and Smith, J. U. (2005) Similar response of labile and resistant soil organic matter pools to changes in temperature. *Nature*, 433, 57–59.

Fanin, N., Clemmensen, K. E., Lindahl, B. D. et al. (2022) Ericoid shrubs shape fungal communities and suppress organic matter decomposition in boreal forests. *New Phytologist*, 236, 684–697.

Fanin, N., Lin, D., Freschet, G. T. et al. (2021) Home-field advantage of litter decomposition: from the phyllosphere to the soil. *New Phytologist*, 231, 1353–1358.

FAO (2015) *The Status of the World's Soil Resources*. Food and Agriculture Organization of the United Nations, Rome.

Farrar, J., Hawes, M., Jones, D. and Lindow, S. (2003) How roots control the flux of carbon to the rhizosphere. *Ecology*, 84, 827–837.

Farrell, M., Hill, P. W., Wanniarachchi, S. D. et al. (2011) Rapid peptide metabolism: a major component of soil nitrogen cycling. *Global Biogeochemical Cycles*, 25, GB3014.

Farrer, E. C., Porazinska, D. L., Spasojevic, M. J. et al. (2019) Soil microbial networks shift across a high-elevation successional gradient. *Frontiers in Microbiology*, 10, 2887.

Fenner, N., Freeman, C., Lock, M. A. et al. (2007) Interactions between elevated CO_2 and warming could amplify DOC exports from peatland catchments. *Environmental Science and Technology*, 41, 3146–3152.

Fickling, N. W., Abbott, C. A., Brame, J. E. et al. (2024) Light–dark cycles may influence in situ soil bacterial networks and diurnally-sensitive taxa. *Ecology and Evolution*, 14, e11018.

Fierer, N. (2017) Embracing the unknown: disentangling the complexities of the soil microbiome. *Nature Reviews Microbiology*, 15, 579–590.

Fierer, N. and Jackson, R. B. (2006) The diversity and biogeography of soil bacterial communities. *Proceedings of the National Academy of Sciences*, 103, 626–631.

Fierer, N., Bradford, M. A. and Jackson, R. B. (2007a) Toward an ecological classification of soil bacteria. *Ecology*, 88, 1354–1364.

Fierer, N., Breitbart, M., Nulton, J. et al. (2007b) Metagenomic and small-subunit rRNA analyses reveal the genetic diversity of bacteria, archaea, fungi, and viruses in soil. *Applied and Environmental Microbiology*, 73, 7059–7066.

Fierer, N., Schimel, J. P., Cates, R. G. and Zou, J. (2001) The influence of balsam poplar tannin fractions on carbon and nitrogen dynamics in Alaskan taiga floodplain soils. *Soil Biology & Biochemistry*, 33, 1827–1839.

Fierer, N., Schimel, J. P. and Holden, P. A. (2003) Variations in microbial community composition through two soil depth profiles. *Soil Biology and Biochemistry*, 35, 167–176.

Fierer, N., Strickland, M. S., Liptzin, D. et al. (2009) Global patterns in belowground communities. *Ecology Letters*, 12, 1238–1249.

Findlay, S., Carreiro, M., Krischik, V. and Jones, C. G. (1996) Effects of damage to living plants on leaf litter quality. *Ecological Applications*, 6, 269–275.

Finlay, B. J. (2002) Global dispersal of free-living microbial eukaryote species. *Science*, 296, 1061–1063.

Finlay, B. J., Esteban, G. F., Olmo, J. L. and Tyler, P. A. (1999) Global distribution of free-living microbial species. *Ecography*, 22, 138–144.

Finlay, B. J., Esteban, G. F., Clarke, K. J. and Olmo, J. L. (2001) Biodiversity of terrestrial protozoa appears homogeneous across local and global spatial scales. *Protist*, 152, 355–366.

Finzi, A. C., DeLucia, E. H., Hamilton J. G., et al. (2002) The nitrogen budget of a pine forest under free-air CO_2 enrichment. *Oecologia*, 132, 567–578.

Fischer, E. M., Sippel, S. and Knutti, R. (2021) Increasing probability of record-shattering climate extremes. *Nature Climate Change*, 11, 689–695.

Fitzpatrick, C. R., Copeland, J., Wang, P. W. et al. (2018) Assembly and ecological function of the root microbiome across angiosperm plant species. *Proceedings of the National Academy of Sciences*, 115, E1157–E1165.

Fontaine, S., Mariotti, A. and Abbadie, L. (2003) The priming effect of organic matter: a question of microbial competition? *Soil Biology and Biochemistry*, 35, 837–843.

Fonte, S. J., Hsieh, M. and Mueller, N. D. (2023) Earthworms contribute significantly to global food production. *Nature Communications*, 14, 5713.

Fornara, D. A. and Tilman, D. (2008) Plant functional composition influences rates of soil carbon and nitrogen accumulation. *Journal of Ecology*, 96, 314–322.

Fragoso, C., Brown, G. G., Patrón, J. C. et al. (1997) Agricultural intensification, soil biodiversity and agroecosystem function in the tropics: the role of earthworms. *Applied Soil Ecology*, 6, 17–35.

Fragoso, C., Kanyonyo, J., Moreno, A. et al. (1999) A survey of tropical earthworms: taxonomy, biogeography and environmental plasticity. In: *Earthworm Management in Tropical Agroecosystems* (Lavelle, P., Brussaard, L., Hendrix P., eds), pp. 1–26. CAB International, Wallingford.

Francioli, D., Cid, G., Kanukollu, S. et al. (2021) Flooding causes dramatic compositional shifts and depletion of putative beneficial bacteria on the spring wheat microbiota. *Frontiers in Microbiology*, 12, 2021.

Franco, A. L. C., Gherardi, L. A., De Tomasel, C. M. et al. (2019a) Drought suppresses soil predators and promotes root herbivores in mesic, but not in xeric grasslands. *Proceedings of the National Academy of Sciences*, 116, 12883–12888.

Franco, A. L. C., Sobral, B. W., Silva, A. L. and Wall, D. H. (2019b) Amazonian deforestation and soil biodiversity. *Conservation Biology*, 33, 590–600.

Frank, D. A. (2008) Evidence for top predator control of a grazing system. *Oikos*, 117, 1718–1724.

Frank, D. A. and McNaughton, S. J. (1992) The ecology of plants, large mammalian herbivores, and drought in Yellowstone National Park. *Ecology*, 73, 2043–2058.

Frank, D. A., Wallen, R. L. and White, P. J. (2016) Ungulate control of grassland production: grazing intensity and ungulate species composition in Yellowstone Park. *Ecosphere*, 7, e01603.

Franken, O., Huizinga, M., Ellers, J. and Berg, M. P. (2018) Heated communities: large inter-and intraspecific variation in heat tolerance across trophic levels of a soil arthropod community. *Oecologia*, 186, 311–322.

Frankland, J. C. (1982) Biomass and nutrient cycling by decomposer basidiomycetes. In: *Decomposer Basidiomycetes: Their Biology and Ecology* (Frankland, J. C., Hedger, J. N. and Swift, M. J., eds), pp. 241–261. Cambridge University Press, Cambridge.

Freeman, C., Fenner, N., Ostle, N. J. et al. (2004) Export of dissolved organic carbon from peatlands under elevated carbon dioxide levels. *Nature*, 430, 195–198.

Frelich, L. E., Hale, C. M., Scheu, S. et al. (2006) Earthworm invasion into previously earthworm-free temperate and boreal forests. *Biological Invasions*, 8, 1235–1245.

Freschet, G. T., Cornelissen, J. H. C., Van Logtestijn, R. S. P. and Aerts, R. (2010) Substantial nutrient resorption from leaves, stems and roots in a subarctic flora: what is the link with other resource economics traits? *New Phytologist*, 186, 879–889.

Freschet, G. T., Roumet, C., Comas, L. H. et al. (2021) Root traits as drivers of plant and ecosystem functioning: current understanding, pitfalls and future research needs. *New Phytologist*, 232, 1123–1158.

Frey, S. D. (2019) Mycorrhizal fungi as mediators of soil organic matter dynamics. *Annual Review of Ecology, Evolution, and Systematics*, 50, 237–259.

Frey, S. D., Knorr, M., Parrent, J. L. and Simpson, R. T. (2004) Chronic nitrogen enrichment affects the structure and function of the soil microbial community in temperate hardwood and pine forests. *Forest Ecology and Management*, 196, 159–171.

Friedlingstein, P. (2015) Carbon cycle feedbacks and future climate change. *Philosophical Transactions of the Royal Society A: Mathematical, Physical and Engineering Sciences*, 373, 20140421.

Friedlingstein, P., O'Sullivan, M., Jones, M. W. et al. (2020) Global carbon budget 2020. *Earth System Science Data*, 12, 3269–3340.

Friggens, N. L., Hester, A. J., Mitchell, R. J. et al. (2020) Tree planting in organic soils does not result in net carbon sequestration on decadal timescales. *Global Change Biology*, 26, 5178–5188.

Fritz, H. and Loison, A. (2006) Large herbivores across biomes. In: *Large Herbivore Ecology, Ecosystem Dynamics and Conservation* (Danell, K., Duncan, P., Bergström, R. and Pastor, J., eds), pp. 19–49. Cambridge University Press. Cambridge.

Frost, C. J. and Hunter, M. D. (2008) Insect herbivores and their frass affect *Quercus rubra* leaf quality and initial stages of subsequent litter decomposition. *Oikos*, 117, 13–22.

Frost, C. J. and Hunter, M. D. (2007) Recycling of nitrogen in herbivore feces: plant recovery, herbivore assimilation, soil retention, and leaching losses. *Oecologia*, 151, 42–53.

Fry, E. L., Johnson, G. N., Hall, A. L. et al. (2018a) Drought neutralises plant–soil feedback of two mesic grassland forbs. *Oecologia*, 186, 1113–1125.

Fry, E. L., Pilgrim, E. S., Tallowin, J. R. B. et al. (2017) Plant, soil and microbial controls on grassland diversity restoration: a long-term, multi-site mesocosm experiment. *Journal of Applied Ecology*, 54, 1320–1330.

Fry, E. L., Savage, J., Hall, A. L. et al. (2018b) Soil multifunctionality and drought resistance are determined by plant structural traits in restoring grassland. *Ecology*, 99, 2260–2271.

Fryxell, J. M. and Sinclair, A. R. E. (1988) Causes and consequences of migration by large herbivores. *Trends in Ecology & Evolution*, 3, 237–241.

Fukami, T. (2015) Historical contingency in community assembly: integrating niches, species pools, and priority effects. *Annual Review of Ecology, Evolution, and Systematics*, 46, 1–23.

Fulthorpe, R. R., Roesch, L. F. W., Riva, A. and Triplett, E. W. (2008) Distantly sampled soils carry few species in common. *The ISME Journal*, 2, 901–910.

Furey, G. N. and Tilman, D. (2021) Plant biodiversity and the regeneration of soil fertility. *Proceedings of the National Academy of Sciences*, 118, e2111321118.

Fusco, E. J., Finn, J. T., Balch, J. K. et al. (2019) Invasive grasses increase fire occurrence and frequency across US ecoregions. *Proceedings of the National Academy of Sciences*, 116, 23594–23599.

Galloway, J. N., Townsend, A. R., Erisman, J. W. et al. (2008) Transformation of the nitrogen cycle: recent trends, questions, and potential solutions. *Science*, 320, 889–892.

Gange, A. C. and Brown, V. K. (1989) Effects of root herbivory by an insect on a foliar-feeding species, mediated through changes in the host plant. *Oecologia*, 81, 38–42.

Gange, A. C., Brown, V. K. and Sinclair, G. S. (1993) Vesicular-arbuscular mycorrhizal fungi: a determinant of plant community structure in early succession. *Functional Ecology*, 7, 616–622.

Gao, Y. and Cabrera Serrenho, A. (2023) Greenhouse gas emissions from nitrogen fertilizers could be reduced by up to one-fifth of current levels by 2050 with combined interventions. *Nature Food*, 4, 170–178.

Garland, G., Edlinger, A., Banerjee, S. et al. (2021) Crop cover is more important than rotational diversity for soil multifunctionality and cereal yields in European cropping systems. *Nature Food*, 2, 28–37.

Gavazov, K., Ingrisch, J., Hasibeder, R. et al. (2017) Winter ecology of a subalpine grassland: effects of snow removal on soil respiration, microbial structure and function. *Science of the Total Environment*, 590, 316–324.

Geisen, S. (2016) The bacterial–fungal energy channel concept challenged by enormous functional versatility of soil protists. *Soil Biology and Biochemistry*, 102, 22–25.

Geisen, S., Mitchell, E. A. D., Adl, S. et al. (2018) Soil protists: a fertile frontier in soil biology research. *FEMS Microbiology Reviews*, 42, 293–323.

George, P. B., Lallias, D., Creer, S. et al. (2019) Divergent national-scale trends of microbial and animal biodiversity revealed across diverse temperate soil ecosystems. *Nature Communications*, 10, 1107.

Georgiou, K., Jackson, R. B., Vindušková, O. et al. (2022) Global stocks and capacity of mineral-associated soil organic carbon. *Nature Communications*, 13, 3797.

Gerrish, G. and Mueller-Dombois, D. (1980) Behaviour of native and non-native plants in tropical rainforests on Ohau, Hawaiian Islands. *Phytocoenologia*, 8, 237–295.

Gerrits, G. M., Waenink, R., Aradottir, A. L. et al. (2023) Synthesis on the effectiveness of soil translocation for plant community restoration. *Journal of Applied Ecology*, 60, 714–724.

Geyer, K. M. and Barrett, J. E. (2019) Unimodal productivity–diversity relationships among bacterial communities in a simple polar soil ecosystem. *Environmental Microbiology*, 21, 2523–2532.

Gießelmann, U. C., Martins, K. G., Brändle, M. et al. (2011) Lack of home-field advantage in the decomposition of leaf litter in the Atlantic Rainforest of Brazil. *Applied Soil Ecology*, 49, 5–10.

Gilbert, K. J., Fahey, T. J., Maerz, J. C. et al. (2014) Exploring carbon flow through the root channel in a temperate forest soil food web. *Soil Biology and Biochemistry*, 76, 45–52.

Gill, J. L. (2014) Ecological impacts of the late Quaternary megaherbivore extinctions. *New Phytologist*, 201, 1163–1169.

Gill, J. L., Williams, J. W., Jackson, S. T. et al. (2009) Pleistocene megafaunal collapse, novel plant communities, and enhanced fire regimes in North America. *Science*, 326, 1100–1103.

Gill, R. W. (1969) Soil microarthropod abundance following old-field litter manipulation. *Ecology*, 50, 805–816.

Gillingham, A. G., Syers, J. K. and Gregg, P. E. H. (1980) Phosphorus uptake and return in grazed sheep pastures. II. Above-ground components of the phosphorus cycle. *New Zealand Journal of Agricultural Research*, 23, 323–330.

Gillison, A. N., Jones, D. T., Susilo, F. X. and Bignell, D. E. (2003) Vegetation indicates diversity of soil macroinvertebrates: a case study with termites along a land-use intensification gradient in lowland Sumatra. *Organisms Diversity & Evolution*, 3, 111–126.

Golluscio, R. A., Austin, A. T., García Martínez, G. C. et al. (2009) Sheep grazing decreases organic carbon and nitrogen pools in the Patagonian steppe: combination of direct and indirect effects. *Ecosystems*, 12, 686–697.

Gongalsky, K. B. and Persson, T. (2013) Recovery of soil macrofauna after wildfires in boreal forests. *Soil Biology and Biochemistry*, 57, 182–191.

Gongalsky, K. B., Zaitsev, A. S., Korobushkin, D. I. et al. (2021) Forest fire induces short-term shifts in soil food webs with consequences for carbon cycling. *Ecology Letters*, 24, 438–450.

Gordon, H., Haygarth, P. M. and Bardgett, R. D. (2008) Drying and rewetting effects on soil microbial community composition and nutrient leaching. *Soil Biology and Biochemistry*, 40, 302–311.

Gordon, W. S. and Jackson, R. B. (2000) Nutrient concentrations in fine roots. *Ecology*, 81, 275–280.

Gossner, M. M., Lewinsohn, T. M., Kahl, T. et al. (2016) Land-use intensification causes multitrophic homogenization of grassland communities. *Nature*, 540, 266–269.

Gould, I. J., Quinton, J. N., Weigelt, A. et al. (2016) Plant diversity and root traits benefit physical properties key to soil function in grasslands. *Ecology Letters*, 19, 1140–1149.

Grace, J. B., Anderson T. M., Smith, M. D. et al. (2007) Does species diversity limit productivity in natural grassland communities? *Ecology Letters*, 10, 680–689.

Graham, E. B., Camargo, A. P., Wu, R. et al. (2024) A global atlas of soil viruses reveals unexplored biodiversity and potential biogeochemical impacts. *Nature Microbiology*, 9, 1873–1883.

Grayston, S. J., Shenquiang, W., Campbell, C. D. and Edwards, A. C. (1998) Selective influence of plant species on microbial diversity in the rhizosphere. *Soil Biology and Biochemistry*, 30, 369–378.

Griffin-Nolan, R. J., Blumenthal, D. M., Collins, S. L. et al. (2019) Shifts in plant functional composition following long-term drought in grasslands. *Journal of Ecology*, 107, 2133–2148.

Griffiths, B. S., Welschen, R., Van Arendonk, J. J. C. M. and Lambers, H. (1992) The effect of nitrate-nitrogen on bacteria and bacterial-feeding fauna in the rhizosphere of different grass species. *Oecologia*, 91, 253–259.

Griffiths, H. M., Ashton, L. A., Evans, T. A. et al. (2019) Termites can decompose more than half of deadwood in tropical rainforest. *Current Biology*, 29, R118–R119.

Griffiths, H. M., Bardgett, R. D., Louzada, J. and Barlow, J. (2016) The value of trophic interactions for ecosystem function: dung beetle communities influence seed burial and seedling recruitment in tropical forests. *Proceedings of the Royal Society B*, 283, 20161634.

Grigulis, K., Lavorel, S., Krainer, U. et al. (2013) Relative contributions of plant traits and soil microbial properties to mountain grassland ecosystem services. *Journal of Ecology*, 101, 47–57.

Grime, J. P. (1973) Competitive exclusion in herbaceous vegetation. *Nature*, 242, 344–347.

Grime, J. P. (1979) *Plant Strategies and Vegetation Processes*. John Wiley, Chichester.

Grime, J. P. (1998) Benefits of plant diversity to ecosystems: immediate, filter and founder effects. *Journal of Ecology*, 86, 902–910.

Grime, J. P., Fridley, J. D., Askew, A. P. et al. (2008) Long-term resistance to simulated climate change in an infertile grassland. *Proceedings of the National Academy of Sciences*, 105, 10028–10032.

Grime, J. P., Mackey, J. M. L., Hillier, S. H. and Read, D. J. (1987) Floristic diversity in a model system using experimental microcosms. *Nature*, 328, 420–422.

Groß, V., Reinhard, A., Petters, S. et al. (2023) Adding complexity to soil food webs: myxobacteria have broad predation spectra with bacteria, yeasts and filamentous fungi in vitro. *European Journal of Soil Biology*, 117, 103508.

Gu, B., Zhang, X., Lam, S. K. et al. (2023) Cost-effective mitigation of nitrogen pollution from global croplands. *Nature*, 613, 77–84.

Guerra, C. A., Berdugo, M., Eldridge, D. J. et al. (2022) Global hotspots for soil nature conservation. *Nature*, 610, 693–698.

Guerra, C. A., Heintz-Buschart, A., Sikorski, J. et al. (2020) Blind spots in global soil biodiversity and ecosystem function research. *Nature Communications*, 11, 3870.

Guitian, R. and Bardgett, R. D. (2000) Plant and soil microbial responses to defoliation in temperate seminatural grassland. *Plant and Soil*, 220, 271–277.

Gundale, M. J. (2002) Influence of exotic earthworms on the soil organic horizon and the rare fern *Botrychium mormo*. *Conservation Biology*, 16, 1555–1561.

Gundale, M. J., Deluca, T. H. and Nordin, A. (2011) Bryophytes attenuate anthropogenic nitrogen inputs in boreal forests. *Global Change Biology*, 17, 2743–2753.

Gundale, M. J., Wardle, D. A. and Nilsson, M. C. (2010) Vascular plant removal effects on biological N-fixation vary across a boreal forest island gradient. *Ecology*, 91, 1704–1714.

Gundersen, H., Leinaas, H. P. and Thaulow, C. (2014) Surface structure and wetting characteristics of Collembola cuticles. *PLoS ONE*, 9, e86783.

Guo, L. B. and Gifford, R. M. (2002) Soil carbon stocks and land use change: a meta-analysis. *Global Change Biology*, 8, 345–360.

Guo, X., Gao, Q., Yuan, M. et al. (2020) Gene-informed decomposition model predicts lower soil carbon loss due to persistent microbial adaptation to warming. *Nature Communications*, 11, 4897.

Gutknecht, J. L. M., Field, C. B. and Balser, T. C. (2012) Microbial communities and their responses to simulated global change fluctuate greatly over multiple years. *Global Change Biology*, 18, 2256–2269.

Haddaway, N. R., Hedlund, K., Jackson, L. E. et al. (2017) How does tillage intensity affect soil organic carbon? A systematic review. *Environmental Evidence*, 6, 30.

Hagerman, S. H., Sakakibara, S.M. and Durall D. M. (2001) The potential for woody understory plants to provide refuge for ectomycorrhizal inoculum at an interior Douglas-fir forest after clear-cut logging. *Canadian Journal of Forest Research*, 31, 711–721.

Hall, S. J., Trujillo, J., Nakase, D. et al. (2013) Legacies of prehistoric agricultural practices within plant and soil properties across an arid ecosystem. *Ecosystems*, 16, 1273–1293.

Hamilton, E. W. and Frank, D. A. (2001) Can plants stimulate soil microbes and their own nutrient supply? Evidence from a grazing tolerant grass. *Ecology*, 82, 2397–2402.

Hamilton, E. W., Frank, D. A., Hinchey, P. M. and Murray, T. R. (2008) Defoliation induces root exudation and triggers positive rhizospheric feedbacks in a temperate grassland. *Soil Biology and Biochemistry*, 40, 2865–2873.

Han, G., Hao, X., Zhao, M. et al. (2008) Effect of grazing intensity on carbon and nitrogen in soil and vegetation in a meadow steppe in Inner Mongolia. *Agriculture, Ecosystems & Environment*, 125, 21–32.

Han, X., Li, Y., Li, Y. et al. (2022) Soil inoculum identity and rate jointly steer microbiomes and plant communities in the field. *ISME Communications*, 2, 59.

Handa, I. T., Aerts, R., Berendse, F. et al. (2014) Consequences of biodiversity loss for litter decomposition across biomes. *Nature*, 509, 218–221.

Handa, I. T., Harmsen, R. and Jefferies, R. L. (2002) Patterns of vegetation change and the recovery potential of degraded areas in a coastal marsh system of the Hudson Bay lowlands. *Journal of Ecology*, 90, 86–99.

Handley, W. R. C. (1954) *Mull and Mor in Relation to Forest Soils*. Her Majesty's Stationery Office, London.

Handley, W. R. C. (1961) Further evidence for the importance of residual leaf protein complexes in litter decomposition and the supply of nitrogen for plant growth. *Plant and Soil*, 15, 37–73.

Hanley, M. E., Trofmov, S. and Taylor, G. (2004) Species-level effects more important than functional group-level responses to elevated CO_2: evidence from simulated turves. *Functional Ecology*, 18, 304–313.

Hanlon, R. D. G. and Anderson, J. M. (1979) The effects of Collembola grazing on microbial activity in decomposing leaf litter. *Oecologia*, 38, 93–99.

Hannula, S. E., Morriën, E., De Hollander, M. et al. (2017) Shifts in rhizosphere fungal community during secondary succession following abandonment from agriculture. *The ISME Journal*, 11, 2294–2304.

Hansen, R. A. (2000) Effect of habitat complexity and composition on a diverse litter microarthropod assemblage. *Ecology*, 81, 1120–1132.

Hanski, I. and Cambefort, Y. (1991) *Dung Beetle Ecology*. Princeton University Press, Princeton, NJ.

Harris, R. B. (2010) Rangeland degradation on the Qinghai-Tibetan plateau: a review of the evidence of its magnitude and causes. *Journal of Arid Environments*, 74, 1–12.

Harris, W. V. (1964) *Termites: Their Recognition and Control*. Longmans, London.

Harrison, J. G. and Griffin, E. A. (2020) The diversity and distribution of endophytes across biomes, plant phylogeny and host tissues: how far have we come and where do we go from here? *Environmental Microbiology*, 22, 2107–2123.

Harrison, K. A., Bol, R. and Bardgett, R. D. (2007) Preferences for uptake of different nitrogen forms by coexisting plant species and soil microbes. *Ecology*, 88, 989–999.

Harrison, K. A., Bol, R. and Bardgett, R. D. (2008) Do plant species with different growth strategies vary in their ability to compete with soil microbes for chemical forms of nitrogen? *Soil Biology and Biochemistry*, 40, 228–237.

Harsch, M. A., Hulme, P. E., McGlone, M. S. and Duncan, R. P. (2009) Are treelines advancing? A global meta-analysis of treeline response to climate warming. *Ecology Letters*, 12, 1040–1049.

Hart, M. M., Antunes, P. M., Chaudhary, V. B. and Abbott, L. K. (2018) Fungal inoculants in the field. *Functional Ecology*, 32, 126–135.

Hartley, I. P., Hill, T. C., Chadburn, S. E. and Hugelius, G. (2021) Temperature effects on carbon storage are controlled by soil stabilisation capacities. *Nature Communications*, 12, 6713.

Hartley, I. P., Hopkins, D. W., Garnett, M. H. et al. (2008) Soil microbial respiration in arctic soil does not acclimate to temperature. *Ecology Letters*, 11, 1092–1100.

Hartley, S. E. and Jones, C. G. (1997) Plant chemistry and herbivory, or why is the world green? In: *Plant Ecology* (Crawley, M. J., ed.), pp. 284–324. Blackwell Science, Oxford.

Hartmann, M. and Six, J. (2023) Soil structure and microbiome functions in agroecosystems. *Nature Reviews Earth and Environment*, 4, 4–18.

Hartnett, D. C. and Wilson, G. W. T. (1999) Mycorrhizae influence plant community structure and diversity in tallgrass prairie. *Ecology*, 80, 1187–1195.

Hättenschwiler, S. and Vitousek, P. M. (2000) The role of polyphenols in terrestrial ecosystems nutrient cycling. *Trends in Ecology and Evolution*, 15, 238–243.

Hättenschwiler, S., Coq, S., Barantal, S. and Handa, I. T. (2011) Leaf traits and decomposition in tropical rainforests: revisiting some commonly held views and towards a new hypothesis. *New Phytologist*, 189, 950–965.

Hawkes, C. V. and Keitt, T. H. (2015) Resilience vs. historical contingency in microbial responses to environmental change. *Ecology Letters*, 18, 612–625.

Hawkes, C. V., Kivlin, S. N., Rocca, J. D. et al. (2011) Fungal community responses to precipitation. *Global Change Biology*, 17, 1637–1645.

Hawkins, H. J., Cargill, R. I. M., Van Nuland, M. E. et al. (2023) Mycorrhizal mycelium as a global carbon pool. *Current Biology*, 33, R560–R573.

Hawlena, D. and Schmitz, O. J. (2010) Herbivore physiological response to predation risk and implications for ecosystem nutrient dynamics. *Proceedings of the National Academy of Sciences*, 107, 15503–15507.

Hawlena, D. and Zaguri, M. (2016) Fear and below-ground food-webs. *Soil Biology and Biochemistry*, 102, 26–28.

Hawlena, D., Strickland, M. S., Bradford, M. A. and Schmitz, O. J. (2012) Fear of predation slows plant-litter decomposition. *Science*, 336, 1434–1438.

Haygarth, P. M., Jarvie, H. P., Powers, S. M. et al. (2014) Sustainable phosphorus management and the need for a long-term perspective: the legacy hypothesis. *Environmental Science & Technology*, 48, 8417–8419.

Haynes, G. (2012) Elephants (and extinct relatives) as earth-movers and ecosystem engineers. *Geomorphology*, 157, 99–107.

Haystead, A., Malajczuk, N. and Grove, T.S. (1988) Underground transfer of nitrogen between pasture plants infected with arbuscular mycorrhizal fungi. *New Phytologist* 108, 417–423.

Hazard, C. and Johnson, D. (2018) Does genotypic and species diversity of mycorrhizal plants and fungi affect ecosystem function? *New Phytologist*, 220, 1122–1128.

He, L., Rodrigues, J. L. M., Soudzilovskaia, N. A. et al. (2020) Global biogeography of fungal and bacterial biomass carbon in topsoil. *Soil Biology and Biochemistry*, 151, 108024.

He, N. P., Zhang, Y. H., Yu, Q. et al. (2011) Grazing intensity impacts soil carbon and nitrogen storage of continental steppe. *Ecosphere*, 2, 1–10.

He, Z. Y., Hu, H. W., Nguyen, B. A. T. et al. (2024) Distribution of soil tardigrades as revealed by molecular identification across a large-scale area of Australia. *Soil Biology and Biochemistry*, 196, 109506.

He, Z., Xu, M., Deng, Y. et al. (2010) Metagenomic analysis reveals a marked divergence in the structure of belowground microbial communities at elevated CO_2. *Ecology Letters*, 13, 564–575.

Heath, J., Ayres, E., Possell, M. et al. (2005) Rising atmospheric CO_2 reduces sequestration of root-derived soil carbon. *Science*, 309, 1711–1713.

Hedin, L. O., Armesto, J. J. and Johnson, A. H. (1995) Patterns of nutrient loss from unpolluted, old-growth temperate forests: evaluation of biogeochemical theory. *Ecology*, 76, 493–509.

Hedin, L. O., Brookshire, E. N. J., Menge, D. N. L. and Barron, A. R. (2009) The nitrogen paradox in tropical forest ecosystems. *Annual Review of Ecology, Evolution and Systematics*, 40, 613–635.

Hedlund, K., Boddy, L. and Preston, C. M. (1991) Mycelial responses of the soil fungus, *Mortierella isabellina*, to grazing by *Onychiurus armatus* (Collembola). *Soil Biology and Biochemistry*, 23, 361–366.

Heemsbergen, D. A., Berg, M. P., Loreau, M. et al. (2004) Biodiversity effects on soil processes explained by interspecific functional dissimilarity. *Science*, 306, 1019–1020.

Hempson, G. P., Archibald, S. and Bond, W. J. (2015) A continent-wide assessment of the form and intensity of large mammal herbivory in Africa. *Science*, 350, 1056–1061.

Hempson, G. P., Archibald, S. and Bond, W. J. (2017) The consequences of replacing wildlife with livestock in Africa. *Scientific Reports*, 7, 17196.

Hendrix, P. F., Parmelee, R. W., Crossley, D. A. et al. (1986) Detritus food webs in conventional and no-tillage agroecosystems. *Bioscience*, 36, 374–380.

Henneron, L., Cros, C., Picon-Cochard, C. et al. (2020a) Plant economic strategies of grassland species control soil carbon dynamics through rhizodeposition. *Journal of Ecology*, 108, 528–545.

Henneron, L., Kardol, P., Wardle, D. A. et al. (2020b) Rhizosphere control of soil nitrogen cycling: a key component of plant economic strategies. *New Phytologist*, 228, 1269–1282.

Henry, H. A. L. and Jefferies, R. L. (2003) Plant amino acid uptake, soluble N turnover and microbial N capture in soils of a grazed Arctic salt marsh. *Journal of Ecology*, 91, 627–636.

Herdler, S., Kreuzer, K., Scheu, S. and Bonkowski, M. (2008) Interactions between arbuscular mycorrhizal fungi (*Glomus intraradices*, Glomeromycota) and amoebae (*Acanthamoeba castellanii*, Protozoa) in the rhizosphere of rice (*Oryza sativa*). *Soil Biology and Biochemistry*, 40, 660–668.

Herman, R. P., Provencio, K. R., Herrera-Matos, J. and Torrez, R. J. (1995) Resource islands predict the distribution of heterotrophic bacteria in Chihuahuan Desert soils. *Applied and Environmental Microbiology*, 61, 1816–1821.

Herridge, D. F., Peoples, M. B. and Boddey, R. M. (2008) Global inputs of biological nitrogen fixation in agricultural systems. *Plant and Soil*, 311, 1–18.

Hestrin, R., Kan, M., Lafler, M. et al. (2022) Plant-associated fungi support bacterial resilience following water limitation. *The ISME Journal*, 16, 2752–2762.

Hetrick, B. A. D., Wilson, G. W. T. and Cox, T. S. (1993) Mycorrhizal dependence of modern wheat cultivars and ancestors: a synthesis. *Canadian Journal of Botany*, 71, 512–518.

Hewins, D. B., Lyseng, M. P., Schoderbek, D. F. et al. (2018) Grazing and climate effects on soil organic carbon concentration and particle-size association in northern grasslands. *Scientific Reports*, 8, 1336.

Hickling, R., Roy, D. B., Hill, J. K. et al. (2006) The distributions of a wide range of taxonomic groups are expanding polewards. *Global Change Biology*, 12, 450–455.

Hill, P. W., Farrar, J., Roberts, P. et al. (2011a) Vascular plant success in a warming Antarctic may be due to efficient nitrogen acquisition. *Nature Climate Change*, 1, 50–53.

Hill, P. W., Quilliam, R.S., DeLuca, T. H. et al. (2011b) Acquisition and assimilation of nitrogen as peptide-bound and D-enantiomers of amino acids by wheat. *PLoS ONE*, 6, e19220.

Hiltner, L. (1904) Über neuere Erfahrungen und Probleme auf dem Gebiete der Bodenbakteriologie unter besonderer Berücksichtigung der Gründüngung und Brache. *Arbeiten der Deutschen Landwirtschaftlichen Gesellschaft*, 98, 59–78.

Hodda, M. and Wanless, F. R. (1994) Nematodes from an English chalk grassland—species distributions. *Nematologica*, 40, 116–132.

Hodge, A. and Storer, K. (2015) Arbuscular mycorrhiza and nitrogen: implications for individual plants through to ecosystems. *Plant and Soil*, 386, 1–19.

Hodge, A., Campbell, C. D. and Fitter, A. H. (2001) An arbuscular mycorrhizal fungus accelerates decomposition and acquires nitrogen directly from organic material. *Nature*, 413, 297–299.

Hodge, A., Stewart, J., Robinson, D. et al. (1998) Root proliferation, soil fauna and plant nitrogen capture from nutrient-rich patches in soil. *New Phytologist*, 139, 479–494.

Hodkinson, I. D., Coulson, S. J., Harrison, J. and Webb, N. R. (2001) What a wonderful web they weave: spiders, nutrient capture and early ecosystem development in the high Arctic—some counter-intuitive ideas on community assembly. *Oikos*, 95, 349–352.

Hoeppner, S. S. and Dukes, J. S. (2012) Interactive responses of old-field plant growth and composition to warming and precipitation. *Global Change Biology*, 18, 1754–1768.

Hokka, V., Mikola, J., Vestberg, M. and Setälä, H. (2004) Interactive effects of defoliation and an AM fungus on plants and soil organisms in experimental legume–grass communities. *Oikos*, 106, 73–84.

Hol, W. G., De Boer, W., Termorshuizen, A. J. et al. (2010) Reduction of rare soil microbes modifies plant–herbivore interactions. *Ecology Letters*, 13, 292–301.

Holden, S. R., Gutierrez, A. and Treseder, K. K. (2013) Changes in soil fungal communities, extracellular enzyme activities, and litter decomposition across a fire chronosequence in Alaskan boreal forests. *Ecosystems*, 16, 34–46.

Hölldobler, B. and Wilson, E. O. (1990) *The Ants*. Springer, Berlin.

Holmstrup, M., Ehlers, B. K., Slotsbo, S. et al. (2018) Functional diversity of Collembola is reduced in soils subjected to short-term, but not long-term, geothermal warming. *Functional Ecology*, 32, 1304–1316.

Holmstrup, M., Sjursen, H., Ravn, H. and Bayley, M. (2001) Dehydration tolerance and water vapour absorption in two species of soil-dwelling Collembola by accumulation of sugars and polyols. *Functional Ecology*, 15, 647–653.

Hong, S., Yin, G., Piao, S. et al. (2020) Divergent responses of soil organic carbon to afforestation. *Nature Sustainability*, 3, 694–700.

Hou, E., Luo, Y., Kuang, Y. et al. (2020) Global meta-analysis shows pervasive phosphorus limitation of aboveground plant production in natural terrestrial ecosystems. *Nature Communications*, 11, 637.

Houben, D., Daoulas, G., Faucon, M. P. and Dulaurent, A. M. (2020) Potential use of mealworm frass as a fertilizer: impact on crop growth and soil properties. *Scientific Reports*, 10, 4659.

Howison, R. A., Olff, H., Van de Koppel, J. and Smit, C. (2017) Biotically driven vegetation mosaics in grazing ecosystems: the battle between bioturbation and biocompaction. *Ecological Monographs*, 87, 363–378.

Hu, H. Y., Li, H., Hao, M. M. et al. (2021a) Nitrogen fixation and crop productivity enhancements co-driven by intercrop root exudates and key rhizosphere bacteria. *Journal of Applied Ecology*, 58, 2243–2255.

Hu, J., Zhou, S., Tie, L. et al. (2022) Effects of nitrogen addition on soil faunal abundance: a global meta-analysis. *Global Ecology and Biogeography*, 31, 1655–1666.

Hu, L., Robert, C. A. M., Cadot, S. et al. (2018) Root exudate metabolites drive plant–soil feedbacks on growth and defense by shaping the rhizosphere microbiota. *Nature Communications*, 9, 2738.

Hu, S., Chapin, F. S., Firestone, M. K. et al. (2001) Nitrogen limitation of microbial decomposition in a grassland under elevated CO_2. *Nature*, 409, 188–191.

Hu, W., Ran, J., Dong, L. et al. (2021b) Aridity-driven shift in biodiversity–soil multifunctionality relationships. *Nature Communications*, 12, 5350.

Hu, W., Schmidt, S. K., Sommers, P. et al. (2021c) Multiple-trophic patterns of primary succession following retreat of a high-elevation glacier. *Ecosphere*, 12, e03400.

Hudson, J. M. G. and Henry, G. H. R. (2010) High Arctic plant community resists 15 years of experimental warming. *Journal of Ecology*, 98, 1035–1041.

Hungate, B. A., Marks, J. C., Power, M. E. et al. (2021) The functional significance of bacterial predators. *mBbio*, 12, e00466-21.

Hunt, H. W., Coleman, D. C., Ingham, E. R. et al. (1987) The detrital food web in a shortgrass prairie. *Biology and Fertility of Soils*, 3, 57–68.

Hunt, H. W., Ingham, E. R., Coleman, D. C. et al. (1988) Nitrogen limitation of production and decomposition in prairie, mountain meadow, and pine forest. *Ecology*, 69, 1009–1016.

Huston, M. (1979) A general hypothesis of species diversity. *The American Naturalist*, 113, 81–101.

Huston, M. A. (1994) *Biological Diversity. The Coexistence of Species on Changing Landscapes*. Cambridge University Press, Cambridge.

Hyodo, F., Inoue, T., Azuma, J. I. et al. (2000) Role of the mutualistic fungus in lignin degradation of the soil fungus-growing termite *Macrotermes gilvus* (Isoptera: Macrotermitinae). *Soil Biology and Biochemistry*, 32, 653–658.

Ilmarinen, K., Mikola, J., Nieminen, M. and Vestberg, M. (2005) Does plant growth phase determine the response of plants and soil organisms to defoliation? *Soil Biology and Biochemistry*, 37, 433–443.

Inderjit, Pollock, J. L., Callaway, R. M. and Holben, W. (2008) Phytotoxic effects of (±)-catechin in vitro, in soil, and in the field. *PLoS ONE*, 3, e2536.

Ingham, R. E., Trofymow, J. A., Ingham, E. R. and Coleman, D. C. (1985) Interactions of bacteria, fungi, and their nematode grazers: effects on nutrient cycling and plant growth. *Ecological Monographs*, 55, 119–140.

Insam, H. and Haselwandter, K. (1989) Metabolic quotient of the soil microflora in relation to plant succession. *Oecologia*, 79, 174–178.

in't Zandt, D., Herben, T., Van den Brink, A. et al. (2021) Species abundance fluctuations over 31 years are associated with plant–soil feedback in a species-rich mountain meadow. *Journal of Ecology*, 109, 1511–1523.

in't Zandt, D., Van den Brink, A., de Kroon, H. and Visser, E. J. W. (2019) Plant–soil feedback is shut down when nutrients come to town. *Plant and Soil*, 439, 541–551.

IPBES (2019) *Summary for policymakers of the global assessment report on biodiversity and ecosystem services of the Intergovernmental Science-Policy Platform on Biodiversity and Ecosystem Services* (Díaz, S., Settele, J., Brondízio, E., et al., eds). IPBES Secretariat, Bonn.

IPCC (2021) *Climate Change 2021: The Physical Science Basis. Contribution of Working Group I to the Sixth Assessment Report of the Intergovernmental Panel on Climate Change* (Masson-Delmotte, V., Zhai, P., Pirani, A., eds). Cambridge University Press, Cambridge.

Isbell, F., Adler, P. R., Eisenhauer, N. et al. (2017) Benefits of increasing plant diversity in sustainable agroecosystems. *Journal of Ecology*, 105, 871–879.

Isbell, F., Craven, D., Connolly, J. et al. (2015) Biodiversity increases the resistance of ecosystem productivity to climate extremes. *Nature*, 526, 574–577.

Jackson, R. B., Lajtha, K., Crow, S. E. et al. (2017) The ecology of soil carbon: pools, vulnerabilities, and biotic and abiotic controls. *Annual Review of Ecology, Evolution, and Systematics*, 48, 419–445.

Jaeger, C. H., Monson, R. K., Fisk, M. C. and Schmidt, S. K. (1999) Seasonal partitioning of nitrogen by plants and soil microorganisms in an alpine ecosystem. *Ecology*, 80, 1883–1891.

Jansa, J. and Hodge, A. (2021) Swimming, gliding, or hyphal riding? On microbial migration along the arbuscular mycorrhizal hyphal highway and functional consequences thereof. *New Phytologist*, 230, 14–16.

Janzen, D. H. (1970) Herbivores and the number of tree species in tropical forests. *The American Naturalist*, 104, 501–528.

Jefferies, R. L. and Rockwell, R. F. (2002) Foraging geese, vegetation loss and soil degradation in an Arctic salt marsh. *Applied Vegetation Science*, 5, 7–16.

Jeffery, S., Verheijen, F. G. A., Van der Velde, M. and Bastos, A. C. (2011) A quantitative review of the effects of biochar application to soils on crop productivity using meta-analysis. *Agriculture, Ecosystems and Environment*, 144, 175–187.

Jenkinson, D. S., Adams, D. E and Wild, A. (1991) Model estimates of CO_2 emissions from soil in response to global warming. *Nature*, 351, 304–306.

Jenny, H. (1941) *Factors of Soil Formation: A System of Quantitative Pedology*. Dover Publications, New York.

Jewell, P. L., Käuferle, D., Güsewell, S. et al. (2007) Redistribution of phosphorus by cattle on a traditional mountain pasture in the Alps. *Agriculture, Ecosystems and Environment*, 122, 377–386.

Ji, X., Liu, M., Yang, J. and Feng, F. (2022) Meta-analysis of the impact of freeze–thaw cycles on soil microbial diversity and C and N dynamics. *Soil Biology and Biochemistry*, 168, 108608.

Jia, S., Wang, X., Yuan, Z. et al. (2020) Tree species traits affect which natural enemies drive the Janzen–Connell effect in a temperate forest. *Nature Communications*, 11, 286.

Jiang, F., Zhang, L., Zhou, J. et al. (2021) Arbuscular mycorrhizal fungi enhance mineralisation of organic phosphorus by carrying bacteria along their extraradical hyphae. *New Phytologist*, 230, 304–315.

Jiang, M., Crous, K.Y., Carrillo, Y. et al. (2024) Microbial competition for phosphorus limits the CO_2 response of a mature forest. *Nature*, 630, 660–665.

Jochum, M., Fischer, M., Isbell, F. et al. (2020) The results of biodiversity–ecosystem functioning experiments are realistic. *Nature Ecology and Evolution*, 4, 1485–1494.

Johnson, D. J., Clay, K. and Phillips, R. P. (2018) Mycorrhizal associations and the spatial structure of an old-growth forest community. *Oecologia*, 186, 195–204.

Johnson, S. N. and Rasmann, S. (2015) Root-feeding insects and their interactions with organisms in the rhizosphere. *Annual Review of Entomology*, 60, 517–535.

Jolly, W. M., Cochrane, M. A., Freeborn, P. H. et al. (2015) Climate-induced variations in global wildfire danger from 1979 to 2013. *Nature Communications*, 6, 7537.

Jonasson, S., Michelsen, A., Schmidt, I. K. and Nielsen, E. V. (1999) Responses in microbes and plants to changed temperature, nutrient, and light regimes in the arctic. *Ecology*, 80, 1828–1843.

Jones, D. L. and Kielland, K. (2002) Soil amino acid turnover dominates the nitrogen flux in permafrost-dominated taiga forest soils. *Soil Biology and Biochemistry*, 34, 209–219.

Jones, D. L., Hodge, A. and Kuzyakov, Y. (2004) Plant and mycorrhizal regulation of rhizodeposition. *New Phytologist*, 163, 459–480.

Jones, D. T., Susilo, F. X., Bignell, D. E. et al. (2008) Termite assemblage collapse along a land-use intensification gradient in lowland central Sumatra, Indonesia. *Journal of Applied Ecology*, 40, 380–391.

Jonsson, L. M., Nilsson, M. C., Wardle, D. A. and Zackrisson, O. (2001) Context dependent effects of ectomycorrhizal species richness on tree seedling productivity. *Oikos*, 93, 353–364.

Jørgensen, H. B., Johansson, T., Canbäck, B. et al. (2005) Selective foraging of fungi by collembolans in soil. *Biology Letters*, 1, 243–246.

Jousset, A. (2012) Ecological and evolutive implications of bacterial defences against predators. *Environmental Microbiology*, 14, 1830–1843.

Jover, L. F., Effler, T. C., Buchan, A. et al. (2014) The elemental composition of virus particles: implications for marine biogeochemical cycles. *Nature Reviews Microbiology*, 12, 519–528.

Jurgensen, M. F., Finér, L., Domisch, T. et al. (2008) Organic mound-building ants: their impact on soil properties in temperate and boreal forests. *Journal of Applied Entomology*, 132, 266–275.

Kabir, Z. (2005) Tillage or no-tillage: impact on mycorrhizae. *Canadian Journal of Plant Science*, 85, 23–29.

Kahmen, A., Renker, A., Unsicker, S. B. and Buchmann, N. (2006) Niche complementarity for nitrogen: an explanation for the biodiversity and ecosystem functioning relationship. *Ecology*, 87, 1244–1255.

Kaisermann, A., De Vries, F. T., Griffiths, R. I. and Bardgett, R. D. (2017) Legacy effects of drought on plant–soil feedbacks and plant–plant interactions. *New Phytologist*, 215, 1413–1424.

Kampichler, C., Kandeler, E., Bardgett, R. D. et al. (1998) Impact of elevated atmospheric CO_2 concentration on soil microbial biomass and activity in a complex, weedy field model ecosystem. *Global Change Biology*, 4, 335–346.

Kantzas, E. P., Val Martin, M., Lomas, M. R. et al. (2022) Substantial carbon drawdown potential from enhanced rock weathering in the United Kingdom. *Nature Geoscience*, 15, 382–389.

Kaplan, I. and Denno, R. F. (2007) Interspecific interactions in phytophagous insects revisited: a quantitative assessment of competition theory. *Ecology Letters*, 10, 977–994.

Kaplan, I., Sardanelli, S., Rehill, B. J. and Denno, R. F. (2011) Toward a mechanistic understanding of competition in vascular-feeding herbivores: an empirical test of the sink competition hypothesis. *Oecologia*, 166, 627–636.

Kardol, P., Bezemer, T. M. and Van der Putten, W. H. (2006) Temporal variation in plant–soil feedback controls succession. *Ecology Letters*, 9, 1080–1088.

Kardol, P., Cornips, N. J., Van Kempen, M. M. et al. (2007) Microbe-mediated plant–soil feedback causes historical contingency effects in plant community assembly. *Ecological Monographs*, 77, 147–162.

Karst, J., Jones, M. D. and Hoeksema, J. D. (2023) Positive citation bias and overinterpreted results lead to misinformation on common mycorrhizal networks in forests. *Nature Ecology and Evolution*, 7, 501–511.

Kaufmann, R. (2001) Invertebrate succession on an alpine glacier foreland. *Ecology*, 82, 2261–2278.

Kaye, J. P. and Hart, S. C. (1997) Competition for nitrogen between plants and soil microorganisms. *Trends in Ecology and Evolution*, 12, 139–143.

Keane, B. J., Hartley, I. P., Taylor, C.R. et al. (2023) Grassland responses to elevated CO_2 determined by plant–microbe competition for phosphorus. *Nature Geoscience*, 16, 704–709.

Kearns, P. J. and Shade, A. (2018) Trait-based patterns of microbial dynamics in dormancy potential and heterotrophic strategy: case studies of resource-based and post-press succession. *The ISME Journal*, 12, 2575–2581.

Keesing, F., Ostfeld, R. S., Okanga, S. et al. (2018) Consequences of integrating livestock and wildlife in an African savanna. *Nature Sustainability*, 1, 566–573.

Keesstra, S. D., Bouma, J., Wallinga, J. et al. (2016) The significance of soils and soil science towards realization of the UN sustainable development goals (SDGs). *Soil*, 2, 111–128.

Kielland, K. (1994) Amino acid absorption by arctic plants: implication for plant nutrition and nitrogen cycling. *Ecology*, 75, 2373–2383.

Kiers, E. T., Duhamel, M., Beesetty, Y. et al. (2011) Reciprocal rewards stabilize cooperation in the mycorrhizal symbiosis. *Science*, 333, 880–882.

Kimuyu, D. M., Sensenig, R. L., Riginos, C. et al. (2014) Native and domestic browsers and grazers reduce fuels, fire temperatures, and acacia ant mortality in an African savanna. *Ecological Applications*, 24, 741–749.

King, K. L. and Hutchinson, K. J. (1976) The effects of sheep stocking intensity on the abundance and distribution of mesofauna in pastures. *Journal of Applied Ecology*, 13, 41–55.

King, K. L., Hutchinson, K. J. and Greenslade, P. (1976) The effects of sheep numbers on associations of Collembola in sown pastures. *Journal of Applied Ecology*, 13, 731–739.

Kirschbaum, M. U. F. (2006) The temperature dependence of organic-matter decomposition—still a topic of debate. *Soil Biology and Biochemistry*, 38, 2510–2518.

Kivlin, S. N., Mann, M. A., Lynn, J. S. et al. (2022) Grass species identity shapes communities of root and leaf fungi more than elevation. *ISME Communications*, 2, 25.

Klein, T., Siegwolf, R. T. W. and Körner, C. (2016) Belowground carbon trade among tall trees in a temperate forest. *Science*, 352, 342–344.

Klironomos, J. N. (2002) Feedback with soil biota contributes to plant rarity and invasiveness in communities. *Nature*, 417, 67–70.

Klironomos, J. N. (2003) Variation in plant response to native and exotic arbuscular mycorrhizal fungi. *Ecology*, 84, 2292–2301.

Klironomos, J. N., Widden, P. and Deslandes, I. (1992) Feeding preferences of the collembolan *Folsomia candida* in relation to microfungal successions on decaying litter. *Soil Biology and Biochemistry*, 24, 685–692.

Klumpp, K., Fontaine, S., Attard, E. et al. (2009) Grazing triggers soil carbon loss by altering plant roots and their control on soil microbial community. *Journal of Ecology*, 97, 876–885.

Klumpp, K., Soussana, J. F. and Falcimagne, R. (2007) Effects of past and current disturbance on carbon cycling in grassland mesocosms. *Agriculture, Ecosystems and Environment*, 121, 59–73.

Knight, C. G., Nicolitch, O., Griffiths, R. I. et al. (2024) Soil microbiomes show consistent and predictable responses to extreme events. *Nature*, 636, 690–696.

Knight, D., Elliot, P. W., Anderson, J. M. and Scholefield, D. (1992) The role of earthworms in managed, permanent pastures in Devon, England. *Soil Biology and Biochemistry*, 24, 1511–1517.

Knorr, M. A., Contosta, A. R., Morrison, E. W. et al. (2024) Unexpected sustained soil carbon flux in response to simultaneous warming and nitrogen enrichment compared with single factors alone. *Nature Ecology and Evolution*, 8, 2277–2285.

Knorr, W., Prentice, I. C., House, J. I. and Holland, E. A. (2005) Long-term sensitivity of soil carbon turnover to warming. *Nature*, 433, 298–301.

Kohler, F., Gillet, F., Reust, S. et al. (2006) Spatial and seasonal patterns of cattle habitat use in a mountain wooded pasture. *Landscape Ecology*, 21, 281–295.

Kohout, P., Sýkorová, Z., Bahram, M. et al. (2011) Ericaceous dwarf shrubs affect ectomycorrhizal fungal community of the invasive *Pinus strobus* and native *Pinus sylvestris* in a pot experiment. *Mycorrhiza*, 21, 403–412.

Koller, R., Rodriguez, A., Robin, C. et al. (2013a) Protozoa enhance foraging efficiency of arbuscular mycorrhizal fungi for mineral nitrogen from organic matter in soil to the benefit of host plants. *New Phytologist*, 199, 203–211.

Koller, R., Scheu, S., Bonkowski, M. and Robin, C. (2013b) Protozoa stimulate N uptake and growth of arbuscular mycorrhizal plants. *Soil Biology and Biochemistry*, 65, 204–210.

Koltz, A. M., Classen, A. T. and Wright, J. P. (2018) Warming reverses top-down effects of predators on belowground ecosystem function in Arctic tundra. *Proceedings of the National Academy of Sciences*, 115, E7541–E7549.

Kong, D., Ma, C., Zhang, Q. et al. (2014) Leading dimensions in absorptive root trait variation across 96 subtropical forest species. *New Phytologist*, 203, 863–872.

Kong, D., Wang, J., Wu, H. et al. (2019) Nonlinearity of root trait relationships and the root economics spectrum. *Nature Communications*, 10, 2203.

Korb, J. and Linsenmair, K. E. (2002) Evaluation of predation risk in the collectively foraging termite *Macrotermes bellicosus*. *Insectes Sociaux*, 49, 264–269.

Körner, C. (1999) *Alpine Plant Life: Functional Plant Ecology of High Mountain Ecosystems*. Springer, Berlin.

Köster, K., Berninger, F., Lindén, A. et al. (2014) Recovery in fungal biomass is related to decrease in soil organic matter turnover time in a boreal fire chronosequence. *Geoderma*, 235–236, 74–82.

Koziol, L. and Bever, J. D. (2017) The missing link in grassland restoration: arbuscular mycorrhizal fungi inoculation increases plant diversity and accelerates succession. *Journal of Applied Ecology*, 54, 1301–1309.

Koziol, L., Bauer, J. T., Duell, E. B. et al. (2022) Manipulating plant microbiomes in the field: native mycorrhizae advance plant succession and improve native plant restoration. *Journal of Applied Ecology*, 59, 1976–1985.

Kraamwinkel, C. T., Beaulieu, A., Dias, T. et al. (2021) Planetary limits to soil degradation. *Communications Earth and Environment*, 2, 249.

Kramer-Walter, K. R., Bellingham, P. J., Millar, T. R. et al. (2016) Root traits are multidimensional: specific root length is independent from root tissue density and the plant economic spectrum. *Journal of Ecology*, 104, 1299–1310.

Kreibich, H., Van Loon, A. F., Schröter, K. et al. (2022) The challenge of unprecedented floods and droughts in risk management. *Nature*, 608, 80–86.

Kristensen, J. Å., Rousk, J. and Metcalfe, D. B. (2020) Below-ground responses to insect herbivory in ecosystems with woody plant canopies: a meta-analysis. *Journal of Ecology*, 108, 917–930.

Kristensen, J. A., Svenning, J. C., Georgiou, K. and Malhi, Y. (2022) Can large herbivores enhance ecosystem carbon persistence? *Trends in Ecology and Evolution*, 37, 117–128.

Kulkarni, O. S., Mazumder, M., Kini, S. et al. (2024) Volatile methyl jasmonate from roots triggers host-beneficial soil microbiome biofilms. *Nature Chemical Biology*, 20, 473–483.

Kulmatiski, A., Beard, K. H. and Heavilin, J. (2012) Plant–soil feedbacks provide an additional explanation for diversity–productivity relationships. *Proceedings of the Royal Society B: Biological Sciences*, 279, 3020–3026.

Kulmatiski, A., Beard, K. H., Stevens, J. R. and Cobbold, S. M. (2008) Plant–soil feedbacks: a meta-analytical review. *Ecology Letters*, 11, 980–992.

Kuzyakov, Y. (2010) Priming effects: interactions between living and dead organic matter. *Soil Biology & Biochemistry*, 42, 1363–1371.

Kuzyakov, Y. and Mason-Jones, K. (2018) Viruses in soil: nano-scale undead drivers of microbial life, biogeochemical turnover and ecosystem functions. *Soil Biology and Biochemistry*, 127, 305–317.

Labouyrie, M., Ballabio, C., Romero, F. et al. (2023) Patterns in soil microbial diversity across Europe. *Nature Communications*, 14, 3311.

Laganiere, J., Angers, D. A. and Pare, D. (2010) Carbon accumulation in agricultural soils after afforestation: a meta-analysis. *Global Change Biology*, 16, 439–453.

Lagerström, A., Nilsson, M. C., Zackrisson, O. and Wardle, D. A. (2007) Ecosystem input of nitrogen through biological fixation in feather mosses during ecosystem retrogression. *Functional Ecology*, 21, 1027–1033.

Lal, R. (2004) Soil carbon sequestration to mitigate climate change. *Geoderma*, 123, 1–22.

Laliberté, E., Grace, J. B., Huston, M. A. et al. (2013) How does pedogenesis drive plant diversity? *Trends in Ecology and Evolution*, 28, 331–340.

Laliberté, E., Turner, B. L., Costes, T. et al. (2012) Experimental assessment of nutrient limitation along a 2-million-year dune chronosequence in the south-western Australia biodiversity hotspot. *Journal of Ecology*, 100, 631–642.

LaManna, J. A., Walton, M. L., Turner, B. L. and Myers, J. A. (2016) Negative density dependence is stronger in resource-rich environments and diversifies communities when stronger for common but not rare species. *Ecology Letters*, 19, 657–667.

Lang, B., Rall, B. C., Scheu, S. and Brose, U. (2014) Effects of environmental warming and drought on size-structured soil food webs. *Oikos*, 123, 1224–1233.

Lange, M., Eisenhauer, N., Chen, H. and Gleixner, G. (2023) Increased soil carbon storage through plant diversity strengthens with time and extends into the subsoil. *Global Change Biology*, 29, 2627–2639.

Lange, M., Eisenhauer, N., Sierra, C. A. et al. (2015) Plant diversity increases soil microbial activity and soil carbon storage. *Nature Communications*, 6, 6707.

Langley, J. A., McKinley, D. C., Wolf, A. A. et al. (2009) Priming depletes soil carbon and releases nitrogen in a scrub-oak ecosystem exposed to elevated CO_2. *Soil Biology and Biochemistry*, 41, 54–60.

Lankau, R. A. (2011) Resistance and recovery of soil microbial communities in the face of *Alliaria petiolata* invasions. *New Phytologist*, 189, 536–548.

Lankau, R. A., Nuzzo, V., Spyreas, G. and Davis, A. S. (2009) Evolutionary limits ameliorate the negative impact of an invasive plant. *Proceedings of the National Academy of Sciences*, 106, 15362–15367.

Lau, J. A. and Lennon, J. T. (2012) Rapid responses of soil microorganisms improve plant fitness in novel environments. *Proceedings of the National Academy of Sciences*, 109, 14058–14062.

Laughlin, D. C. (2011) Nitrification is linked to dominant leaf traits rather than functional diversity. *Journal of Ecology*, 99, 1091–1099.

Lavallee, J. M., Soong, J. L. and Cotrufo, M. F. (2020) Conceptualizing soil organic matter into particulate and mineral-associated forms to address global change in the 21st century. *Global Change Biology*, 26, 261–273.

Lavelle, P. and Martin, A. (1992) Small-scale and large-scale effects of endogeneic earthworms on soil organic matter dynamics in soil and the humid tropics. *Soil Biology and Biochemistry*, 24, 1491–1498.

Lavelle, P. and Pashanasi, B. (1989) Soil macrofauna and land management in Peruvian Amazonia (Yurimaguas, Loreto). *Pedobiologia*, 33, 283–291.

Lavelle, P., Bignell, D., Lepage, M. and Wolters, V. (1997) Soil function in a changing world: the role of invertebrate ecosystem engineers. *European Journal of Soil Biology*, 33, 159–193.

Lavelle, P., Lattaud, C., Trigo, D. and Barois, I. (1995) Mutualism and biodiversity in soils. *Plant and Soil*, 170, 23–33.

Lavelle, P., Mathieu, J., Spain, A. et al. (2022) Soil macroinvertebrate communities: a world-wide assessment. *Global Ecology and Biogeography*, 31, 1261–1276.

Lazcano, C., Zhu-Barker, X. and Decock, C. (2021) Effects of organic fertilizers on the soil microorganisms responsible for N_2O emissions: a review. *Microorganisms*, 9, 983.

Leadley, P. W., Niklaus, P. A., Stocker, R. and Körner, C. (1999) A field study of the effects of elevated CO_2 on plant biomass and community structure in a calcareous grassland. *Oecologia*, 118, 39–49.

Ledgard, S. F. (1991) Transfer of fixed nitrogen from white clover to associated grasses in swards grazed by dairy cows, estimated using ^{15}N methods. *Plant and Soil*, 131, 215–223.

Lee, K. (1985) *Earthworms: Their Ecology and Relationships with Soils and Land Use.* Academic Press, New York.

Lee, M. R., Bernhardt, E. S., Van Bodegom, P. M. et al. (2017) Invasive species' leaf traits and dissimilarity from natives shape their impact on nitrogen cycling: a meta-analysis. *New Phytologist*, 213, 128–139.

Leff, J. W., Bardgett, R. D., Wilkinson, A. et al. (2018) Predicting the structure of soil communities from plant community taxonomy, phylogeny, and traits. *The ISME Journal*, 12, 1794–1805.

Leff, J., Jones, S. E. and Prober, S. M. (2015) Consistent responses of soil microbial communities to elevated nutrient inputs in grasslands across the globe, *Proceedings of the National Academy of Sciences*, 112, 10967–10972.

Lehmann, A., Zheng, W. and Rillig, M. C. (2017) Soil biota contributions to soil aggregation. *Nature Ecology and Evolution*, 1, 1828–1835.

Lehmann, A., Zheng, W., Ryo, M. et al. (2020a) Fungal traits important for soil aggregation. *Frontiers in Microbiology*, 10, 2904.

Lehmann, J. and Kleber, M. (2015) The contentious nature of soil organic matter. *Nature*, 528, 60–68.

Lehmann, J., Bossio, D. A., Kögel-Knabner, I. and Rillig, M. C. (2020b) The concept and future prospects of soil health. *Nature Reviews Earth and Environment*, 1, 544–553.

Lehmann, J., Cowie, A., Masiello, C. A. et al. (2021) Biochar in climate change mitigation. *Nature Geosciences*, 14, 883–892.

Leigh, J., Hodge, A. and Fitter, A. H. (2009) Arbuscular mycorrhizal fungi can transfer substantial amounts of nitrogen to their host plant from organic material. *New Phytologist*, 181, 199–207.

Leishman, M. R., Haslehurst, T., Ares, A. and Baruch, Z. (2007) Leaf trait relationships of native and invasive plants: community- and global-scale comparisons. *New Phytologist*, 176, 635–643.

Lekberg, Y., Gibbons, S. M., Rosendahl, S. and Ramsey, P. W. (2013) Severe plant invasions can increase mycorrhizal fungal abundance and diversity. *The ISME Journal*, 7, 1424–1433.

Lenoir, L., Persson, T., Bengtsson, J. et al. (2007) Bottom-up or top-down control in forest soil microcosms? Effects of soil fauna on fungal biomass and C/N mineralization. *Biology and Fertility of Soils*, 43, 281–294.

Lensing, J. R. and Wise, D. H. (2006) Predicted climate change alters the indirect effect of predators on an ecosystem process. *Proceedings of the National Academy of Sciences*, 103, 15502–15505.

Leonard, M. A. (1984) Observations on the influence of culture conditions on the fungal preference of *Folsomia candida* (Collembola; Isotomidae). *Pedobiologia*, 26, 361–367.

Le Roux, E., Kerley, G. I. H. and Cromsigt, J. P. G. M. (2018) Megaherbivores modify trophic cascades triggered by fear of predation in an African savanna ecosystem. *Current Biology*, 28, 2493–2499.

Le Roux, E., Van Veenhuisen, L. S., Kerley, G. I. and Cromsigt, J. P. G. M. (2020) Animal body size distribution influences the ratios of nutrients supplied to plants. *Proceedings of the National Academy of Sciences*, 117, 22256–22263.

Lewis, S. L., Wheeler, C. E., Mitchard, E. T. A. and Koch, A. (2019) Restoring natural forests is the best way to remove atmospheric carbon. *Nature*, 568, 25–28.

Li, B., Li, Y. Y., Wu, H. M. et al. (2016) Root exudates drive interspecific facilitation by enhancing nodulation and N_2 fixation. *Proceedings of the National Academy of Sciences*, 113, 6496–6501.

Li, C., Chen, X., Jia, Z. et al. (2024a) Meta-analysis reveals the effects of microbial inoculants on the biomass and diversity of soil microbial communities. *Nature Ecology and Evolution*, 8, 1270–1284.

Li, G., Liu, T., Whalen, J. K. and Wei, Z. (2024b) Nematodes: an overlooked tiny engineer of plant health. *Trends in Plant Science*, 29, 52–63.

Li, J., Wang, J., Liu, H., Macdonald, C. A. and Singh, B. K. (2023) Microbial inoculants with higher capacity to colonize soils improved wheat drought tolerance. *Microbial Biotechnology*, 16, 2131–2144.

Li, L., Tilman, D., Lambers, H. and Zhang, F. S. (2014) Plant diversity and overyielding: insights from belowground facilitation of intercropping in agriculture. *New Phytologist*, 203, 63–69.

Li, S. P., Wang, P., Chen, Y. et al. (2020a) Island biogeography of soil bacteria and fungi: similar patterns, but different mechanisms. *The ISME Journal*, 14, 1886–1896.

Li, X., De Boer, W., Zhang, Y., Ding, C. et al. (2018a) Suppression of soil-borne *Fusarium* pathogens of peanut by intercropping with the medicinal herb *Atractylodes lancea*. *Soil Biology and Biochemistry*, 116, 120–130.

Li, X., Risch, A. C., Sanders, D. et al. (2021a) A facilitation between large herbivores and ants accelerates litter decomposition by modifying soil microenvironmental conditions. *Functional Ecology*, 35, 1822–1832.

Li, X., Wang, T., Chang, S. X. et al. (2020b) Biochar increases soil microbial biomass but has variable effects on microbial diversity: a meta-analysis. *Science of the Total Environment*, 749, 141593.

Li, X., Zhong, Z., Sanders, D. et al. (2018b) Reciprocal facilitation between large herbivores and ants in a semi-arid grassland. *Proceedings of the Royal Society B*, 285, 20181665.

Li, X., Zhu, H., Geisen, S. et al. (2020c) Agriculture erases climate constraints on soil nematode communities across large spatial scales. *Global Change Biology*, 26, 919–930.

Li, X. F., Wang, Z. G., Bao, X. G. et al. (2021b) Long-term increased grain yield and soil fertility from intercropping. *Nature Sustainability*, 4, 943–950.

Li, Y., Ma, J., Yu, Y. et al. (2022) Effects of multiple global change factors on soil microbial richness, diversity and functional gene abundances: a meta-analysis. *Science of the Total Environment*, 815, 152737.

Liang, C., Schimel, J. P. and Jastrow, J. D. (2017) The importance of anabolism in microbial control over soil carbon storage. *Nature Microbiology*, 2, 17105.

Liang, J., Crowther, T. W., Picard, N. et al. (2016) Positive biodiversity–productivity relationship predominant in global forests. *Science*, 354, aaf8957.

Liao, C., Peng, R., Luo, Y. et al. (2008) Altered ecosystem carbon and nitrogen cycles by plant invasion: a meta-analysis. *New Phytologist*, 177, 706–714.

Liao, H., Li, H., Duan, C. S. et al. (2022) Response of soil viral communities to land use changes. *Nature Communications*, 13, 6027.

Liiri, M., Setälä, H., Haimi, J. et al. (2002) Relationship between soil microarthropod species diversity and plant growth does not change when the system is disturbed. *Oikos*, 96, 137–149.

Lindberg, N., Engtsson, J. B. and Persson, T. (2002) Effects of experimental irrigation and drought on the composition and diversity of soil fauna in a coniferous stand. *Journal of Applied Ecology*, 39, 924–936.

Lipson, D. A. and Schmidt, S. K. (2004) Seasonal changes in an alpine bacterial community in the Colorado Rocky Mountains. *Applied Environmental Microbiology*, 70, 2867–2879.

Liu, H., Mi, Z., Lin, L. I. et al. (2018a) Shifting plant species composition in response to climate change stabilizes grassland primary production. *Proceedings of the National Academy of Sciences*, 115, 4051–4056.

Liu, L., Zhu, K., Wurzburger, N. and Zhang, J. (2020) Relationships between plant diversity and soil microbial diversity vary across taxonomic groups and spatial scales. *Ecosphere*, 11, e02999.

Liu, M., Xu, X, Wanek, W. et al. (2025) Nitrogen availability in soil controls uptake of different nitrogen forms by plants. *New Phytologist*, 245, 1450–1467.

Liu, N., Kan, H. M., Yang, G. W. and Zhang, Y. J. (2015a) Changes in plant, soil, and microbes in a typical steppe from simulated grazing: explaining potential change in soil C. *Ecological Monographs*, 85, 269–286.

Liu, Q., Qiao, N., Xu, X. et al. (2016) Nitrogen acquisition by plants and microorganisms in a temperate grassland. *Scientific Reports*, 6, 22642.

Liu, S., García-Palacios, P., Tedersoo, L. et al. (2022) Phylotype diversity within soil fungal functional groups drives ecosystem stability. *Nature Ecology and Evolution*, 6, 900–909.

Liu, X., Burslem, D. F. R. P., Taylor, J. D. et al. (2018b) Partitioning of soil phosphorus among arbuscular and ectomycorrhizal trees in tropical and subtropical forests. *Ecology Letters*, 21, 713–723.

Liu, X., Etienne, R. S., Liang, M. et al. (2015b) Experimental evidence for an intraspecific Janzen–Connell effect mediated by soil biota. *Ecology*, 96, 662–671.

Liu, X., Trogisch, S., He, J. S. et al. (2018c) Tree species richness increases ecosystem carbon storage in subtropical forests. *Proceedings of the Royal Society B*, 285, 20181240.

Liu, Y. and He, F. (2021) Warming intensifies soil pathogen negative feedback on a temperate tree. *New Phytologist*, 231, 2297–2307.

Lobry De Bruyn, L. and Conacher, A. J. (1994) The effect of ant biopores on water infiltration in soils in undisturbed bushland and in farmland in a semi-arid environment. *Pedobiologia*, 38, 193–207.

Longepierre, M., Widmer, F., Keller, T. et al. (2021) Limited resilience of the soil microbiome to mechanical compaction within four growing seasons of agricultural management. *ISME Communications*, 1, 44.

Loreau, M. and De Mazancourt, C. (2013) Biodiversity and ecosystem stability: a synthesis of underlying mechanisms. *Ecology Letters*, 16, 106–115.

Loreau, M. and Hector, A. (2001) Partitioning selection and complementarity in biodiversity experiments. *Nature*, 412, 72–76.

Lori, M., Symnaczik, S., Mäder, P. et al. (2017) Organic farming enhances soil microbial abundance and activity—a meta-analysis and meta-regression. *PLoS ONE*, 12, e0180442.

Lovett, G. M. and Ruesink, A. E. (1995) Carbon and nitrogen mineralization from decomposing gypsy moth frass. *Oecologia*, 104, 133–138.

Lu, C. and Tian, H. (2017) Global nitrogen and phosphorus fertilizer use for agriculture production in the past half century: shifted hot spots and nutrient imbalance. *Earth System Science Data*, 9, 181–192.

Lu, M., Zhou, X., Yang, Q. et al. (2013) Responses of ecosystem carbon cycle to experimental warming: a meta-analysis. *Ecology*, 94, 726–738.

Lu, X., Taylor, A. E., Myrold, D. D. and Neufeld, J. D. (2020) Expanding perspectives of soil nitrification to include ammonia-oxidizing archaea and comammox bacteria. *Soil Science Society of America Journal*, 84, 287–302.

Lubbers, I. M., Berg, M. P., De Deyn, G. B. et al. (2020) Soil fauna diversity increases CO_2 but suppresses N_2O emissions from soil. *Global Change Biology*, 26, 1886–1898.

Lubbers, I. M., Pulleman, M. M. and Van Groenigen, J. W. (2017) Can earthworms simultaneously enhance decomposition and stabilization of plant residue carbon? *Soil Biology and Biochemistry*, 105, 12–24.

Lundberg, D. S., Lebeis, S. L., Paredes, S. H. et al. (2012) Defining the core *Arabidopsis thaliana* root microbiome. *Nature*, 488, 86–90.

Luo, G., Li, L., Friman, V. P. et al. (2018) Organic amendments increase crop yields by improving microbe-mediated soil functioning of agroecosystems: a meta-analysis. *Soil Biology and Biochemistry*, 124, 105–115.

Luo, S., Png, G. K., Ostle, N. J. et al. (2023) Grassland degradation-induced declines in soil fungal complexity reduce fungal community stability and ecosystem multifunctionality. *Soil Biology and Biochemistry*, 176, 108865.

Luo, Y., Sherry, R., Zhou, X. and Wan, S. (2009) Terrestrial carbon-cycle feedback to climate warming: experimental evidence on plant regulation and impacts of biofuel feedstock harvest. *GCB Bioenergy*, 1, 62–74.

Luo, Y., Su, B., Currie, W. S. et al. (2004) Progressive nitrogen limitation responses to rising atmospheric carbon dioxide. *BioScience*, 54, 731–739.

Luo, Y., Wan, S. and Hui, D. (2001) Acclimatization of soil respiration to warming in tall grass prairie. *Nature*, 413, 622–625.

Lustenhouwer, N., Maynard, D. S., Bradford, M. A. et al. (2020) A trait-based understanding of wood decomposition by fungi. *Proceedings of the National Academy of Sciences*, 117, 11551–11558.

Lutz, S., Bodenhausen, N., Hess, J. et al. (2023) Soil microbiome indicators can predict crop growth response to large-scale inoculation with arbuscular mycorrhizal fungi. *Nature Microbiology*, 8, 2277–2289.

Lynch, M. D. J. and Neufeld, J. D. (2015) Ecology and exploration of the rare biosphere. *Nature Reviews Microbiology*, 13, 217–229.

Ma, X., Geng, Q., Zhang, H. et al. (2021) Global negative effects of nutrient enrichment on arbuscular mycorrhizal fungi, plant diversity and ecosystem multifunctionality. *New Phytologist*, 229, 2957–2969.

Ma, Y., Woolf, D., Fan, M. et al. (2023) Global crop production increase by soil organic carbon. *Nature Geoscience*, 16, 1159–1165.

Ma, Z., Guo, D., Xu, X. et al. (2018) Evolutionary history resolves global organization of root functional traits. *Nature*, 555, 94–97.

MacArthur, R. H. and Wilson, E. O. (1967) *The Theory of Island Biogeography*. Princeton University Press, Princeton, NJ.

Maestre, F. T., Bautista, S., Cortina, J. and Bellot, J. (2001) Potential for using facilitation by grasses to establish shrubs on a semiarid degraded steppe. *Ecological Applications*, 11, 1641–1655.

Maestre, F. T., Delgado-Baquerizo, M., Jeffries, T. C. et al. (2015) Increasing aridity reduces soil microbial diversity and abundance in global drylands. *Proceedings of the National Academy of Sciences*, 112, 15684–15689.

Mahé, F., de Vargas, C., Bass, D. et al. (2017) Parasites dominate hyperdiverse soil protist communities in Neotropical rainforests. *Nature Ecology and Evolution*, 1, 91.

Maherali, H. and Klironomos, J. N. (2007) Influence of phylogeny on fungal community assembly and ecosystem functioning. *Science*, 316, 1746–1748.

Malhi, Y., Doughty, C. E., Galetti, M. et al. (2016) Megafauna and ecosystem function from the Pleistocene to the Anthropocene. *Proceedings of the National Academy of Sciences*, 113, 838–846.

Malik, A. A., Chowdhury, S., Schlager, V. et al. (2016) Soil fungal:bacterial ratios are linked to altered carbon cycling. *Frontiers in Microbiology*, 7, 1247.

Malmström, A. (2010) The importance of measuring fire severity—evidence from microarthropod studies. *Forest Ecology and Management*, 260, 62–70.

Maltz, M. R. and Treseder, K. K. (2015) Sources of inocula influence mycorrhizal colonization of plants in restoration projects: a meta-analysis. *Restoration Ecology*, 23, 625–634.

Mangan, S. A., Schnitzer, S. A., Herre, E. A. et al. (2010) Negative plant–soil feedback predicts tree-species relative abundance in a tropical forest. *Nature*, 466, 752–755.

Maraun, M., Martens, H., Migge, S. et al. (2003) Adding to 'the enigma of soil animal diversity': fungal feeders and saprophagous soil invertebrates prefer similar food substrates. *European Journal of Soil Biology*, 39, 85–95.

Maraun, M., Schatz, H. and Scheu, S. (2007) Awesome or ordinary? Global diversity patterns of oribatid mites. *Ecography*, 30, 209–216.

Mariotte, P. (2014) Do subordinate species punch above their weight? Evidence from above-and belowground. *New Phytologist*, 203, 16–21.

Mariotte, P., Canarini, A. and Dijkstra, F. A. (2017) Stoichiometric N:P flexibility and mycorrhizal symbiosis favour plant resistance against drought. *Journal of Ecology*, 105, 958–967.

Mariotte, P., Vandenberghe, C., Meugnier, C. et al. (2013) Subordinate plant species impact on soil microbial communities and ecosystem functioning in grasslands: findings from a removal experiment. *Perspectives in Plant Ecology, Evolution and Systematics*, 15, 77–85.

Maron, J. L., Marler, M., Klironomos, J. N. and Cleveland, C. C. (2011) Soil fungal pathogens and the relationship between plant diversity and productivity. *Ecology Letters*, 14, 36–41.

Maron, J. L., Smith, A. L., Ortega, Y. K. et al. (2016) Negative plant–soil feedbacks increase with plant abundance, and are unchanged by competition. *Ecology*, 97, 2055–2063.

Martin, P. A., Fisher, L., Pérez-Izquierdo, L. et al. (2024) Meta-analysis reveals that the effects of precipitation change on soil and litter fauna in forests depend on body size. *Global Change Biology*, 30, e17305.

Martin, P. A., Newton, A. C. and Bullock, J. M. (2017) Impacts of invasive plants on carbon pools depend on both species' traits and local climate. *Ecology*, 98, 1026–1035.

Martín-Robles, N., Lehmann, A., Seco, E. et al. (2018) Impacts of domestication on the arbuscular mycorrhizal symbiosis of 27 crop species. *New Phytologist*, 218, 322–334.

Mason, R. E., Craine, J. M., Lany, N. K. et al. (2022) Evidence, causes, and consequences of declining nitrogen availability in terrestrial ecosystems. *Science*, 376, eabh3767.

Masters, G. J., Brown, V. K. and Gange, A. C. (1993) Plant mediated interactions between aboveground and belowground insect herbivores. *Oikos*, 66, 148–151.

Matthews, J. A. (1992) *The Ecology of Recently Deglaciated Terrain: A Geoecological Approach to Glacier Forelands and Primary Succession*. Cambridge University Press, Cambridge.

Mawarda, P. C., Le Roux, X., Van Elsas, J. D. and Salles, J. F. (2020) Deliberate introduction of invisible invaders: a critical appraisal of the impact of microbial inoculants on soil microbial communities. *Soil Biology and Biochemistry*, 148, 107874.

Mawdsley, J. L. and Bardgett, R. D. (1997) Continuous defoliation of perennial ryegrass (*Lolium perenne*) and white clover (*Trifolium repens*) and associated changes in the composition and activity of the microbial population of an upland grassland soil. *Biology and Fertility of Soils*, 24, 52–58.

Mayor, J. R., Sanders, N. J., Classen, A. T. et al. (2017) Elevation alters ecosystem properties across temperate treelines globally. *Nature*, 542, 91–95.

Mazel, F., Malard, L., Niculita-Hirzel, H. et al. (2022) Soil protist function varies with elevation in the Swiss Alps. *Environmental Microbiology*, 24, 1689–1702.

McCormack, S. A., Ostle, N., Bardgett, R. D. et al. (2013) Biochar in bioenergy cropping systems: impacts on soil faunal communities and linked ecosystem processes. *GCB Bioenergy*, 5, 81–95.

McDaniel, M. D., Grandy, A. S., Tiemann, L. K. and Weintraub, M. N. (2014) Crop rotation complexity regulates the decomposition of high and low quality residues. *Soil Biology and Biochemistry*, 78, 243–254.

McIntoch, P. D., Allen, R. B. and Scott, N. (1997) Effects of exclosure and management on biomass and soil nutrient pools in seasonally dry high county, New Zealand. *Journal of Environmental Management*, 51, 169–186.

McKane, R. B., Johnson, L. C., Shaver, G. R. et al. (2002) Resource-based niches provide a basis for plant species diversity and dominance in arctic tundra. *Nature*, 413, 68–71.

McNaughton, S. J. (1983) Serengeti grassland ecology: the role of composite environmental factors and contingency in community organization. *Ecological Monographs*, 53, 291–320.

McNaughton, S. J. (1985) Ecology of a grazing system: the Serengeti. *Ecological Monographs*, 55, 259–294.

McNaughton, S. J., Banyikwa, F. F. and McNaughton, M. M. (1997) Promotion of the cycling of diet-enhancing nutrients by African grazers. *Science*, 278, 1798–1800.

McNaughton, S. J., Oesterheld, M., Frank, D. A. et al. (1989) Ecosystem-level patterns of primary productivity and herbivory in terrestrial habitats. *Nature*, 341, 142–144.

McSherry, M. E. and Ritchie, M. E. (2013) Effects of grazing on grassland soil carbon: a global review. *Global Change Biology*, 19, 1347–1357.

Medina-Roldán, E. and Bardgett, R. D. (2011) Plant and soil responses to defoliation: a comparative study of grass species with contrasting life history strategies. *Plant and Soil*, 344, 377–388.

Medina-Roldán, E., Paz-Ferreiro, J. and Bardgett, R. D. (2012) Grazing exclusion affects soil and plant communities, but has no impact on soil carbon storage in an upland grassland. *Agriculture, Ecosystems & Environment*, 149, 118–123.

Meier, I. C., Finzi, A. C. and Phillips, R. P. (2017) Root exudates increase N availability by stimulating microbial turnover of fast-cycling N pools. *Soil Biology and Biochemistry*, 106, 119–128.

Meinhardt, K. A. and Gehring, C. A. (2012) Disrupting mycorrhizal mutualisms: a potential mechanism by which exotic tamarisk outcompetes native cottonwoods. *Ecological Applications*, 22, 532–549.

Meisner, A., De Deyn, G. B., De Boer, W. and Van der Putten, W. H. (2013) Soil biotic legacy effects of extreme weather events influence plant invasiveness. *Proceedings of the National Academy of Sciences*, 110, 9835–9838.

Melguizo-Ruiz, N., Jiménez-Navarro, G., De Mas, E. et al. (2020) Field exclusion of large soil predators impacts lower trophic levels and decreases leaf-litter decomposition in dry forests. *Journal of Animal Ecology*, 89, 334–346.

Melillo, J. M., Frey, S. D., DeAngelis, K. M. et al. (2017) Long-term pattern and magnitude of soil carbon feedback to the climate system in a warming world. *Science*, 358, 101–105.

Melillo, J. M., Steudler, P. A., Aber, J. D. et al. (2002) Soil warming and carbon-cycle feedbacks to the climate system. *Science*, 298, 2173–2176.

Mendes, R., Kruijt, M., De Bruijn, I. et al. (2011) Deciphering the rhizosphere microbiome for disease-suppressive bacteria. *Science*, 332, 1097–1100.

Menge, D. N. L. and Hedin, L. O. (2009) Nitrogen fixation in different biogeochemical niches along a 120,000-year chronosequence in New Zealand. *Ecology*, 90, 2190–2201.

Menzel, A., Hempel, S., Klotz, S. et al. (2017) Mycorrhizal status helps explain invasion success of alien plant species. *Ecology*, 98, 92–102.

Metcalfe, D. B. (2017) Microbial change in warming soils. *Science*, 358, 41–42.

Midolo, G., Alkemade, R., Schipper, A. M. et al. (2019) Impacts of nitrogen addition on plant species richness and abundance: a global meta-analysis. *Global Ecology and Biogeography*, 28, 398–413.

Mikola, J., Setälä, H., Virkajärvi, P. et al. (2009) Defoliation and patchy nutrient return drive grazing effects on plant and soil properties in a dairy cow pasture. *Ecological Monographs*, 79, 221–244.

Milchunas, D. G. and Lauenroth, W. K. (1993) Quantitative effects of grazing on vegetation and soils over a global range of environments. *Ecological Monographs*, 63, 327–366.

Milcu, A. and Manning, P. (2011) All size classes of soil fauna and litter quality control the acceleration of litter decay in its home environment. *Oikos*, 120, 1366–1370.

Miller, A. E. and Bowman, W. D. (2002) Variation in nitrogen-15 natural abundance and nitrogen uptake traits among co-occurring alpine species: do species partition by nitrogen form? *Oecologia*, 130, 609–616.

Miller, A. E. and Bowman, W. D. (2003) Alpine plants show species-level differences in the uptake of organic and inorganic nitrogen. *Plant and Soil*, 250, 283–292.

Minasny, B., Malone, B. P., McBratney, A. B. et al. (2017) Soil carbon 4 per mille. *Geoderma*, 292, 59–86.

Moinet, G. Y., Hijbeek, R., Van Vuuren, D. P. and Giller, K. E. (2023) Carbon for soils, not soils for carbon. *Global Change Biology*, 29, 2384–2398.

Mommer, L., Cotton, T. E. A., Raaijmakers, J. M. et al. (2018) Lost in diversity: the interactions between soil-borne fungi, biodiversity and plant productivity. *New Phytologist*, 218, 542–553.

Mondal, S. and Chakraborty, D. (2022) Global meta-analysis suggests that no-tillage favourably changes soil structure and porosity. *Geoderma*, 405, 115443.

Mondal, S., Chakraborty, D., Bandyopadhyay, K. et al. (2020) A global analysis of the impact of zero-tillage on soil physical condition, organic carbon content, and plant root response. *Land Degradation & Development*, 31, 557–567.

Monk, J. D. and Schmitz, O. J. (2022) Landscapes shaped from the top down: predicting cascading predator effects on spatial biogeochemistry. *Oikos*, 2022, e08554.

Montanarella, L., Pennock, D. J., McKenzie, N. et al. (2016) World's soils are under threat. *Soil*, 2, 79–82.

Montgomery, D. R. (2007) Soil erosion and agricultural sustainability. *Proceedings of the National Academy of Sciences*, 104, 13268–13272.

Moore, J. C. and de Ruiter, P. C. (2000) Invertebrates in detrital food webs along gradients of productivity. In: *Invertebrates as Webmasters in Ecosystems* (Coleman, D. C. and Hendrix, P. F., eds), pp. 161–184. CAB International, Wallingford.

Moore, J. C. and Hunt, W. H. (1988) Resource compartmentation and the stability of real ecosystems. *Nature*, 333, 261–263.

Moore, J. C., McCann, K., Setälä, H. and de Ruiter, P. C. (2003) Top-down is bottom-up: does predation in the rhizosphere regulate aboveground dynamics? *Ecology*, 84, 846–857.

Moore, J. C., St. John, T. V. and Coleman, D. C. (1985) Ingestion of vesicular-arbuscular mycorrhizal hyphae and spores by soil microarthropods. *Ecology*, 66, 1979–1981.

Moreau, D., Bardgett, R. D., Finlay, R. D. et al. (2019) A plant perspective on nitrogen cycling in the rhizosphere. *Functional Ecology*, 33, 540–552.

Mori, A. S., Dee, L. E., Gonzalez, A. et al. (2021) Biodiversity–productivity relationships are key to nature-based climate solutions. *Nature Climate Change*, 11, 543–550.

Morriën, E. and Van der Putten, W. H. (2013) Soil microbial community structure of range-expanding plant species differs from co-occurring natives. *Journal of Ecology*, 101, 1093–1102.

Morriën, E., Duyts, H. and Van der Putten, W. H. (2012) Effects of native and exotic range-expanding plant species on taxonomic and functional composition of nematodes in the soil food web. *Oikos*, 121, 181–190.

Morriën, E., Hannula, S. E., Snoek, L. B. et al. (2017) Soil networks become more connected and take up more carbon as nature restoration progresses. *Nature Communications*, 8, 14349.

Morugán-Coronado, A., Pérez-Rodríguez, P., Insolia, E. et al. (2022) The impact of crop diversification, tillage and fertilization type on soil total microbial, fungal and bacterial abundance: a worldwide meta-analysis of agricultural sites. *Agriculture, Ecosystems and Environment*, 329, 107867.

Moutinho, P., Nepstad, D. C. and Davidson, E. A. (2003) Influence of leaf-cutting ant nests on secondary forest growth and soil properties in Amazonia. *Ecology*, 84, 1265–1276.

Mueller, K. E., Blumenthal, D. M., Carrillo, Y. et al. (2016) Elevated CO_2 and warming shift the functional composition of soil nematode communities in a semiarid grassland. *Soil Biology and Biochemistry*, 103, 46–51.

Müller L. M. and Bahn, M. (2022) Drought legacies and ecosystem responses to subsequent drought. *Global Change Biology*, 28, 5086–5103.

Müller, P. E. (1884) Forest soil studies: on mull and mor in oak-forests and on heaths with some physical and chemical investigations of forest and heath soils by CFA Tuxen. *Tidsskr. Skovbrug Copenhagen*, 7(1).

Munakata, R., Larbat, R., Duriot, L. et al. (2019) Polyphenols from plant roots: an expanding biological frontier. *Recent Advances in Polyphenol Research*, 6, 207–236.

Murali, G., Iwamura, T., Meiri, S. and Roll, U. (2023) Future temperature extremes threaten land vertebrates. *Nature*, 615, 461–467.

Myers-Smith, I. H. and Hik, D. S. (2018) Climate warming as a driver of tundra shrubline advance. *Journal of Ecology*, 106, 547–560.

Myers-Smith, I. H., Forbes, B. C., Wilmking, M. et al. (2011) Shrub expansion in tundra ecosystems: dynamics, impacts and research priorities. *Environmental Research Letters*, 6, 045509.

Nadelhoffer, K. J., Aber, J. D. and Melillo, J. M. (1985) Fine roots, net primary production, and soil nitrogen availability: a new hypothesis. *Ecology*, 66, 1377–1390.

Näsholm, T., Ekblad, A., Nordin, A. et al. (1998) Boreal forest plants take up organic nitrogen. *Nature*, 392, 914–916.

Näsholm, T., Huss-Danell, K. and Högberg, P. (2000) Uptake of organic nitrogen in the field by four agriculturally important plant species. *Ecology*, 81, 1155–1161.

Näsholm, T., Huss-Danell, K. and Högberg, P. (2001) Uptake of glycine by field grown wheat. *New Phytologist*, 150, 59–63.

Näsholm, T., Kielland, K. and Ganeteg, U. (2009) Uptake of organic nitrogen by plants. *New Phytologist*, 182, 31–48.

Navarro-Cano, J. A., Verdú, M. and Goberna, M. (2018) Trait-based selection of nurse plants to restore ecosystem functions in mine tailings. *Journal of Applied Ecology*, 55, 1195–1206.

Neff, J. C., Chapin, F. S. and Vitousek, P. M. (2003) Breaks in the cycle: dissolved organic nitrogen in terrestrial ecosystems. *Frontiers in Ecology and Environment*, 1, 205–211.

Nemergut, D. R., Anderson, S. P., Cleveland, C. C. et al. (2007) Microbial community succession in an unvegetated, recently deglaciated soil. *Microbial Ecology*, 53, 110–122.

Nemergut, D. R., Knelman, J. E., Ferrenberg, S. et al. (2016) Decreases in average bacterial community rRNA operon copy number during succession. *The ISME Journal*, 10, 1147–1156.

Nerva, L., Sandrini, M., Moffa, L. et al. (2022) Breeding toward improved ecological plant–microbiome interactions. *Trends in Plant Science*, 27, 1134–1143.

Neuenkamp, L., Prober, S. M., Price, J. N. et al. (2019) Benefits of mycorrhizal inoculation to ecological restoration depend on plant functional type, restoration context and time. *Fungal Ecology*, 40, 140–149.

Newell, K. (1984a) Interaction between two decomposer basidiomycetes and a collembolan under Sitka spruce: grazing and its potential effects on fungal distribution and litter decomposition. *Soil Biology and Biochemistry*, 16, 235–239.

Newell, K. (1984b) Interaction between two decomposer basidiomycetes and a collembolan under Sitka spruce: distribution, abundance and selective grazing. *Soil Biology and Biochemistry*, 16, 227–233.

Newton, P. C. D., Clark, H., Bell, C. C. et al. (1995) Plant growth and soil processes in temperate grassland communities at elevated CO_2. *Journal of Biogeography*, 22, 235–240.

Nichols, E., Spector, S., Louzada, J. et al. (2008) Ecological functions and ecosystem services provided by Scarabaenae dung beetles. *Biological Conservation*, 141, 1461–1474.

Nicolas, A. M., Sieradzki, E. T., Pett-Ridge, J. et al. (2023) A subset of viruses thrives following microbial resuscitation during rewetting of a seasonally dry California grassland soil. *Nature Communications*, 14, 5835.

Nie, M., Lu, M., Bell, J. et al. (2013) Altered root traits due to elevated CO_2: a meta-analysis. *Global Ecology and Biogeography*, 22, 1095–1105.

Nielsen, U. N. (2019) *Soil Fauna Assemblages*. Cambridge University Press, Cambridge.

Nielsen, U. N., Ayres, E., Wall, D. H. and Bardgett, R. D. (2011) Soil biodiversity and carbon cycling: a review and synthesis of studies examining diversity–function relationships. *European Journal of Soil Science*, 62, 105–116.

Nielsen, U. N., Ayres, E., Wall, D. H. et al. (2014) Global-scale patterns of assemblage structure of soil nematodes in relation to climate and ecosystem properties. *Global Ecology and Biogeography*, 23, 968–978.

Niklaus, P. A., Leadley, P. W., Schmid, B. and Körner, C. (2001) A long-term field study on biodiversity × elevated CO_2 interactions in grassland. *Ecological Monographs*, 71, 341–356.

Nordin, A., Högberg, P. and Näsholm, T. (2001) Soil nitrogen form and plant nitrogen uptake along a boreal forest productivity gradient. *Oecologia*, 129, 125–132.

Norén, B. and Odham, G. (1973) Antagonistic effects of *Myxococcus xanthus* on fungi: II. Isolation and characterization of inhibitory lipid factors. *Lipids*, 8, 573–583.

Northup, R. R., Yu, Z., Dahlgren, R. A. and Vogt, K. A. (1995) Polyphenol control on nitrogen release from pine litter. *Nature*, 377, 227–229.

Nottingham, A. T., Turner, B. L., Winter, K. et al. (2013) Root and arbuscular mycorrhizal mycelial interactions with soil microorganisms in lowland tropical forest. *FEMS Microbiology Ecology*, 85, 37–50.

Nykanen, H. and Koricheva, J. (2004) Damage-induced changes in woody plants and their effects on insect herbivore performance: a meta-analysis. *Oikos*, 104, 247–268.

Ochoa-Hueso, R., Collins, S. L., Delgado-Baquerizo, M. et al. (2018) Drought consistently alters the composition of soil fungal and bacterial communities in grasslands from two continents. *Global Change Biology*, 24, 2818–2827.

Ohtonen, R., Fritze, H., Pennanen, T. et al. (1999) Ecosystem properties and microbial community changes in primary succession on a glacier forefront. *Oecologia*, 119, 239–246.

Olff, H. and Ritchie, M. E. (1998) Effects of herbivores on grassland plant diversity. *Trends in Ecology and Evolution*, 13, 261–265.

Olff, H., Huisman, J. and Van Tooren, B. F. (1993) Species dynamics and nutrient accumulation during early primary succession in coastal sand dunes. *Journal of Ecology*, 81, 693–706.

Olff, H., Ritchie, M. E. and Prins, H. H. T. (2002) Global environmental controls of diversity in large herbivores. *Nature*, 415, 901–904.

Oliverio, A. M., Geisen, S., Delgado-Baquerizo, M. et al. (2020) The global-scale distributions of soil protists and their contributions to belowground systems. *Science Advances*, 6, eaax8787.

Olofsson, J. and Post, E. (2018) Effects of large herbivores on tundra vegetation in a changing climate, and implications for rewilding. *Philosophical Transactions of the Royal Society B: Biological Sciences*, 373, 20170437.

Olofsson, J., Oksanen, L., Callaghan, T. et al. (2009) Herbivores inhibit climate-driven shrub expansion on the tundra. *Global Change Biology*, 15, 2681–2693.

Olsen, S. L., Sandvik, S. M. and Totland, Ø. (2013) Influence of two N-fixing legumes on plant community properties and soil nutrient levels in an alpine ecosystem. *Arctic, Antarctic, and Alpine Research*, 45, 363–371.

Oram, N. J., Ingrisch, J., Bardgett, R. D. et al. (2023) Drought intensity alters productivity, carbon allocation and plant nitrogen uptake in fast versus slow grassland communities. *Journal of Ecology*, 111, 1681–1699.

Orgiazzi, A., Bardgett, R. D., Barrios, E. et al. (2016) *Global Soil Biodiversity Atlas*. European Commission, Luxembourg.

Ortiz-Álvarez, R., Fierer, N., de los Ríos, A. et al. (2018) Consistent changes in the taxonomic structure and functional attributes of bacterial communities during primary succession. *The ISME Journal*, 12, 1658–1667.

Orwin, K. H., Buckland, S. M., Johnson, D. et al. (2010) Linkages of plant traits to soil properties and the functioning of temperate grassland. *Journal of Ecology*, 98, 1074–1083.

Ostfeld, R.S. and Keesing, F. (2004) Oh the locusts sang, then they dropped dead. *Science*, 306, 1488–1489.

Owen, A. G. and Jones, D. L. (2001) Competition for amino acids between wheat roots and rhizosphere microorganisms and the role of amino acids in plant N acquisition. *Soil Biology and Biochemistry*, 33, 651–657.

Packer, A. and Clay, K. (2000) Soil pathogens and spatial patterns of seedling mortality in a temperate tree. *Nature*, 404, 278–281.

Padilla, F. M. and Pugnaire, F. I. (2006) The role of nurse plants in the restoration of degraded environments. *Frontiers in Ecology and the Environment*, 4, 196–202.

Panikov, N. S., Flanagan, P. W., Oechel, W. C. et al. (2006) Microbial activity in soils frozen to below −39°C. *Soil Biology and Biochemistry*, 38, 785–794.

Parker, T. C., Sadowsky, J., Dunleavy, H. et al. (2017) Slowed biogeochemical cycling in sub-arctic birch forest linked to reduced mycorrhizal growth and community change after a defoliation event. *Ecosystems*, 20, 316–330.

Parker, T. C., Subke, J. A. and Wookey, P. A. (2015) Rapid carbon turnover beneath shrub and tree vegetation is associated with low soil carbon stocks at a subarctic treeline. *Global Change Biology*, 21, 2070–2081.

Parkes, J. P., Easdale, T. A., Williamson, W. M. and Forsyth, D. M. (2015) Causes and consequences of ground disturbance by feral pigs (*Sus scrofa*) in a lowland New Zealand conifer–angiosperm forest. *New Zealand Journal of Ecology*, 39, 34–42.

Parkinson, D., Visser, S. and Whittaker, J. B. (1979) Effects of collembolan grazing on fungal colonization of leaf litter. *Soil Biology and Biochemistry*, 11, 529–535.

Parmenter, R. R. and MacMahon, J. A. (2009) Carrion decomposition and nutrient cycling in a semiarid shrub-steppe ecosystem. *Ecological Monographs*, 79, 637–661.

Pastor, J., Dewey, B., Naiman, R. J. et al. (1993) Moose browsing and soil fertility in the boreal forests of Isle Royale National Park. *Ecology*, 74, 467–480.

Paterson, E. and Sim, A. (2000) Effect of nitrogen supply and defoliation on loss of organic compounds from roots of *Festuca rubra*. *Journal of Experimental Botany*, 51, 1449–1457.

Paul, E. A. and Clark, F. E. (1996) *Soil Microbiology and Biochemistry*. Academic Press, San Diego.

Paul, E. and Frey, S. (2023) *Soil Microbiology, Ecology and Biochemistry*. 5th Edition. Elsevier, Amsterdam.

Pausch, J., Holz, M., Zhu, B. and Cheng, W. (2024) Rhizosphere priming promotes plant nitrogen acquisition by microbial necromass recycling. *Plant, Cell & Environment*, 47, 1987–1996.

Paustian, K., Lehmann, J., Ogle, S. et al. (2016) Climate-smart soils. *Nature*, 532, 49–57.

Peay, K. G., Kennedy, P. G. and Talbot, J. M. (2016) Dimensions of biodiversity in the earth mycobiome. *Nature Reviews Microbiology*, 14, 434–447.

Peh, K. S. H., Lewis, S. L. and Lloyd, J. (2011) Mechanisms of monodominance in diverse tropical tree-dominated systems. *Journal of Ecology*, 99, 891–898.

Pellegrini, A. F. A., Harden, J., Georgiou, K. et al. (2022) Fire effects on the persistence of soil organic matter and long-term carbon storage. *Nature Geoscience*, 15, 5–13.

Pellegrini, A. F. A., Hobbie, S. E., Reich, P. B. et al. (2020) Repeated fire shifts carbon and nitrogen cycling by changing plant inputs and soil decomposition across ecosystems. *Ecological Monographs*, 90, e01409.

Peltzer, D. A., Bellingham, P. J. and Kurokawa, H. (2009) Punching above their weight: low-biomass non-native plant species alter properties during primary succession. *Oikos*, 118, 1001–1014.

Peltzer, D. A., Wardle, D. A., Allison, V. J. et al. (2010) Understanding ecosystem retrogression. *Ecological Monographs*, 80, 509–529.

Peng, Y., Peñuelas, J., Vesterdal, L. et al. (2022) Responses of soil fauna communities to the individual and combined effects of multiple global change factors. *Ecology Letters*, 25, 1961–1973.

Peoples, M. B., Brockwell, J., Hunt, J. R. et al. (2012) Factors affecting the potential contributions of N_2 fixation by legumes in Australian pasture systems. *Crop and Pasture Science*, 63, 759–786.

Perakis, S. S. and Hedin, L. O. (2002) Nitrogen loss from unpolluted South American forests mainly via dissolved organic compounds. *Nature*, 415, 416–419.

Peralta, A. L., Sun, Y., McDaniel, M. D. and Lennon, J. T. (2018) Crop rotational diversity increases disease suppressive capacity of soil microbiomes. *Ecosphere*, 9, e02235.

Perkins-Kirkpatrick, S. E. and Lewis, S. C. (2020) Increasing trends in regional heatwaves. *Nature Communications*, 11, 3357.

Perry, L. G., Thelen, G. C., Ridenour, W. M. et al. (2007) Concentrations of the allelochemical (±)-catechin in *Centaurea maculosa* soils. *Journal of Chemical Ecology*, 33, 2337–2344.

Persson, J. and Näsholm, T. (2001) Amino acid uptake: a widespread ability among boreal forest plants. *Ecology Letters*, 4, 434–438.

Petersen, H. and Luxton, M. (1982) A comparative analysis of soil fauna populations and their role in decomposition processes. *Oikos*, 39, 288–388.

Petters, S., Groß, V., Söllinger, A., Pichler, M. et al. (2021) The soil microbial food web revisited: predatory myxobacteria as keystone taxa? *The ISME Journal*, 15, 2665–2675.

Philippot, L., Chenu, C., Kappler, A. et al. (2023) The interplay between microbial communities and soil properties. *Nature Reviews Microbiology*, 22, 226–239.

Philippot, L., Raaijmakers, J. M., Lemanceau, P. and Van der Putten, W. H. (2013) Going back to the roots: the microbial ecology of the rhizosphere. *Nature Reviews Microbiology*, 11, 789–799.

Phillips, H. R. P., Cameron, E. K., Eisenhauer, N. et al. (2024) Global changes and their environmental stressors have a significant impact on soil biodiversity—a meta-analysis. *iScience*, 27(9).

Phillips, H. R. P., Guerra, C. A., Bartz, M. L. C. et al. (2019) Global distribution of earthworm diversity. *Science*, 366, 480–485.

Phillips, R. P., Brzostek, E. and Midgley, M. G. (2013) The mycorrhizal-associated nutrient economy: a new framework for predicting carbon–nutrient couplings in temperate forests. *New Phytologist*, 199, 41–51.

Phillips, R. P., Finzi, A. C. and Bernhardt, E. S. (2011) Enhanced root exudation induces microbial feedbacks to N cycling in a pine forest under long-term CO_2 fumigation. *Ecology Letters*, 14, 187–194.

Phillips, R. P., Meier, I. C., Bernhardt, E. S. et al. (2012) Roots and fungi accelerate carbon and nitrogen cycling in forests exposed to elevated CO_2. *Ecology Letters*, 15, 1042–1049.

Phoenix, G. K., Johnson, D. A., Muddimer, S. P. et al. (2020) Niche differentiation and plasticity in soil phosphorus acquisition among co-occurring plants. *Nature Plants*, 6, 349–354.

Phoenix, G. K., Leake, J. R., Read, D. J. et al. (2005) Accumulation of pollutant nitrogen in calcareous and acidic grasslands: evidence from N flux and ^{15}N tracer studies. *Water, Air, & Soil Pollution: Focus*, 4, 159–167.

Pickard, J. (1984) Exotic plants on Lord Howe Island: distribution in space and time, 1853–1981. *Journal of Biogeography*, 11, 181–208.

Pimm, S. L., Jenkins, C. N., Abell, R. et al. (2014) The biodiversity of species and their rates of extinction, distribution, and protection. *Science*, 344, 1246752.

Piñeiro, G., Paruelo, J. M., Jobbágy, E. G. et al. (2009) Grazing effects on belowground C and N stocks along a network of cattle exclosures in temperate and subtropical grasslands of South America. *Global Biogeochemical Cycles*, 23(2).

Piñeiro, G., Paruelo, J. M. and Oesterheld, M. (2006) Potential long-term impacts of livestock introduction on carbon and nitrogen cycling in grasslands of southern South America. *Global Change Biology*, 12, 1267–1284.

Pollierer, M. M., Langel, R., Körner, C. et al. (2007) The underestimated importance of belowground carbon input for forest soil animal food webs. *Ecology Letters*, 10, 729–736.

Ponge, J. F. (2003) Humus forms in terrestrial ecosystems: a framework to biodiversity. *Soil Biology and Biochemistry*, 35, 935–945.

Ponsard, S and Arditi, R. (2000) What can stable isotopes ($\delta^{15}N$ and $\delta^{13}C$) tell about the food web of soil macroinvertebrates? *Ecology*, 81, 852–864.

Poorter, H., Knopf, O., Wright, I. J. et al. (2022) A meta-analysis of responses of C_3 plants to atmospheric CO_2: dose–response curves for 85 traits ranging from the molecular to the whole-plant level. *New Phytologist*, 233, 1560–1596.

Popovici, I. (1989) Soil nematode communities in the Carpathian beech forests of Romania. *Studia Universitatis Babescedilla-bolyai, Biologia*, 34, 38–44.

Porazinska, D. L., Bardgett, R. D., Blaauw, M. B. et al. (2003) Relationships at the aboveground–belowground interface: plants, soil biota, and soil processes. *Ecological Monographs*, 73, 377–395.

Porter, E. M., Bowman, W. D., Clark, C. M. et al. (2013) Interactive effects of anthropogenic nitrogen enrichment and climate change on terrestrial and aquatic biodiversity. *Biogeochemistry*, 114, 93–120.

Post, E. and Pedersen, C. (2008) Opposing plant community responses to warming with and without herbivores. *Proceedings of the National Academy of Sciences*, 105, 12353–12358.

Post, E., Peterson, R. O., Stenseth, N. C. and McLaren B. E. (1999) Ecosystem consequences of wolf behavioral response to climate. *Nature*, 401, 905–907.

Potapov, A. M., Beaulieu, F., Birkhofer, K. et al. (2022) Feeding habits and multifunctional classification of soil-associated consumers from protists to vertebrates. *Biological Reviews*, 97, 1057–1117.

Potapov, A. M., Drescher, J., Darras, K. et al. (2024) Rainforest transformation reallocates energy from green to brown food webs. *Nature*, 627, 116–122.

Potapov, A. M., Guerra, C. A., Van den Hoogen, J. et al. (2023) Globally invariant metabolism but density–diversity mismatch in springtails. *Nature Communications*, 14, 674.

Potapov, A. M., Klarner, B., Sandmann, D. et al. (2019) Linking size spectrum, energy flux and trophic multifunctionality in soil food webs of tropical land-use systems. *Journal of Animal Ecology*, 88, 1845–1859.

Powell, J. R. and Rillig, M. C. (2018) Biodiversity of arbuscular mycorrhizal fungi and ecosystem function. *New Phytologist*, 220, 1059–1075.

Prairie, A. M., King, A. E. and Cotrufo, M. F. (2023) Restoring particulate and mineral-associated organic carbon through regenerative agriculture. *Proceedings of the National Academy of Sciences*, 120, e2217481120.

Press, M. C. and Phoenix, G. K (2005) Impacts of parasitic plants on natural communities. *New Phytologist*, 166, 737–751.

Pressler, Y., Moore, J. C. and Cotrufo, M. F. (2019) Belowground community responses to fire: meta-analysis reveals contrasting responses of soil microorganisms and mesofauna. *Oikos*, 128, 309–327.

Prober, S. M., Leff, J. W., Bates, S. T. et al. (2015) Plant diversity predicts beta but not alpha diversity of soil microbes across grasslands worldwide. *Ecology Letters*, 18, 85–95.

Prommer, J., Walker, T. W. N., Wanek, W. et al. (2020) Increased microbial growth, biomass, and turnover drive soil organic carbon accumulation at higher plant diversity. *Global Change Biology*, 26, 669–681.

Prosser, J. I. and Nicol, G. W. (2008) Relative contributions of archaea and bacteria to aerobic ammonia oxidation in the environment. *Environmental Microbiology*, 10, 2931–2941.

Pugnaire, F. I., Haase, P., Puigdefábregas, J. et al. (1996) Facilitation and succession under the canopy of a leguminous shrub, *Retama sphaerocarpa*, in a semi-arid environment in south-east Spain. *Oikos*, 76, 455–464.

Pugnaire, F. I., Morillo, J. A., Peñuelas, J. et al. (2019) Climate change effects on plant–soil feedbacks and consequences for biodiversity and functioning of terrestrial ecosystems. *Science Advances*, 5, eaaz1834.

Qin, S., Chen, L., Fang, K. et al. (2019) Temperature sensitivity of SOM decomposition governed by aggregate protection and microbial communities. *Science Advances*, 5, eaau1218.

Qin, Y., Payne, R., Gu, Y. et al. (2017) Short-term response of testate amoebae to wildfire. *Applied Soil Ecology*, 116, 64–69.

Qiu, L., Zhang, Q., Zhu, H. et al. (2021) Erosion reduces soil microbial diversity, network complexity and multifunctionality. *The ISME Journal*, 15, 2474–2489.

Qu, X., Li, X., Bardgett, R. D. et al. (2024) Deforestation impacts soil biodiversity and ecosystem services worldwide. *Proceedings of the National Academy of Sciences*, 121, e2318475121.

Quested, H. M., Cornelissen, J. H. C., Press, M. C. et al. (2003a) Decomposition of sub-arctic plants with differing nitrogen economies: a functional role for hemiparasites. *Ecology*, 84, 3209–3221.

Quested, H. M., Press, M. C. and Callaghan, T. V. (2003b) Litter of the hemiparasite *Bartsia alpina* enhances plant growth: evidence for a functional role in nutrient cycling. *Oecologia*, 135, 606–614.

Quested, H. M., Press, M. C., Callaghan, T. V. and Cornelissen, J. H. C. (2002) The hemiparasite angiosperm *Bartsia alpina* has the potential to accelerate decomposition in sub-arctic communities. *Oecologia*, 130, 88–95.

Quinton, J. N., Govers, G., Van Oost, K. and Bardgett, R. D. (2010) The impact of agricultural soil erosion on biogeochemical cycling. *Nature Geoscience*, 3, 311–314.

Raab, T. K., Lipson, D. A. and Monson, R. K. (1996) Non-mycorrhizal uptake of amino acids by roots of the alpine sedge *Kobresia myosuroides*: implications for the alpine nitrogen cycle. *Oecologia*, 108, 488–494.

Raab, T. K., Lipson, D. A. and Monson, R. K. (1999) Soil amino acid utilization among species of the Cyperacea: plant and soil processes. *Ecology*, 80, 2408–2419.

Ramirez, K. S., Craine, J. M. and Fierer, N. (2010) Nitrogen fertilization inhibits soil microbial respiration regardless of the form of nitrogen applied. *Soil Biology and Biochemistry*, 42, 2336–2338.

Ramirez, K. S., Craine, J. M. and Fierer, N. (2012) Consistent effects of nitrogen amendments on soil microbial communities and processes across biomes. *Global Change Biology*, 18, 1918–1927.

Ranjard, L., Dequiedt, S., Prévost-Bouré, N. C. et al. (2013) Turnover of soil bacterial diversity driven by wide-scale environmental heterogeneity. *Nature Communications*, 4, 1434.

Ratering, S. and Conrad, R. (1998) Effects of short-term drainage and aeration on the production of methane in submerged rice soil. *Global Change Biology*, 4, 397–407.

Ray, D.K., Ramankutty, N., Mueller, N.D. et al. (2012) Recent patterns of crop yield growth and stagnation. *Nature Communications*, 3, 1293.

Read, D. J. (1991) Mycorrhizas in ecosystems. *Experimentia*, 47, 376–391.

Read, D. J. (1994) Plant–microbe mutualisms and community structure. In: *Biodiversity and Ecosystem Function* (Schulze, E. D. and Mooney, H. A., eds), pp. 181–209. Springer, Berlin.

Read, D. J. and Perez-Moreno, J. (2003) Mycorrhizas and nutrient cycling in ecosystems—a journey towards relevance? *New Phytologist*, 157, 475–492.

Read, D. J., Haggar, J., Magkourilou, E. et al. (2024) Photosynthate transfer from an autotrophic orchid to conspecific heterotrophic protocorms through a common mycorrhizal network. *New Phytologist*, 243, 398–406.

Reeder, J. D. and Schuman, G. E. (2002) Influence of livestock grazing on C sequestration in semi-arid mixed-grass and short-grass rangelands. *Environmental Pollution*, 116, 457–463.

Reich, P. B. (2014) The world-wide 'fast–slow' plant economics spectrum: a traits manifesto. *Journal of Ecology*, 102, 275–301.

Reichstein, M., Bahn, M., Ciais, P. et al. (2013) Climate extremes and the carbon cycle. *Nature*, 500, 287–295.

Reider, K. E. and Schmidt, S. K. (2021) Vicuña dung gardens at the edge of the cryosphere. *Ecology*, 102, e03228.

Reinhart, K. O. and Callaway, R. M. (2004) Soil biota facilitate exotic Acer invasions in Europe and North America. *Ecological Applications*, 14, 1737–1745.

Reinhart, K. O., Bauer, J. T., McCarthy-Neumann, S. et al. (2021) Globally, plant–soil feedbacks are weak predictors of plant abundance. *Ecology and Evolution*, 11, 1756–1768.

Reinhart, K. O., Packer, A., Van der Putten, W. H. and Clay, K. (2003) Plant–soil biota interactions and spatial distribution of black cherry in its native and invasive ranges. *Ecology Letters*, 6, 1046–1050.

Reinhart, K. O., Tytgat, T., Van der Putten, W. H. and Clay, K. (2010) Virulence of soil-borne pathogens and invasion by *Prunus serotina*. *New Phytologist*, 186, 484–495.

Rhoades, D. F. (1985) Offensive–defensive interactions between herbivores and plants: their relevance in herbivore population dynamics and ecological theory. *American Naturalist*, 125, 205–238.

Richardson, A. E. and Simpson, R. J. (2011) Soil microorganisms mediating phosphorus availability update on microbial phosphorus. *Plant Physiology*, 156, 989–996.

Richardson, A. E., Barea, J. M., McNeill, A. M. and Prigent-Combaret, C. (2009) Acquisition of phosphorus and nitrogen in the rhizosphere and plant growth promotion by microorganisms. *Plant and Soil*, 321, 305–339.

Rietkerk, M., Dekker, S. C., de Ruiter, P. C. and Van de Koppel, J. (2004) Self-organized patchiness and catastrophic shifts in ecosystems. *Science*, 305, 1926–1929.

Rillig, M. C. and Mummey, D. L. (2006) Mycorrhizas and soil structure. *New Phytologist*, 171, 41–53.

Rillig, M. C., Ryo, M., Lehmann, A. et al. (2019) The role of multiple global change factors in driving soil functions and microbial biodiversity. *Science*, 366, 886–890.

Rinnan, R., Stark, S. and Tolvanen, A. (2009) Responses of vegetation and soil microbial communities to warming and simulated herbivory in a subarctic heath. *Journal of Ecology*, 97, 788–800.

Ripple, W. J. and Beschta, R. L. (2004) Wolves and the ecology of fear: can predation risk structure ecosystems? *BioScience*, 54, 755–766.

Risch, A. C., Frossard, A., Schütz, M. et al. (2020) Effects of elk and bison carcasses on soil microbial communities and ecosystem functions in Yellowstone, USA. *Functional Ecology*, 34, 1933–1944.

Risch, A. C., Ochoa-Hueso, R., Van der Putten, W. H. et al. (2018) Size-dependent loss of aboveground animals differentially affects grassland ecosystem coupling and functions. *Nature Communications*, 9, 3684.

Risch, A. C., Schütz, M., Vandegehuchte, M. L. et al. (2015) Aboveground vertebrate and invertebrate herbivore impact on net N mineralization in subalpine grasslands. *Ecology*, 96, 3312–3322.

Ritchie, M. E., Tilman, D. and Knops, J. M. H. (1998) Herbivore effects on plant and nitrogen dynamics in oak savanna. *Ecology*, 79, 165–177.

Ritz, K. and Young. I. M. (2004) Interactions between soil structure and fungi. *Mycologist*, 18, 52–59.

Roberts, M. S. and Cohan, F. M. (1995) Recombination and migration rates in natural populations of *Baccillus subtilus* and *Baccillus mojavensis*. *Evolution*, 49, 1081–1094.

Robinson, C. H., Ineson, P., Piearce, T. G. and Parrington, J. (1996) Effects of earthworms on cation and phosphate mobilisation in limed peat soils under *Picea sitchensis*. *Forest Ecology and Management*, 86, 253–258.

Robinson, C. H., Ineson, P., Piearce, T. G. and Rowland, A. P. (1992) Nitrogen mobilization by earthworms in limed peat soils under *Picea sitchensis*. *Journal of Applied Ecology*, 29, 226–237.

Rockström, J., Steffen, W., Noone, K. et al. (2009) A safe operating space for humanity. *Nature*, 461, 472–475.

Rodrigues, J. L., Pellizari, V. H., Mueller, R. et al. (2013) Conversion of the Amazon rainforest to agriculture results in biotic homogenization of soil bacterial communities. *Proceedings of the National Academy of Sciences*, 110, 988–993.

Rodríguez-Echeverría, S. and Pérez-Fernández, M. A. (2003) Soil fertility and herb facilitation mediated by *Retama sphaerocarpa*. *Journal of Vegetation Science*, 14, 807–814.

Roesch, L. F. W., Fulthorpe, R. R., Riva, A. et al. (2007) Pyrosequencing enumerates and contrasts soil microbial diversity. *The ISME Journal*, 1, 283–290.

Rohatyn, S., Yakir, D., Rotenberg, E. and Carmel, Y. (2022) Limited climate change mitigation potential through forestation of the vast dryland regions. *Science*, 377, 1436–1439.

Rolo, V., Rivest, D., Maillard, É. and Moreno, G. (2023) Agroforestry potential for adaptation to climate change: a soil-based perspective. *Soil Use and Management*, 39, 1006–1032.

Romero, F., Labouyrie, M., Orgiazzi, A. et al. (2024) Soil health is associated with higher primary productivity across Europe. *Nature Ecology and Evolution*, 8, 1847–1855.

Rooney, N., McCann, K., Gellner, G. and Moore, J. C. (2006) Structural asymmetry and the stability of diverse food webs. *Nature*, 442, 265–269.

Ross, D. J., Newton, P. C. D. and Tate, K. R. (2004) Elevated [CO_2] effects on herbage and soil carbon and nitrogen pools and mineralization in a species-rich, grazed pasture on a seasonally dry sand. *Plant and Soil*, 260, 183–196.

Roy, S. and Bagchi, S. (2022) Large mammalian herbivores and the paradox of soil carbon in grazing ecosystems: role of microbial decomposers and their enzymes. *Ecosystems*, 25, 976–988.

Roy, S., Naidu, D. G. T. and Bagchi, S. (2023) Functional substitutability of native herbivores by livestock for soil carbon stock is mediated by microbial decomposers. *Global Change Biology*, 29, 2141–2155.

Rubin, R. L., Van Groenigen, K. J. and Hungate, B. A. (2017) Plant growth promoting rhizobacteria are more effective under drought: a meta-analysis. *Plant and Soil*, 416, 309–323.

Rule, S., Brook, B. W., Haberle, S. G. et al. (2012) The aftermath of megafaunal extinction: ecosystem transformation in Pleistocene Australia. *Science*, 335, 1483–1486.

Rumpel, C., Amiraslani, F., Chenu, C. et al. (2020) The 4p1000 initiative: opportunities, limitations and challenges for implementing soil organic carbon sequestration as a sustainable development strategy. *Ambio*, 49, 350–360.

Rumpf, S. B., Gravey, M., Brönnimann, O. et al. (2022) From white to green: snow cover loss and increased vegetation productivity in the European Alps. *Science*, 376, 1119–1122.

Ryan, M. H., Graham, J. H., Morton, J. B. and Kirkegaard, J. A. (2019) Research must use a systems agronomy approach if management of the arbuscular mycorrhizal symbiosis is to contribute to sustainable intensification. *The New Phytologist*, 222, 1176–1178.

Saetre, P. and Bååth, E. (2000) Spatial variation and patterns of soil microbial community structure in a mixed spruce–birch stand. *Soil Biology & Biochemistry*, 30, 909–917.

Salamon, J. A., Schaefer, M., Alphei, J. et al. (2004) Effects of plant diversity on Collembola in an experimental grassland ecosystem. *Oikos*, 106, 51–60.

Salomon, M. J., Demarmels, R., Watts-Williams, S. J. et al. (2022) Global evaluation of commercial arbuscular mycorrhizal inoculants under greenhouse and field conditions. *Applied Soil Ecology*, 169, 104225.

Salvagiotti, F., Cassman, K. G., Specht, J. E. et al. (2008) Nitrogen uptake, fixation and response to fertilizer N in soybeans: a review. *Field Crops Research*, 108, 1–13.

Sangabriel-Conde, W., Negrete-Yankelevich, S., Maldonado-Mendoza, I. E. et al. (2014) Native maize landraces from Los Tuxtlas, Mexico show varying mycorrhizal dependency for P uptake. *Biology and Fertility of Soils*, 50, 405–414.

Sankaran, M. (2019) Droughts and the ecological future of tropical savanna vegetation. *Journal of Ecology*, 107, 1531–1549.

Sankaran, M., Hanan, N. P., Scholes, R. J. et al. (2005) Determinants of woody cover in African savannas. *Nature*, 438, 846–849.

Santana, M. M. and Gonzalez, J. M. (2015) High temperature microbial activity in upper soil layers. *FEMS Microbiology Letters*, 362, fnv182.

Santonja, M., Fernandez, C., Proffit, M. et al. (2017) Plant litter mixture partly mitigates the negative effects of extended drought on soil biota and litter decomposition in a Mediterranean oak forest. *Journal of Ecology*, 105, 801–815.

Santos, P. F., Phillips, J. and Whitford, W. G. (1981) The role of mites and nematodes in early stages of buried litter decomposition in a desert. *Ecology*, 63, 664–669.

Sasaki, T., Konno, M., Hasegawa, Y. et al. (2019) Role of mycorrhizal associations in tree spatial distribution patterns based on size class in an old-growth forest. *Oecologia*, 189, 971–980.

Satchell, J. E. (1967) Lumbricidae. In: *Soil Biology* (Burges, A. and Raw, F., eds), pp. 259–322. Academic Press, London.

Schädler, M., Jung, G., Brandl, R. and Auge, H. (2004) Secondary succession is influenced by belowground insect herbivory on a productive site. *Oecologia*, 138, 242–252.

Schadt, C. W., Martin, A. P., Lipson, D. A. and Schmidt, S. K. (2003) Seasonal dynamics of previously unknown fungal lineages in tundra soils. *Science*, 301, 1359–1361.

Schaefer, I. and Caruso, T. (2019) Oribatid mites show that soil food web complexity and close aboveground–belowground linkages emerged in the early Paleozoic. *Communications Biology*, 2, 387.

Scherber, C., Eisenhauer, N., Weisser, W. W. et al. (2010) Bottom-up effects of plant diversity on multitrophic interactions in a biodiversity experiment. *Nature*, 468, 553–556.

Scheu, S. and Falca, M. (2000) The soil food web of two beech forests (*Fagus sylvatica*) of contrasting humus type: stable isotope analysis of macro- and a mesofauna-dominated community. *Oecologia*, 123, 285–296.

Scheu, S. and Parkinson, D. (1994) Effects of invasion of an aspen forest (Canada) by *Dendrobaena octaedra* (Lumbricidae) on plant growth. *Ecology*, 75, 2348–2361.

Scheu, S. and Schulz, E. (1996) Secondary succession, soil formation and development of a diverse community of oribatids and saprophagous soil macroinvertebrates. *Biological Conservation*, 5, 235–250.

Scheu, S. and Simmerling, F. (2004) Growth and reproduction of fungal feeding Collembola as affected by fungal species, melanin and mixed diets. *Oecologia*, 139, 347–353.

Scheu, S., Theenhaus, A. and Jones, T. H. (1999) Links between the detritivore and the herbivore system: effects of earthworms and Collembola on plant growth and aphid development. *Oecologia*, 119, 541–551.

Schimel, J. P. and Chapin, F. S. (1996) Tundra plant uptake of amino acid and NH_4^+ nitrogen in situ: plants compete well for amino acid N. *Ecology*, 77, 2142–2147.

Schimel, J. P., Bilbrough, C. and Welker, J. M. (2004) Increased snow depth affects microbial activity and nitrogen mineralization in two Arctic tundra communities. *Soil Biology and Biochemistry*, 36, 217–227.

Schimel, J. P., Cates, R. G. and Ruess, R. (1998) The role of balsam poplar secondary chemicals in controlling soil nutrient dynamics through succession in the Alaskan taiga. *Biogeochemistry*, 42, 221–234.

Schmidt, H. P., Kammann, C., Hagemann, N. et al. (2021) Biochar in agriculture—a systematic review of 26 global meta-analyses. *GCB Bioenergy*, 13, 1708–1730.

Schmidt, M. W. I., Torn, M. S., Abiven, S. et al. (2011) Persistence of soil organic matter as an ecosystem property. *Nature*, 478, 49–56.

Schmidt, O., Clements, R. O. and Donaldson, G. (2003) Why do cereal–legume intercrops support large earthworm populations? *Applied Soil Ecology*, 22, 181–190.

Schmidt, S. K., Reed, S. C., Nemergut, D. R. et al. (2008) The earliest stages of ecosystem succession in high-elevation (5000 metres above sea level), recently deglaciated soils. *Proceedings of the Royal Society B*, 275, 2793–2802.

Schmitz, O. J. and Leroux, S. J. (2020) Food webs and ecosystems: linking species interactions to the carbon cycle. *Annual Review of Ecology, Evolution, and Systematics*, 51, 271–295.

Schmitz, O. J., Hawlena, D. and Trussell, G. C. (2010) Predator control of ecosystem nutrient dynamics. *Ecology Letters*, 13, 1199–1209.

Schnitzer, S. A., Klironomos, J. N., Hillerislambers, J. et al. (2011) Soil microbes drive the classic plant diversity–productivity pattern. *Ecology*, 92, 296–303.

Schnürer, J. Clarholm, M. and Rosswall, T. (1986) Fungi, bacteria and protozoa in soil from four arable cropping systems. *Biology and Fertility of Soils*, 2, 119–126.

Schrama, M. and Bardgett, R. D. (2016) Grassland invasibility varies with drought effects on soil functioning. *Journal of Ecology*, 104, 1250–1258.

Schrama, M., Heijning, P., Bakker, J. P. et al. (2013) Herbivore trampling as an alternative pathway for explaining differences in nitrogen mineralization in moist grasslands. *Oecologia*, 172, 231–243.

Schrama, M., Quist, C.W., Arjen de Groot, G. et al. (2023) Cessation of grazing causes biodiversity loss and homogenization of soil food webs. *Proceedings of the Royal Society B*, 290, 20231345.

Schuman, G. E., Reeder, J. D., Manley, J. T. et al. (1999) Impact of grazing management on the carbon and nitrogen balance of a mixed-grass rangeland. *Ecological Applications*, 9, 65–71.

Schuur, E. A. G., Bracho, R., Celis, G. et al. (2021) Tundra underlain by thawing permafrost persistently emits carbon to the atmosphere over 15 years of measurements. *Journal of Geophysical Research: Biogeosciences*, 126, e2020JG006044.

Schweitzer, J. A., Bailey, J. K., Hart, S. C. et al. (2005) The interaction of plant genotype and herbivory decelerate leaf litter decomposition and alter nutrient dynamics. *Oikos*, 110, 133–145.

Seagle, S. W. (2003) Can ungulates foraging in multiple-use landscapes alter forest nitrogen budgets? *Oikos*, 103, 230–234.

Semchenko, M., Leff, J. W., Lozano, Y. M. et al. (2018) Fungal diversity regulates plant–soil feedbacks in temperate grassland. *Science Advances*, 4, eaau4578.

Setälä, H. (1995) Growth of birch and pine seedlings in relation to grazing by soil fauna on ectomycorrhizal fungi. *Ecology*, 76, 1844–1851.

Setälä, H. and Huhta, V. (1991) Soil fauna increase *Betula pendula* growth: laboratory experiments with coniferous forest floor. *Ecology*, 72, 665–671.

Shangguan, W., Gong, P., Liang, L. et al. (2014) Soil diversity as affected by land use in China: consequences for soil protection. *The Scientific World Journal*, 2014, 913852

Sharma, S. B., Sayyed, R. Z., Trivedi, M. H. and Gobi, T. A. (2013) Phosphate solubilizing microbes: sustainable approach for managing phosphorus deficiency in agricultural soils. *SpringerPlus*, 2, 587.

Sharpley, A. N. and Syers, J. K. (1976) Potential role of earthworm casts for the phosphorus enrichment of run-off waters. *Soil Biology and Biochemistry*, 8, 341–346.

Sharpley, A. N., Jarvie, H. P., Buda, A. R. et al. (2013) Phosphorus legacy: overcoming the effects of past management practices to mitigate future water quality impairment. *Journal of Environmental Quality*, 42, 1308–1326.

Sharpley, A. N., Syers, J. K. and Springett, J. A. (1979) Effect of surface-casting earthworms on the transport of phosphorus and nitrogen in surface runoff from pasture. *Soil Biology and Biochemistry*, 11, 459–462.

Shepherd, V. E. and Chapman, C. A. (1998) Dung beetles as secondary seed dispersers: impact on seed predation and germination. *Journal of Tropical Ecology*, 14, 199–215.

Shrestha, G. and Stahl, P. D. (2008) Carbon accumulation and storage in semi-arid sagebrush steppe: effects of long-term grazing exclusion. *Agriculture, Ecosystems and Environment*, 125, 173–181.

Siebert, J., Sünnemann, M., Auge, H. et al. (2019) The effects of drought and nutrient addition on soil organisms vary across taxonomic groups, but are constant across seasons. *Scientific Reports*, 9, 639.

Siepel, H. and Van de Bund, C. F. (1988) The influence of management practices on the microarthropod community of grassland. *Pedobiologia*, 31, 339–354.

Simard, S. W., Beiler, K. J., Bingham, M. A. et al. (2012) Mycorrhizal networks: mechanisms, ecology and modelling. *Fungal Biology Reviews*, 26, 39–60.

Simard, S. W., Perry, D. A., Jones, M. D. et al. (1997) Net transfer of carbon between ectomycorrhizal tree species in the field. *Nature*, 388, 579–582.

Simberloff, D., Martin, J. L., Genovesi, P. et al. (2013) Impacts of biological invasions: what's what and the way forward. *Trends in Ecology and Evolution*, 28, 58–66.

Simmons, B. L., Wall, D. H., Adams, B. J. et al. (2009) Long-term experimental warming reduces soil nematode populations in the McMurdo Dry Valleys, Antarctica. *Soil Biology and Biochemistry*, 41, 2052–2060.

Sitters, J. and Venterink, H. O. (2018) A stoichiometric perspective of the effect of herbivore dung on ecosystem functioning. *Ecology and Evolution*, 8, 1043–1046.

Sitters, J., Cherif, M., Egelkraut, D. et al. (2019) Long-term heavy reindeer grazing promotes plant phosphorus limitation in arctic tundra. *Functional Ecology*, 33, 1233–1242.

Sitters, J., Kimuyu, D. M., Young, T. P. et al. (2020a) Negative effects of cattle on soil carbon and nutrient pools reversed by megaherbivores. *Nature Sustainability*, 3, 360–366.

Sitters, J., Te Beest, M., Cherif, M. et al. (2017) Interactive effects between reindeer and habitat fertility drive soil nutrient availabilities in arctic tundra. *Ecosystems*, 20, 1266–1277.

Sitters, J., Wubs, E. R. J., Bakker, E. S. et al. (2020b) Nutrient availability controls the impact of mammalian herbivores on soil carbon and nitrogen pools in grasslands. *Global Change Biology*, 26, 2060–2071.

Sitvarin, M. I. and Rypstra, A. L. (2014) Fear of predation alters soil carbon dioxide flux and nitrogen content. *Biology Letters*, 10, 20140366.

Sitvarin, M. I., Romanchek, C. and Rypstra, A. L. (2015) Nonconsumptive predator–prey interactions: sensitivity of the detritivore *Sinella curviseta* (Collembola: Entomobryidae) to cues of predation risk from the spider *Pardosa milvina* (Araneae: Lycosidae). *Environmental Entomology*, 44, 349–355.

Six, J., Frey, S. D., Thiet, R. K. and Batten, K. M. (2006) Bacterial and fungal contributions to carbon sequestration in agroecosystems. *Soil Science Society of America Journal*, 70, 555–569.

Six, J., Ogle, S. M., Breidt, F. J. et al. (2004) The potential to mitigate global warming with no-tillage management is only realized when practised in the long term. *Global Change Biology*, 10, 155–160.

Skinner, C., Gattinger, A., Krauss, M. et al. (2019) The impact of long-term organic farming on soil-derived greenhouse gas emissions. *Scientific Reports*, 9, 1702.

Smith, M. E., Vico, G., Costa, A. et al. (2023) Increasing crop rotational diversity can enhance cereal yields. *Communications Earth and Environment*, 4, 89.

Smith, S. E. and Read, D. J. (2010) *Mycorrhizal Symbiosis*. 3rd Edition. Academic Press, New York.

Smith, S. E. and Smith, F. A. (2011) Roles of arbuscular mycorrhizas in plant nutrition and growth: new paradigms from cellular to ecosystem scales. *Annual Review of Plant Biology*, 62, 227–250.

Smykla, J., Iakovenko, N., Devetter, M. and Kaczmarek, Ł. (2012) Diversity and distribution of tardigrades in soils of Edmonson Point (Northern Victoria Land, continental Antarctica). *Czech Polar Reports*, 2, 61–70.

Sokol, N. W. and Bradford, M. A. (2019) Microbial formation of stable soil carbon is more efficient from belowground than aboveground input. *Nature Geoscience*, 12, 46–53.

Sokol, N. W., Slessarev, E., Marschmann, G. L. (2022) Life and death in the soil microbiome: how ecological processes influence biogeochemistry. *Nature Reviews Microbiology*, 20, 415–430.

Soler, R., Van der Putten, W. H., Harvey, J. A. et al. (2012) Root herbivore effects on aboveground multitrophic interactions: patterns, processes and mechanisms. *Journal of Chemical Ecology*, 38, 755–767.

Song, D., Pan, K., Tariq, A. et al. (2017) Large-scale patterns of distribution and diversity of terrestrial nematodes. *Applied Soil Ecology*, 114, 161–169.

Soong, J. L., Castanha, C., Hicks Pries, C. E. et al. (2021) Five years of whole-soil warming led to loss of subsoil carbon stocks and increased CO_2 efflux. *Science Advances*, 7, eabd1343.

Sørensen, L. I., Kytöviita, M. M., Olofsson, J. and Mikola, J. (2008) Soil feedback on plant growth in a sub-arctic grassland as a result of repeated defoliation. *Soil Biology and Biochemistry*, 40, 2891–2897.

Sørensen, L. I., Mikola, J., Kytöviita, M. M. et al. (2009) Trampling and spatial heterogeneity explain decomposer abundances in a sub-arctic grassland subjected to simulated reindeer grazing. *Ecosystems*, 12, 830–842.

Soussana, J. F., Loiseau, P., Vuichard, N. et al. (2004) Carbon cycling and sequestration opportunities in temperate grasslands. *Soil Use and Management*, 20, 219–230.

Spain, A. V. and McIvor, J. G. (1988) The nature of herbaceous vegetation associated with termitaria in north-eastern Australia. *Journal of Ecology*, 76, 181–191.

Spohn, M., Bagchi, S., Biederman, L. A. et al. (2023) The positive effect of plant diversity on soil carbon depends on climate. *Nature Communications*, 14, 6624.

Springett, J. A. (1992) Distribution of lumbricid earthworms in New Zealand. *Soil Biology and Biochemistry*, 24, 1377–1381.

Srour, A. Y., Ammar, H. A., Subedi, A. et al. (2020) Microbial communities associated with long-term tillage and fertility treatments in a corn–soybean cropping system. *Frontiers in Microbiology*, 11, 1363.

Standen, V. (1984) Production and diversity of enchytraeids, earthworms and plants in fertilized hay meadow plots. *Journal of Applied Ecology*, 21, 293–312.

Stanescu, S. and Maherali, H. (2017) Arbuscular mycorrhizal fungi alter the competitive hierarchy among old-field plant species. *Oecologia*, 183, 479–491.

Stanton, N. L. (1979) Patterns of species diversity in the temperate and tropical litter mites. *Ecology*, 60, 295–304.

Stark S., Tuomi, J., Strömmer, R. and Helle, T. (2003) Non-parallel changes in soil microbial carbon and nitrogen dynamics due to reindeer grazing in northern boreal forests. *Ecography*, 26, 51–59.

Stark, S., Wardle, D. A., Ohtonen, R. et al. (2000) The effect of reindeer grazing on decomposition, mineralisation and soil biota in a dry oligotrophic Scots pine forest. *Oikos*, 90, 301–310.

Starr, E. P., Nuccio, E. E., Pett-Ridge, J. et al. (2019) Metatranscriptomic reconstruction reveals RNA viruses with the potential to shape carbon cycling in soil. *Proceedings of the National Academy of Sciences*, 116, 25900–25908.

Steen, A. D., Crits-Christoph, A., Carini, P. et al. (2019) High proportions of bacteria and archaea across most biomes remain uncultured. *The ISME Journal*, 13, 3126–3130.

Ştefan, V. (1977) Soil Enchytraeidae from the Cerna Valley. Fourth Symposium on Soil Biology. Rumanian National Society of Soil Science, Editura Ceres, Bucureşti, pp. 277–283.

Steinbauer, M. J., Grytnes, J. A., Jurasinski, G. et al. (2018) Accelerated increase in plant species richness on mountain summits is linked to warming. *Nature*, 556, 231–234.

Steltzer, H. and Bowman, W. D. (1998) Original Articles: differential influence of plant species on soil N transformations within moist meadow alpine tundra. *Ecosystems*, 1, 464–474.

Sterner, R. W. and Elser, J. J. (2002) *Ecological Stoichiometry: The Biology of Elements from Molecules to the Biosphere*. Princeton University Press, Princeton, NJ.

St. John, M. G., Orwin, K. H. and Dickie, I. A. (2011) No 'home' versus 'away' effects of decomposition found in a grassland–forest reciprocal litter transplant study. *Soil Biology and Biochemistry*, 43, 1482–1489.

Stinson, K. A., Campbell, S. A., Powell, J. R. et al. (2006) Invasive plant suppresses the growth of native tree seedlings by disrupting belowground mutualisms. *PLoS Biology*, 4, e140.

Stockmann, U., Minasny, B. and McBratney, A. B. (2014) How fast does soil grow? *Geoderma*, 216, 48–61.

Strassburg, B. B. N., Iribarrem, A., Beyer, H. L. et al. (2020) Global priority areas for ecosystem restoration. *Nature*, 586, 724–729.

Street, L. E., Garnett, M. H., Subke, J. A. et al. (2020) Plant carbon allocation drives turnover of old soil organic matter in permafrost tundra soils. *Global Change Biology*, 26, 4559–4571.

Strickland, M. S., Lauber, C., Fierer, N. and Bradford, M. A. (2009) Testing the functional significance of microbial community composition. *Ecology*, 90, 441–451.

Subler, S., Baranski, C. M. and Edwards, C. A. (1997) Earthworm additions increased short-term nitrogen availability and leaching in two grain-crop agroecosystems. *Soil Biology and Biochemistry*, 29, 413–421.

Sun, L., Ataka, M., Han, M. et al. (2021a) Root exudation as a major competitive fine-root functional trait of 18 coexisting species in a subtropical forest. *New Phytologist*, 229, 259–271.

Sun, Y., Wang, C., Yang, J. et al. (2021b) Elevated CO_2 shifts soil microbial communities from K- to r-strategists. *Global Ecology and Biogeography*, 30, 961–972.

Sun, Z. K. and He, W. M. (2010) Evidence for enhanced mutualism hypothesis: *Solidago canadensis* plants from regular soils perform better. *PLoS ONE*, 5, e15418.

Suttle, C. A. (2007) Marine viruses—major players in the global ecosystem. *Nature Reviews Microbiology*, 5, 801–812.

Sweeney, C. J., De Vries, F. T., Van Dongen, B. E. and Bardgett, R. D. (2021) Root traits explain rhizosphere fungal community composition among temperate grassland plant species. *New Phytologist*, 229, 1492–1507.

Swift M. J., Heal, O.W. and Anderson, J.M. (1979) *Decomposition in Terrestrial Ecosystems*. Blackwell, Oxford.

Szanser, M., Ilieva-Makulec, K., Kajak, A. et al. (2011) Impact of litter species diversity on decomposition processes and communities of soil organisms. *Soil Biology and Biochemistry*, 43, 9–19.

Talbot, J. M., Bruns, T. D., Taylor, J. W. et al. (2014) Endemism and functional convergence across the North American soil mycobiome. *Proceedings of the National Academy of Sciences*, 111, 6341–6346.

Tamburini, G., Bommarco, R., Wanger, T. C. et al. (2020) Agricultural diversification promotes multiple ecosystem services without compromising yield. *Science Advances*, 6, eaba1715.

Tang, B., Man, J., Lehmann, A. and Rillig, M. C. (2023a) Arbuscular mycorrhizal fungi benefit plants in response to major global change factors. *Ecology Letters*, 26, 2087–2097.

Tang, J., Cheng, H. and Fang, C. (2017) The temperature sensitivity of soil organic carbon decomposition is not related to labile and recalcitrant carbon. *PLoS ONE*, 12, e0186675.

Tang, Y., Winterfeldt, S., Brangarí, A. C. et al. (2023b) Higher resistance and resilience of bacterial growth to drought in grasslands with historically lower precipitation. *Soil Biology and Biochemistry*, 177, 108889.

Tao, F., Huang, Y., Hungate, B. A. et al. (2023) Microbial carbon use efficiency promotes global soil carbon storage. *Nature*, 618, 981–985.

Tao, X., Yang, Z., Feng, J. et al. (2024) Experimental warming accelerates positive soil priming in a temperate grassland ecosystem. *Nature Communications*, 15, 1178.

Tardiff, S. E. and Stanford, J. A. (1998) Grizzly bear digging: effects on subalpine meadow plants in relation to mineral nitrogen availability. *Ecology*, 79, 2219–2228.

Taylor, B. N. and Menge, D. N. L. (2018) Light regulates tropical symbiotic nitrogen fixation more strongly than soil nitrogen. *Nature Plants*, 4, 655–661.

Taylor, D. L., Hollingsworth, T. N., McFarland, J. W. et al. (2014) A first comprehensive census of fungi in soil reveals both hyperdiversity and fine-scale niche partitioning. *Ecological Monographs*, 84, 3–20.

Tedersoo, L., Bahram, M., Põlme, S. et al. (2014) Global diversity and geography of soil fungi. *Science*, 346, 1256688.

Tedersoo, L., Bahram, M. and Zobel, M. (2020) How mycorrhizal associations drive plant population and community biology. *Science*, 367, eaba1223.

Tedersoo, L., Mikryukov, V., Zizka, A. et al. (2022) Global patterns in endemicity and vulnerability of soil fungi. *Global Change Biology*, 28, 6696–6710.

Terasaki Hart, D. E., Yeo, S., Almaraz, M. et al. (2023) Priority science can accelerate agroforestry as a natural climate solution. *Nature Climate Change*, 13, 1179–1190.

Terborgh, J., Lopez, L., Nuñez, P. et al. (2001) Ecological meltdown in predator-free forest fragments. *Science*, 294, 1923–1926.

Terrer, C., Phillips, R. P., Hungate, B. A. et al. (2021) A trade-off between plant and soil carbon storage under elevated CO_2. *Nature*, 591, 599–603.

Teste, F. P., Kardol, P., Turner, B. L. et al. (2017) Plant–soil feedback and the maintenance of diversity in Mediterranean-climate shrublands. *Science*, 355, 173–176.

Thakur, M. P., Milcu, A., Manning, P. et al. (2015) Plant diversity drives soil microbial biomass carbon in grasslands irrespective of global environmental change factors. *Global Change Biology*, 21, 4076-4085.

Thakur, M. P., Phillips, H. R., Brose, U. et al. (2020) Towards an integrative understanding of soil biodiversity. *Biological Reviews*, 95, 350–364.

Thakur, M. P., Sigurðsson, B. D., Sigurðsson, P. and Holmstrup, M. (2023) Warming shifts the biomass distribution of soil microarthropod communities. *Soil Biology and Biochemistry*, 177, 108894.

Thakur, M. P., Van der Putten, W. H., Apon, F. et al. (2021) Resilience of rhizosphere microbial predators and their prey communities after an extreme heat event. *Functional Ecology*, 35, 216–225.

Thaler, E. A., Larsen, I. J. and Yu, Q. (2021) The extent of soil loss across the US Corn Belt. *Proceedings of the National Academy of Sciences*, 118, e1922375118.

Thilakarathna, M. S., McElroy, M. S., Chapagain, T. et al. (2016) Belowground nitrogen transfer from legumes to non-legumes under managed herbaceous cropping systems. A review. *Agronomy for Sustainable Development*, 36, 58.

Thion, C. E., Poirel, J. D., Cornulier, T. et al. (2016) Plant nitrogen-use strategy as a driver of rhizosphere archaeal and bacterial ammonia oxidiser abundance. *FEMS Microbiology Ecology*, 92, fiw091.

Thompson, C. H. (1981) Podzol chronosequences on coastal dunes of eastern Australia. *Nature*, 291, 59–61.

Thorpe, A. S., Thelen, G. C., Diaconu, A. and Callaway, R. M. (2009) Root exudate is allelopathic in invaded community but not in native community: field evidence for the novel weapons hypothesis. *Journal of Ecology*, 97, 641–645.

Tian, D., Xiang, Y., Seabloom, E. et al. (2023) Soil carbon sequestration benefits of active versus natural restoration vary with initial carbon content and soil layer. *Communications Earth and Environment*, 4, 83.

Tian, G., Kang, B. T. and Brussaard, L. (1992) Biological effects of plant residues with contrasting chemical composition under humid tropical conditions—decomposition and nutrient release. *Soil Biology and Biochemistry*, 24, 1051–1060.

Tiemann, L. K., Grandy, A. S., Atkinson, E. E. et al. (2015) Crop rotational diversity enhances belowground communities and functions in an agroecosystem. *Ecology Letters*, 18, 761–771.

Tilman, D. and Downing, J. A. (1994) Biodiversity and stability in grasslands. *Nature*, 367, 363–365.

Tilman, D., Isbell, F. and Cowles, J. M. (2014) Biodiversity and ecosystem functioning. *Annual Review of Ecology, Evolution, and Systematics*, 45, 471–493.

Tisdall, J. M. and Oades, J. M. (1982) Organic matter and water-stable aggregates in soils. *Journal of Soil Science*, 33, 141–163.

Tomasek, A. A., Hondzo, M., Kozarek, J. L. et al. (2019) Intermittent flooding of organic-rich soil promotes the formation of denitrification hot moments and hot spots. *Ecosphere*, 10, e02549.

Torsvik, V., Øvreås, L. and Thingstad, T. F. (2002) Prokaryotic diversity—magnitude, dynamics, and controlling factors. *Science*, 296, 1064–1066.

Totsche, K. U., Amelung, W., Gerzabek, M. H. et al. (2018) Microaggregates in soils. *Journal of Plant Nutrition and Soil Science*, 181, 104–136.

Towne, E. G. (2000) Prairie vegetation and soil nutrient responses to ungulate carcasses. *Oecologia*, 122, 232–239.

Tracy, B. F. and Frank, D. A. (1998) Herbivore influence on soil microbial biomass and N mineralization in a northern grassland ecosystem: Yellowstone National Park. *Oecologia*, 114, 556–562.

Treseder, K. K. (2004) A meta-analysis of mycorrhizal responses to nitrogen, phosphorus, and atmospheric CO_2 in field studies. *New Phytologist*, 164, 347–355.

Triebwasser, D. J., Tharayil, N., Preston, C. M. and Gerard, P. D. (2012) The susceptibility of soil enzymes to inhibition by leaf litter tannins is dependent on the tannin chemistry, enzyme class and vegetation history. *New Phytologist*, 196, 1122–1132.

Trivedi, P., Batista, B. D., Bazany, K. E. and Singh, B. K. (2022) Plant–microbiome interactions under a changing world: responses, consequences and perspectives. *New Phytologist*, 234, 1951–1959.

Tscherko, D., Rustemeier, J., Richter, A. et al. (2003) Functional diversity of the soil microflora in primary succession across two glacier forelands in the Central Alps. *European Journal of Soil Science*, 54, 685–696.

Tsiafouli, M. A., Kallimanis, A. S., Katana, E. et al. (2005) Responses of soil microarthropods to experimental short-term manipulations of soil moisture. *Applied Soil Ecology*, 29, 17–26.

Tsiafouli, M. A., Thébault, E., Sgardelis, S. P. et al. (2015) Intensive agriculture reduces soil biodiversity across Europe. *Global Change Biology*, 21, 973–985.

Tuomi, M., Stark, S., Hoset, K. S. et al. (2019) Herbivore effects on ecosystem process rates in a low-productive system. *Ecosystems*, 22, 827–843.

Tuomi, M., Väisänen, M., Ylänne, H. et al. (2021) Stomping in silence: conceptualizing trampling effects on soils in polar tundra. *Functional Ecology*, 35, 306–317.

Turner, B. L. (2008) Resource partitioning for soil phosphorus: a hypothesis. *Journal of Ecology*, 96, 698–702.

Upson, R., Read, D. J. and Newsham, K. K. (2009) Nitrogen form influences the response of *Deschampsia antarctica* to dark septate root endophytes. *Mycorrhiza*, 20, 1–11.

Vahter, T., Bueno, C. G., Davison, J. et al. (2020) Co-introduction of native mycorrhizal fungi and plant seeds accelerates restoration of post-mining landscapes. *Journal of Applied Ecology*, 57, 1741–1751.

Valencia, E., Maestre, F. T., Le Bagousse-Pinguet, Y. et al. (2015) Functional diversity enhances the resistance of ecosystem multifunctionality to aridity in Mediterranean drylands. *New Phytologist*, 206, 660–671.

Van Breemen, N., Finlay, R., Lundström, U. et al. (2000) Mycorrhizal weathering: a true case of mineral plant nutrition? *Biogeochemistry*, 49, 53–67.

Van den Hoogen, J., Geisen, S., Routh, D. et al. (2019) Soil nematode abundance and functional group composition at a global scale. *Nature*, 572, 194–198.

Van den Putte, A., Govers, G., Diels, J. et al. (2010) Assessing the effect of soil tillage on crop growth: a meta-regression analysis on European crop yields under conservation agriculture. *European Journal of Agronomy*, 33, 231–241.

Van der Heijden, M. G. A., Bardgett, R. D. and Van Straalen, N. M. (2008) The unseen majority: soil microbes as drivers of plant diversity and productivity in terrestrial ecosystems. *Ecology Letters*, 11, 296–310.

Van der Heijden, M. G. A., Klironomos, J. N., Ursic, M. et al. (1998) Mycorrhizal fungal diversity determines plant biodiversity, ecosystem variability and productivity. *Nature*, 396, 69–72.

Van der Heijden, M. G., Martin, F. M., Selosse, M. A. and Sanders, I. R. (2015) Mycorrhizal ecology and evolution: the past, the present, and the future. *New Phytologist*, 205, 1406–1423.

Van der Putten, W. H., Bardgett, R. D., Bever, J. D. et al. (2013) Plant–soil feedbacks: the past, the present and future challenges. *Journal of Ecology*, 101, 265–276.

Van der Putten, W. H., Macel, M. and Visser, M. E. (2010) Predicting species distribution and abundance responses to climate change: why it is essential to include biotic interactions across trophic levels. *Philosophical Transactions of the Royal Society B: Biological Sciences*, 365, 2025–2034.

Van der Putten, W. H., Van Dijk, C. and Peters, B. A. M. (1993) Plant-specific soil-borne diseases contribute to succession in foredune vegetation. *Nature*, 362, 53–56.

Van der Wal, A., Van Veen, J. A., Smant, W. et al. (2006) Fungal biomass development in a chronosequence of land abandonment. *Soil Biology and Biochemistry*, 38, 51–60.

Van der Wal, R. (2006) Do herbivores cause habitat degradation or vegetation state transition? Evidence from the tundra. *Oikos*, 114, 117–186.

Van der Wal, R., Bardgett, R. D., Harrison, K. A. and Stien, A. (2004) Vertebrate herbivores and ecosystem control: cascading effects of faeces on tundra ecosystems. *Ecography*, 27, 242–252.

Van der Wal, R., Sjögersten, S., Woodin, S. J. et al. (2007) Spring feeding by pink-footed geese reduces carbon stocks and sink strength in tundra ecosystems. *Global Change Biology*, 13, 539–545.

Van Gestel, N., Shi, Z., Van Groenigen, K. J. et al. (2018) Predicting soil carbon loss with warming. *Nature*, 554, E4–E5.

Van Groenigen, J. W., Lubbers, I. M., Vos, H. M. J. et al. (2014a) Earthworms increase plant production: a meta-analysis. *Scientific Reports*, 4, 6365.

Van Groenigen, K. J., Qi, X., Osenberg, C. W. et al. (2014b) Faster decomposition under increased atmospheric CO_2 limits soil carbon storage. *Science*, 344, 508–509.

Van Grunsven, R. H. A., Van der Putten, W. H., Bezemer, T. M. et al. (2007) Reduced plant–soil feedback of plant species expanding their range as compared to natives. *Journal of Ecology*, 95, 1050–1057.

Van Kessel, C., Clough, T. and Van Groenigen, J. W. (2009) Dissolved organic nitrogen: an overlooked pathway of nitrogen loss from agricultural systems? *Journal of Environmental Quality*, 38, 393–401.

Van Kessel, C., Venterea, R., Six, J. et al. (2013) Climate, duration, and N placement determine N_2O emissions in reduced tillage systems: a meta-analysis. *Global Change Biology*, 19, 33–44.

Van Kessel, M. A. H. J., Speth, D. R, Albertsen, M. et al. (2015) Complete nitrification by a single microorganism. *Nature*, 528, 555–559.

Van Kleunen, M., Dawson, W., Essl, F. et al. (2015) Global exchange and accumulation of non-native plants. *Nature*, 525, 100–103.

Van Klink, R., Schrama, M., Nolte, S. et al. (2015) Defoliation and soil compaction jointly drive large-herbivore grazing effects on plants and soil arthropods on clay soil. *Ecosystems*, 18, 671–685.

Vannette, R. L., Leopold, D. R. and Fukami, T. (2016) Forest area and connectivity influence root-associated fungal communities in a fragmented landscape. *Ecology*, 97, 2374–2383.

Van Ruijven, J., Ampt, E., Francioli, D. and Mommer, L. (2020) Do soil-borne fungal pathogens mediate plant diversity–productivity relationships? Evidence and future opportunities. *Journal of Ecology*, 108, 1810–1821.

Van Wijnen, H. J., Van der Wal, R. and Bakker, J. P. (1999) The impact of herbivores on nitrogen mineralization rate: consequences for salt-marsh succession. *Oecologia*, 118, 225–231.

Väre, H., Ohtonen, R. and Mikkola, K. (1996) The effect and extent of heavy grazing by reindeer in oligotrophic pine heaths in northeastern Fennoscandia. *Ecography*, 19, 245–253.

Veen, G. F., Freschet, G. T., Ordonez, A. and Wardle, D. A. (2015) Litter quality and environmental controls of home-field advantage effects on litter decomposition. *Oikos*, 124, 187–195.

Vega, F. E., Goettel, M. S., Blackwell, M. et al. (2009) Fungal entomopathogens: new insights on their ecology. *Fungal Ecology*, 2, 149–159.

Veldman, J. W., Overbeck, G. E., Negreiros, D. et al. (2015) Where tree planting and forest expansion are bad for biodiversity and ecosystem services. *BioScience*, 65, 1011–1018.

Verbrigghe, N., Leblans, N. I. W., Sigurdsson, B. D. et al. (2022) Soil carbon loss in warmed subarctic grasslands is rapid and restricted to topsoil. *Biogeosciences*, 19, 3381–3393.

Verchot, L. V., Moutinho, P. R. and Davidson E. A. (2003) Leaf-cutting ant (*Atta sexdens*) and nutrient cycling: deep soil inorganic nitrogen stocks, mineralization, and nitrification in Eastern Amazonia. *Soil Biology and Biochemistry*, 35, 1219–1222.

Verity, P. G. (1991) Feeding in planktonic protozoans: evidence for non-random acquisition of prey. *Journal of Protozoology*, 38, 69–76.

Vilà, M., Espinar, J. L., Hejda, M. et al. (2011) Ecological impacts of invasive alien plants: a meta-analysis of their effects on species, communities and ecosystems. *Ecology Letters*, 14, 702–708.

Violle, C., Navas, M. L., Vile, D. et al. (2007) Let the concept of trait be functional! *Oikos*, 116, 882–892.

Virginia, R. A. and Wall, D. H. (1999) How soils structure communities in the Antarctic Dry Valleys. *BioScience*, 49, 973–983.

Vitousek, P. (2006) Ecosystem science and human–environment interactions in the Hawaiian archipelago. *Journal of Ecology*, 94, 510–521.

Vitousek, P. M. and Walker, L. R. (1989) Biological invasion by *Myrica faya* in Hawai'i: plant demography, nitrogen fixation, ecosystem effects. *Ecological Monographs*, 59, 247–265.

Vitousek, P. M., Menge, D. N. L., Reed, S. C. and Cleveland, C. C. (2013) Biological nitrogen fixation: rates, patterns and ecological controls in terrestrial ecosystems. *Philosophical Transactions of the Royal Society B: Biological Sciences*, 368, 20130119.

Vivanco, L. and Austin, A. T. (2008) Tree species identity alters forest litter decomposition through long-term plant and soil interactions in Patagonia, Argentina. *Journal of Ecology*, 96, 727–736.

Vivanco, L. and Austin, A. T. (2011) Nitrogen addition stimulates forest litter decomposition and disrupts species interactions in Patagonia, Argentina. *Global Change Biology*, 17, 1963–1974.

Vogelsang, K. M., Reynolds, H. L. and Bever, J. D. (2006) Mycorrhizal fungal identity and richness determine the diversity and productivity of a tallgrass prairie system. *New Phytologist*, 172, 554–562.

Volsi, B., Higashi, G. E., Bordin, I. and Telles, T. S. (2022) The diversification of species in crop rotation increases the profitability of grain production systems. *Scientific Reports*, 12, 19849.

Volz, P. (1951) Unterschungen uber die Microfauna des Waldbodens. *Zoologische Jahrbücher Abteilung für Systematik, Geographie und Biologie der Tiere*, 79, 514–566.

Wagg, C., Bender, S. F., Widmer, F. and Van der Heijden, M. G. A. (2014) Soil biodiversity and soil community composition determine ecosystem multifunctionality. *Proceedings of the National Academy of Sciences*, 111, 5266–5270.

Wagg, C., Hautier, Y., Pellkofer, S. et al. (2021) Diversity and asynchrony in soil microbial communities stabilizes ecosystem functioning. *Elife*, 10, e62813.

Wagg, C., Jansa, J., Schmid, B. and Van der Heijden, M. G. A. (2011) Belowground biodiversity effects of plant symbionts support aboveground productivity. *Ecology Letters*, 14, 1001–1009.

Waldrop, M. P. and Harden, J. W. (2008) Interactive effects of wildfire and permafrost on microbial communities and soil processes in an Alaskan black spruce forest. *Global Change Biology*, 14, 2591–2602.

Waldrop, M. P., Zak, D. R. and Sinsabaugh, R. L. (2004) Microbial community response to nitrogen deposition in northern forest ecosystems. *Soil Biology and Biochemistry*, 36, 1443–1451.

Walker, L. R. (1993) Nitrogen fixers and species replacements in primary succession. In: *Primary Succession on Land* (Miles, J. and Walton, D. W. H., eds), pp. 249–272. Blackwell Scientific, Oxford.

Walker, L. R. and Del Moral, R. (2003) *Primary Succession and Ecosystem Rehabilitation*. Cambridge University Press, Cambridge.

Walker, L. R., Wardle, D. A., Bardgett, R. D. and Clarkson, B. D. (2010) The use of chronosequences in studies of ecological succession and soil development. *Journal of Ecology*, 98, 725–736.

Walker, T. W. and Syers, J. K. (1976) The fate of phosphorus during pedogenesis. *Geoderma*, 15, 1–19.

Walker, T. W. N., Kaiser, C., Strasser, F. et al. (2018) Microbial temperature sensitivity and biomass change explain soil carbon loss with warming. *Nature Climate Change*, 8, 885–889.

Wall, D. H., Bradford, M. A., St. John, M. G. et al. (2008) Global decomposition experiment shows soil animal impacts on decomposition are climate-dependent. *Global Change Biology*, 14, 2661–2677.

Wall, D. H., Nielsen, U. N. and Six, J. (2015) Soil biodiversity and human health. *Nature*, 528, 69–76.

Wall, J. W., Skene, K. R. and Neilsen, R. (2002) Nematode community and trophic structure along a sand dune succession. *Biology and Fertility of Soils*, 35, 293–301.

Wallace, R. L. (2002) Rotifers: exquisite metazoans. *Integrative and Comparative Biology*, 42, 660–667.

Wallwork, A., Banin, L. F., Dent, D. H. et al. (2022) Soil carbon storage is related to tree functional composition in naturally regenerating tropical forests. *Functional Ecology*, 36, 3175–3187.

Walther, G. R., Beißner, S. and Burga, C. A. (2005) Trends in the upward shift of alpine plants. *Journal of Vegetation Science*, 16, 541–548.

Wang, B., An, S., Liang, C. et al. (2021) Microbial necromass as the source of soil organic carbon in global ecosystems. *Soil Biology and Biochemistry*, 162, 108422.

Wang, L., Delgado-Baquerizo, M., Wang, D. et al. (2019) Diversifying livestock promotes multidiversity and multifunctionality in managed grasslands. *Proceedings of the National Academy of Sciences*, 116, 6187–6192.

Wang, Q., Zhong, M. and Wang, S. (2012) A meta-analysis on the response of microbial biomass, dissolved organic matter, respiration, and N mineralization in mineral soil to fire in forest ecosystems. *Forest Ecology and Management*, 271, 91–97.

Ward, B. B. (2013) How nitrogen is lost. *Science*, 341, 352–353.

Wardle, D. A. (2002) *Communities and Ecosystems: Linking the Aboveground and Belowground Components*. Princeton University Press, Princeton, NJ.

Wardle, D. A. and Van der Putten, W. H. (2002) Biodiversity, ecosystem functioning and aboveground–belowground linkages. In: *Biodiversity and Ecosystem Functioning* (Loreau, M., Naeem, S. and Inchausti, P., eds), pp. 155–168. Oxford University Press, Oxford.

Wardle, D. A. and Zackrisson, O. (2005) Effects of species and functional group loss on island ecosystem properties. *Nature*, 435, 806–810.

Wardle, D. A., Bardgett, R. D., Callaway, R. M. and Van der Putten, W. H. (2011a) Terrestrial ecosystem responses to species gains and losses. *Science*, 332, 1273–1277.

Wardle, D. A., Bardgett, R. D., Klironomos, J. N. et al. (2004a) Ecological linkages between aboveground and belowground biota. *Science*, 304, 1629–1633.

Wardle, D. A., Bonner, K. I., Barker, G. M. et al. (1999a) Plant removals in perennial grassland: vegetation dynamics, decomposers, soil biodiversity, and ecosystem properties. *Ecological Monographs*, 69, 535–568.

Wardle, D. A., Hyodo, F., Bardgett, R. D. et al. (2011b) Long-term aboveground and belowground consequences of red wood ant exclusion in boreal forest. *Ecology*, 92, 645–656.

Wardle, D. A., Jonsson, M., Kalela-Brundin, M. et al. (2012) Drivers of inter-year variability of plant production and decomposers across contrasting island ecosystems. *Ecology*, 93, 521–531.

Wardle, D. A., Nicholson, K. S., Bonner, K. I. and Yeates, G. W. (1999b) Effects of agricultural intensification on soil-associated arthropod population dynamics, community structure, diversity and temporal variability over a seven-year period. *Soil Biology and Biochemistry*, 31, 1691–1706.

Wardle, D. A., Nilsson, M. C. and Zackrisson, O. (2008) Fire-derived charcoal causes loss of forest humus. *Science*, 320, 629–629.

Wardle, D. A., Walker, L. R. and Bardgett, R. D. (2004b) Ecosystem properties and forest decline in contrasting long-term chronosequences. *Science*, 305, 509–513.

Wardle, D. A., Yeates, G. W., Barker, G. M. et al. (2003a) Island biology and ecosystem functioning in epiphytic soil communities. *Science*, 301, 1717–1720.

Wardle, D. A., Yeates, G. W., Williamson, W. and Bonner, K. I. (2003b) The response of a three trophic level soil food web to the identity and diversity of plant species and functional groups. *Oikos*, 102, 45–56.

Warnock, A. J., Fitter, A. H. and Usher, M. B. (1982) The influence of a springtail *Folsomia candida* (Insecta, Collembola) on the mycorrhizal association of leek *Allium porrum* and the vesicular–arbuscular mycorrhizal endophyte *Glomus fasciculatus*. *New Phytologist*, 90, 285–292.

Webb, D. P. (1977) Regulation of deciduous forest litter decomposition by soil arthropod feces. In: *The Role of Arthropods in Forest Ecosystems* (Mattson, W. J., ed.), pp. 57–69. Springer, Berlin.

Weber, J., King, J. A., Abraham, N. L. et al. (2024) Chemistry–albedo feedbacks offset up to a third of forestation's CO_2 removal benefits. *Science*, 383, 860–864.

Weber, P. and Bardgett, R. D. (2011) Influence of single trees on spatial and temporal patterns of belowground properties in native pine forest. *Soil Biology and Biochemistry*, 43, 1372–1378.

Weemstra, M., Mommer, L., Visser, E. J. W. et al. (2016) Towards a multidimensional root trait framework: a tree root review. *New Phytologist*, 211, 1159–1169.

Wei, X., Shao, M., Gale, W. and Li, L. (2014) Global pattern of soil carbon losses due to the conversion of forests to agricultural land. *Scientific Reports*, 4, 4062.

Weidlich, E. W. A., Nelson, C. R., Maron, J. L. et al. (2021) Priority effects and ecological restoration. *Restoration Ecology*, 29, e13317.

Weidlich, E. W. A., Von Gillhaussen, P., Max, J. F. et al. (2018) Priority effects caused by plant order of arrival affect below-ground productivity. *Journal of Ecology*, 106, 774–780.

Weigelt, A., Bol, R. and Bardgett, R. D. (2005) Preferential uptake of soil nitrogen forms by grassland plant species. *Oecologia*, 142, 627–635.

Weil, R. R. and Brady, N. C. (2017) *The Nature and Properties of Soils*. 15th Edition. Pearson, Upper Saddle River, NJ.

Weis-Fogh, T. (1948) Ecological investigations of mites and collemboles in the soil. Appendix: Description of some new mites (Acari). *Natura Jutlandica*, 1, 139–277.

Weltzin, J. F., Bridgham, S. D., Pastor, J. et al. (2003) Potential effects of warming and drying on peatland plant community composition. *Global Change Biology*, 9, 141–151.

White, R. E. (2005) *Principles and Practice of Soil Science: The Soil as a Natural Resource*. 4th Edition. Wiley-Blackwell, Oxford.

White, S. R., Bork, E. W. and Cahill, J. F. (2014) Direct and indirect drivers of plant diversity responses to climate and clipping across northern temperate grassland. *Ecology*, 95, 3093–3103.

Whitman, T., Whitman, E., Woolet, J. et al. (2019) Soil bacterial and fungal response to wildfires in the Canadian boreal forest across a burn severity gradient. *Soil Biology and Biochemistry*, 138, 107571.

Wilkinson, A., Hill, P. W., Farrar, J. F. et al. (2014) Rapid microbial uptake and mineralization of amino acids and peptides along a grassland productivity gradient. *Soil Biology and Biochemistry*, 72, 75–83.

Wilkinson, A., Hill, P. W., Vaieretti, M. V. et al. (2015) Challenging the paradigm of nitrogen cycling: no evidence of in situ resource partitioning by coexisting plant species in grasslands of contrasting fertility. *Ecology and Evolution*, 5, 275–287.

Williams, A. and De Vries, F. T. (2020) Plant root exudation under drought: implications for ecosystem functioning. *New Phytologist*, 225, 1899–1905.

Williams, A., Langridge, H., Straathof, A. L. et al. (2022) Root functional traits explain root exudation rate and composition across a range of grassland species. *Journal of Ecology*, 110, 21–33.

Williamson, K. E., Fuhrmann, J. J., Wommack, K. E. and Radosevich, M. (2017) Viruses in soil ecosystems: an unknown quantity within an unexplored territory. *Annual Review of Virology*, 4, 201–219.

Willott, S. J., Miller, A. J., Incoll, L. D. and Compton, S. G. (2000) The contribution of rabbits (*Oryctolagus cuniculus* L.) to soil fertility in semi-arid Spain. *Biology and Fertility of Soils*, 31, 379–384.

Wilson, C. H., Strickland, M. S., Hutchings, J. A. et al. (2018) Grazing enhances belowground carbon allocation, microbial biomass, and soil carbon in a subtropical grassland. *Global Change Biology*, 24, 2997–3009.

Wookey, P. A., Aerts, R., Bardgett, R. D. et al. (2009) Ecosystem feedbacks and cascade processes: understanding their role in the responses of arctic and alpine ecosystems to environmental change. *Global Change Biology*, 15, 1153–1172.

Worm, B., Lotze, H. K., Hillebrand, H. and Sommer, U. (2002) Consumer versus resource control of species diversity and ecosystem functioning. *Nature*, 417, 848–851.

Wright, I. J., Reich, P. B., Westoby, M. et al. (2004) The worldwide leaf economics spectrum. *Nature*, 428, 821–827.

Wu, C., Coffield, S. R., Goulden, M. L. et al. (2023) Uncertainty in US forest carbon storage potential due to climate risks. *Nature Geoscience*, 16, 422–429.

Wu, T., Ayres, E., Bardgett, R. D. et al. (2011) Molecular study of worldwide distribution and diversity of soil animals. *Proceedings of the National Academy of Sciences*, 108, 17720–17725.

Wu, W., Kuang, L., Li, Y. et al. (2021) Faster recovery of soil biodiversity in native species mixture than in Eucalyptus monoculture after 60 years afforestation in tropical degraded coastal terraces. *Global Change Biology*, 27, 5329–5340.

Wubs, E. R. J., Van der Putten, W. H., Bosch, M. and Bezemer, T. M. (2016) Soil inoculation steers restoration of terrestrial ecosystems. *Nature Plants*, 2, 16107.

Wubs, E. R. J., Van der Putten, W. H., Mortimer, S. R. et al. (2019) Single introductions of soil biota and plants generate long-term legacies in soil and plant community assembly. *Ecology Letters*, 22, 1145–1151.

Wurzburger, N. and Hendrick, R. L. (2009) Plant litter chemistry and mycorrhizal roots promote a nitrogen feedback in a temperate forest. *Journal of Ecology*, 97, 528–536.

Xia, Y., Liu, J., Chen, C. et al. (2022) The multifunctions and future prospects of endophytes and their metabolites in plant disease management. *Microorganisms*, 10, 1072.

Xu, M., Wang, Y. and Yu, S. (2015) Conspecific negative density dependence decreases with increasing species abundance. *Ecosphere*, 6, 1–11.

Xue, K., Yuan, M. M., Shi, Z. J. et al. (2016) Tundra soil carbon is vulnerable to rapid microbial decomposition under climate warming. *Nature Climate Change*, 6, 595–600.

Yachi, S. and Loreau, M. (1999) Biodiversity and ecosystem productivity in a fluctuating environment: the insurance hypothesis. *Proceedings of the National Academy of Sciences*, 96, 1463–1468.

Yacine, Y., Kuťáková, E., in't Zandt, D. et al. (2024) Between- versus within-species variation in plant–soil feedback relates to different functional traits, but exudate variability is involved at both scales. *Functional Ecology*, 38, 1156–1171.

Yan, D., Li, J., Pei, J. et al. (2017) The temperature sensitivity of soil organic carbon decomposition is greater in subsoil than in topsoil during laboratory incubation. *Scientific Reports*, 7, 5181.

Yan, Y., Yan, R., Chen, J. et al. (2018) Grazing modulates soil temperature and moisture in a Eurasian steppe. *Agricultural and Forest Meteorology*, 262, 157–165.

Yang, H., Koide, R. T. and Zhang, Q. (2016) Short-term waterlogging increases arbuscular mycorrhizal fungal species richness and shifts community composition. *Plant and Soil*, 404, 373–384.

Yang, L. H. (2004) Periodical cicadas as resource pulses in North American forests. *Science*, 306, 1565–1567.

Yang, M., Zhang, Y., Qi, L. et al. (2015) Plant–plant–microbe mechanisms involved in soil-borne disease suppression on a maize and pepper intercropping system. *PLoS ONE*, 9, e115052.

Yang, S., Zheng, Q., Yuan, M. et al. (2019a) Long-term elevated CO_2 shifts composition of soil microbial communities in a Californian annual grassland, reducing growth and N utilization potentials. *Science of the Total Environment*, 652, 1474–1481.

Yang, X., Hu, H. W., Yang, G. W. et al. (2023) Crop rotational diversity enhances soil microbiome network complexity and multifunctionality. *Geoderma*, 436, 116562.

Yang, X., Xiong, J., Du, T. et al. (2024) Diversifying crop rotation increases food production, reduces net greenhouse gas emissions, and improves soil health. *Nature Communications*, 15, 198.

Yang, Y., Chen, X., Liu, L. et al. (2022) Nitrogen fertilization weakens the linkage between soil carbon and microbial diversity: a global meta-analysis. *Global Change Biology*, 28, 6446–6461.

Yang, Y., Cheng, H., Gao, H. and An, S. (2020) Response and driving factors of soil microbial diversity related to global nitrogen addition. *Land Degradation and Development*, 31, 190–204.

Yang, Y., Li, T., Wang, Y. et al. (2021) Negative effects of multiple global change factors on soil microbial diversity. *Soil Biology and Biochemistry*, 156, 108229.

Yang, Y., Tilman, D., Furey, G. and Lehman, C. (2019b) Soil carbon sequestration accelerated by restoration of grassland biodiversity. *Nature Communications*, 10, 718.

Yeates, G. W. (1968) An analysis of annual variation of the nematode fauna in dune sand, at Himatangi Beach, New Zealand. *Pedobiologia*, 8, 173–207.

Yeates, G. W. and King, K. L. (1997) Soil nematodes as indicators of the effect of management on grasslands in the New England Tablelands (NSW): comparison of native and improved grasslands. *Pedobiologia*, 41, 526–536.

Yeates, G. W., Bardgett, R. D., Cook, R. et al. (1997) Faunal and microbial diversity in three Welsh grassland soils under conventional and organic management regimes. *Journal of Applied Ecology*, 34, 453–470.

Yeates, G. W., Bongers, T., De Goede, R. G. M. et al. (1993) Feeding habits in soil nematode families and genera—an outline for ecologists. *Journal of Nematology*, 25, 315–331.

Yeates, G. W., Hawke, M. F. and Rijske, W. C. (2000) Changes in soil fauna and soil conditions under *Pinus radiata* agroforestry regimes during a 25-year tree rotation. *Biology and Fertility of Soils*, 31, 391–406.

Yergeau, E., Bokhorst, S., Kang, S. et al. (2012) Shifts in soil microorganisms in response to warming are consistent across a range of Antarctic environments. *The ISME Journal*, 6, 692–702.

Ylänne, H., Kaarlejärvi, E., Väisänen, M. et al. (2020) Removal of grazers alters the response of tundra soil carbon to warming and enhanced nitrogen availability. *Ecological Monographs*, 90, e01396.

Young, T., Okello, B., Kinyua, D. and Palmer, T. M. (1997) KLEE: a long-term multi-species herbivore exclusion experiment in Laikipia, Kenya. *African Journal of Range and Forage Science*, 14, 94–102.

Yu, K., Van den Hoogen, J., Wang, Z. et al. (2022) The biogeography of relative abundance of soil fungi versus bacteria in surface topsoil. *Earth System Science Data*, 14, 4339–4350.

Yue, K., Fornara, D. A., Yang, W. et al. (2017) Influence of multiple global change drivers on terrestrial carbon storage: additive effects are common. *Ecology Letters*, 20, 663–672.

Zaitsev, A. S., Gongalsky, K. B., Malmström, A. et al. (2016) Why are forest fires generally neglected in soil fauna research? A mini-review. *Applied Soil Ecology*, 98, 261–271.

Zak, D. R., Pregitzer, K. S., King, J. S. and Holmes, W. E. (2000) Elevated atmospheric CO_2, fine roots and the response of soil microorganisms: a review and hypothesis. *New Phytologist*, 147, 201–222.

Zak, D. R., Ringelberg, D. B., Pregitzer, K. S. et al. (1996) Soil microbial communities beneath *Populus grandidentata* grown under elevated atmospheric CO_2. *Ecological Applications*, 6, 257–262.

Zanne, A. E., Flores-Moreno, H., Powell, J. R. et al. (2022) Termite sensitivity to temperature affects global wood decay rates. *Science*, 377, 1440–1444.

Zavaleta, E. S., Shaw, M. R., Chiariello, N. R. et al. (2003) Additive effects of simulated climate changes, elevated CO_2, and nitrogen deposition on grassland diversity. *Proceedings of the National Academy of Sciences*, 100, 7650–7654.

Zeller, V., Bardgett, R. D. and Tappeiner, U. (2001) Site and management effects on soil microbial properties of subalpine meadows: a study of land abandonment along a north–south gradient in the European Alps. *Soil Biology and Biochemistry*, 33, 639–649.

Zemunik, G., Turner, B. L., Lambers, H. and Laliberté, E. (2015) Diversity of plant nutrient acquisition strategies increase during long-term ecosystem development. *Nature Plants*, 1, 15050.

Zhang, H. and Forde, B. G. (1998) An Arabidopsis MADS box gene that controls nutrient-induced changes in root architecture. *Science*, 279, 407–409.

Zhang, L., Jing, Y., Chen, C. et al. (2021) Effects of biochar application on soil nitrogen transformation, microbial functional genes, enzyme activity, and plant nitrogen uptake: a meta-analysis of field studies. *GCB Bioenergy*, 13, 1859–1873.

Zhang, L., Xiang, Y., Jing, Y. and Zhang, R. (2019a) Biochar amendment effects on the activities of soil carbon, nitrogen, and phosphorus hydrolytic enzymes: a meta-analysis. *Environmental Science and Pollution Research*, 26, 22990–23001.

Zhang, P., Li, B., Wu, J. and Hu, S. (2019b) Invasive plants differentially affect soil biota through litter and rhizosphere pathways: a meta-analysis. *Ecology Letters*, 22, 200–210.

Zhang, S., Lehmann, A., Zheng, W. et al. (2019c) Arbuscular mycorrhizal fungi increase grain yields: a meta-analysis. *New Phytologist*, *222*, 543–555.

Zhang, Y., Peng, S., Chen, X. and Chen, H. Y. (2022) Plant diversity increases the abundance and diversity of soil fauna: a meta-analysis. *Geoderma*, 411, 115694.

Zhong, L., Bowatte, S., Newton, P. C. D. et al. (2018) An increased ratio of fungi to bacteria indicates greater potential for N_2O production in a grazed grassland exposed to elevated CO_2. *Agriculture, Ecosystems & Environment*, 254, 111–116.

Zhou, G., Zhou, X., He, Y. et al. (2017) Grazing intensity significantly affects belowground carbon and nitrogen cycling in grassland ecosystems: a meta-analysis. *Global Change Biology*, 23, 1167–1179.

Zhou, J., Xue, K., Xie, J. et al. (2012) Microbial mediation of carbon-cycle feedbacks to climate warming. *Nature Climate Change*, 2, 106–110.

Zhou, L., Zhou, X., Shao, J. et al. (2016) Interactive effects of global change factors on soil respiration and its components: a meta-analysis. *Global Change Biology*, 22, 3157–3169.

Zhou, Z., Lu, J. Z., Preiser, J. et al. (2023) Plant roots fuel tropical soil animal communities. *Ecology Letters*, 26, 742–753.

Zhu, B., Gutknecht, J. L. M., Herman, D. J. et al. (2014) Rhizosphere priming effects on soil carbon and nitrogen mineralization. *Soil Biology and Biochemistry*, 76, 183–192.

Zhu, L., Yao, X., Chen, W. et al. (2023) Plastic responses of belowground foraging traits to soil phosphorus-rich patches across 17 coexisting AM tree species in a subtropical forest. *Journal of Ecology*, 111, 830–844.

Zimmer, M. and Topp, W. (1998) Microorganisms and cellulose digestion in the gut of the woodlouse *Porcellio scaber*. *Journal of Chemical Ecology*, 24, 1397–1408.

Zimmer, M. and Topp, W. (2002) The role of coprophagy in nutrient release from feces of phytophagous insects. *Soil Biology and Biochemistry*, 34, 1093–1099.

Zimov, S. A., Chuprynin, V. I., Oreshko, A. P. et al. (1995) Steppe–tundra transition: a herbivore-driven biome shift at the end of the Pleistocene. *The American Naturalist*, 146, 765–794.

Zimov, S. A., Zimov, N. S., Tikhonov, A. N and Chapin, F. S. (2012) Mammoth steppe: a high-productivity phenomenon. *Quaternary Science Reviews*, 57, 26–45.

Zogg, G. P., Zak, D. R., Pregitzer, K. S. and Burton, A. J. (2000) Microbial immobilization and the retention of anthropogenic nitrate in a northern hardwood forest. *Ecology*, 81, 1858–1866.

Zuber, S. M. and Villamil, M. B. (2016) Meta-analysis approach to assess effect of tillage on microbial biomass and enzyme activities. *Soil Biology and Biochemistry*, 97, 176–187.

Zumsteg, A., Luster, J., Göransson, H. et al. (2012) Bacterial, archaeal and fungal succession in the forefield of a receding glacier. *Microbial Ecology*, 63, 552–564.

Zwetsloot, M. J., Kessler, A. and Bauerle, T. L. (2018) Phenolic root exudate and tissue compounds vary widely among temperate forest tree species and have contrasting effects on soil microbial respiration. *New Phytologist*, 218, 530–541.

Index

Tables, figures, and boxes are indicated by an italic *t*, *f*, or *b*.